Undergraduate Lecture Notes in Physics

Series Editors

Neil Ashby, University of Colorado, Boulder, CO, USA

William Brantley, Department of Physics, Furman University, Greenville, SC, USA

Matthew Deady, Physics Program, Bard College, Annandale-on-Hudson, NY, USA

Michael Fowler, Department of Physics, University of Virginia, Charlottesville, VA, USA

Morten Hjorth-Jensen, Department of Physics, University of Oslo, Oslo, Norway

Michael Inglis, Department of Physical Sciences, SUNY Suffolk County Community College, Selden, NY, USA

Barry Luokkala⊕, Department of Physics, Carnegie Mellon University, Pittsburgh, PA, USA

Undergraduate Lecture Notes in Physics (ULNP) publishes authoritative texts covering topics throughout pure and applied physics. Each title in the series is suitable as a basis for undergraduate instruction, typically containing practice problems, worked examples, chapter summaries, and suggestions for further reading.

ULNP titles must provide at least one of the following:

- An exceptionally clear and concise treatment of a standard undergraduate subject.
- A solid undergraduate-level introduction to a graduate, advanced, or non-standard subject.
- A novel perspective or an unusual approach to teaching a subject.

ULNP especially encourages new, original, and idiosyncratic approaches to physics teaching at the undergraduate level.

The purpose of ULNP is to provide intriguing, absorbing books that will continue to be the reader's preferred reference throughout their academic career.

More information about this series at http://www.springer.com/series/8917

Alessandro De Angelis ·
Mário Pimenta · Ruben Conceição

Particle and Astroparticle Physics

Problems and Solutions

 Springer

Alessandro De Angelis
Department of Physics and Astronomy
"Galileo Galilei"
University of Padova
Padova, Italy

Mário Pimenta
Laboratory of Instrumentation
and Experimental Particle Physics
University of Lisboa
Lisboa, Portugal

Ruben Conceição
Laboratory of Instrumentation
and Experimental Particle Physics
University of Lisboa
Lisboa, Portugal

ISSN 2192-4791   ISSN 2192-4805   (electronic)
Undergraduate Lecture Notes in Physics
ISBN 978-3-030-73115-1   ISBN 978-3-030-73116-8   (eBook)
https://doi.org/10.1007/978-3-030-73116-8

This Springer imprint is published by the registered company Springer Nature Switzerland AG
The registered company address is: Gewerbestrasse 11, 6330 Cham, Switzerland

Preface

This book accompanies the textbook *Introduction to Particle and Astroparticle Physics (Multimessenger Astronomy and its Particle Physics Foundations)*, by Alessandro De Angelis and Mário Pimenta, 2nd edition, published by Springer in 2018. Besides providing solutions to all 179 exercises proposed there, it adds 79 new solved problems, for a total of 258 solved problems.

The present book is written in such a way that it can be used independently of its companion textbook, and is addressed in general to undergraduate students in particle and astroparticle physics and to physics teachers who can find here an extensive collection of problems, and be inspired by them. It can be useful to all scientists who want to approach this new physics subject learning by example and are looking for self-training. It preserves the balance in the introduction to particle and astroparticle physics characteristic of the companion textbook, starting from experiment and example.

It is divided into 11 chapters, the same as its companion book. We added to each chapter new exercises as soon as they came out to be useful on the basis of the experience accumulated by teachers using the companion textbook in their courses. Each problem has a title to facilitate the identification of related subjects. The number in parentheses refers to numbering (where available) in the textbook.

A selection of exams from the Astroparticle and Particle Physics courses at IST Lisboa, with solutions, is collected in Appendix. Each exam discusses a particular experimental result or a famous discovery.

Readers are kindly encouraged to inform us of any mistake that they will find, and/or of any better or more straightforward solution.

Lisboa, Portugal
Padova, Italy
April 2021

Alessandro De Angelis
Mário Pimenta
Ruben Conceição

Acknowledgements

Part of the exercises are taken from the examinations at the Particle and Astroparticle Physics courses at Instituto Superior Técnico (IST) of Lisboa, and from selection exams by the Italian National Institute for Nuclear Research (INFN) and by Ph.D. schools across the world.

A special acknowledgement goes to Jorge Romão, who has shared with some of us the responsibility to teach Particle Physics at IST in the last years, and with whom we have discussed many exercises.

A few more colleagues contributed with a substantial number of solutions: we thank in particular P. Abreu, S. Andringa, S. Ansoldi, L. Apolinário, F. Barão, U. Barres de Almeida, D. Bertacca, D. Cannone, M. De Maria, G. Galanti, G. La Mura, C. Lazzaro, R. Lopez Coto, M. Mallamaci, J. Maneira, R. Mirzoyan, R. Ragazzoni, B. Serrano González, R. Shellard, F. Simonetto, B. Tomé.

We thank the teachers using the textbook *Introduction to Particle and Astroparticle Physics: Multimessenger Astronomy and its Particle Physics Foundations* for providing us with useful comments. Part of the problems are taken from tests and exams of their courses.

We thank for useful suggestions and/or smaller contributions our students, and in particular F. Baruffaldi, L. Caloni, S. De Angelis, G. Fardelli, F. Gerardi, A. Granelli, F. Iacob, S. Libanore, A. Marzo, J. Siewyan, A. Spolon, H. Yarar, E. Zatti.

Contents

Chapter 1
Understanding the Universe: Cosmology, Astrophysics, Particles, and Their Interactions

1. *Size of a molecule (1).* Explain how you will be able to find the order of magnitude of the size of a molecule using a drop of oil. Make the experiment and check the result.

Just use a pipette to measure the volume of a drop of oil and let spread it over a water surface. The ratio between the volume of the drop and the oil surface on the water is the width of the oil layer. In a very crude estimation one can consider that the width of the layer is of the order of magnitude of the diameter of one oil molecule; in fact it corresponds to around one hundred atoms radius. Typical values you will obtain with such an experiment are that a drop with a volume of $1\,\mathrm{mm}^3$ will spread over a surface of $1000\,\mathrm{cm}^2$. Thus the order of magnitude of an oil molecule radius is $10^{-6} - 10^{-8}\,\mathrm{cm}$.

2. *Thomson atom (2).* Consider the Thomson atom model applied to a helium atom (the two electrons are in equilibrium inside a homogeneous positive charged sphere of radius $R \sim 10^{-10}\,\mathrm{m}$).

   a. Determine the distance of the electrons to the center of the sphere.
   b. Determine the vibration frequency of the electrons in this model and compare it to the first line of the spectrum of hydrogen.
   a. We start by calculating the electric field produced by a homogeneous positive charged sphere with density ρ, at distance $r < R$, using Gauss' Law, where r is the desired distance between the two electron (see Fig. 1.1). This gives us

   $$E = \frac{1}{3}\frac{\rho}{\varepsilon_0}\frac{r}{2} \tag{1.1}$$

   and positive charge density is

© Springer Nature Switzerland AG 2021

A. De Angelis et al., *Particle and Astroparticle Physics*, Undergraduate Lecture Notes in Physics, https://doi.org/10.1007/978-3-030-73116-8_1

Fig. 1.1 Sketch for the Thomson atom and variables used in this problem

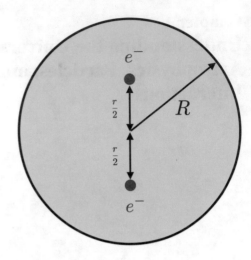

$$\rho = \frac{2e}{4/3\pi R^3} = \frac{3e}{2\pi R^3}.$$ (1.2)

With the above result it is possible to evaluate the amount of positive charge contained in the sphere that involves both electrons as

$$q = \frac{4\pi}{3} \frac{r^3}{8} \rho = \frac{er^3}{4R^3}.$$ (1.3)

The total force in one of the electrons is: $\mathbf{F}_1 = \mathbf{F}_{12} + \mathbf{F}_{1\rho}$, the sum of the force due to the second electron plus the force due to the positive charge distribution. The force between the electrons, having the same charge, is repulsive and is given by

$$F_{12} = \frac{1}{4\pi\varepsilon_0} \frac{e^2}{r^2}.$$ (1.4)

For a system in equilibrium: $\mathbf{F} = 0$. Therefore, using the previous results we can build the following equality

$$\frac{qe}{4\pi\varepsilon_0(r/2)^2} = \frac{e^2}{4\pi\varepsilon_0 r^2}$$ (1.5)

from where we get that the equilibrium condition is $r = R$.

b. In this model the positive charged sphere is fixed and the two electrons can vibrate around the equilibrium position defined above. For simplicity let's consider that both electrons will oscillate longitudinally (along the line defined by the two electrons which includes also the center of the sphere) and be x_1 and x_2 the deviation of each of the electrons from their equilibrium positions (r_0 and $-r_0$ respectively).

The center of mass (CM) of the two electrons will behave as a body with a mass equal to the sum of the mass of the two electrons submitted to the sum of the forces between the charged sphere and each of the electrons:

$$x_{CM} = \frac{x_1 + x_2}{2}$$

$$2\, m_e \ddot{x}_{CM} = -\frac{e^2 (r_0 + x_1)}{4\pi \varepsilon_0 R^3} + \frac{e^2 (r_0 - x_2)}{4\pi \varepsilon_0 R^3}$$

$$\ddot{x}_{CM} = -\frac{e^2}{4\pi \varepsilon_0 R^3 m_e} x_{CM}.$$

The CM of the two electrons behaves then like a harmonic oscillator with an angular frequency:

$$\omega = \sqrt{\frac{e^2}{4\pi \varepsilon_0 R^3 m_e}}. \tag{1.6}$$

The frequency $\nu_{He} = \omega/2\pi \simeq 2.5 \times 10^{15}$ Hz, which can be translated into an energy using $E = h\nu \simeq 10.5\,\text{eV}$. This value is very close to the $n = 2 \rightarrow 1$ transition of the hydrogen atom.

3. *Atom as a box (3).* Consider a simplified model where the hydrogen atom is described by a one dimensional box of length r with the proton at its center and where the electron is free to move around. Compute, taking into account the Heisenberg uncertainty principle, the total energy of the electron as a function of r and determine the value of r for which this energy is minimized.

Let us start by considering that the electron is in a one-dimensional region of size r. In this case the total energy of the electron is given by its kinetic energy plus the Coulomb potential,

$$E = T + V = \frac{p^2}{2m} - \frac{1}{4\pi \varepsilon_0} \frac{qQ}{r} \tag{1.7}$$

where q is the electron charge and Q the proton charge. The quantity p is the electron momentum and m its mass. The Heisenberg uncertainty principle states that the uncertainty in the position of the particle, Δx, and on its momentum, Δp, is related by

$$\Delta x\, \Delta p \geq \frac{\hbar}{2}. \tag{1.8}$$

As the electron can be anywhere within r, then from Eq. 1.8 one has that the electron momentum uncertainty is

Fig. 1.2 Sketch of the evolution of the energy of the electron with the distance from the proton

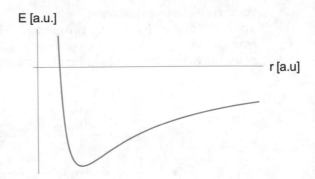

$$\Delta p \geq \frac{\hbar}{2r}.\tag{1.9}$$

Substituting the above condition in Eq. 1.7 one obtains

$$E = \frac{\hbar^2}{4\,mr^2} - \frac{1}{4\pi\varepsilon_0}\frac{e^2}{r}\tag{1.10}$$

where the relation $|q| = Q = e$, being e the absolute value of the electron charge, was also used. The behavior of Eq. 1.10 with r is shown in Fig. 1.2. The distance minimizing the energy can be obtained through

$$\frac{dE}{dr} = 0 \Rightarrow r = \frac{2\hbar\pi\varepsilon_0}{me}.\tag{1.11}$$

Substituting with the appropriate values one gets that the electron should be within $r \simeq 2.6 \times 10^{-11}$ m $= 26$ pm. Experimentally, one has $r_{atom} \sim 30$ pm.

4. *Naming conventions for particles (4).* Write down the symbol, charge and approximate mass for the following particles:

   a. Tau lepton
   b. Anti muon-neutrino
   c. Charm quark
   d. Anti-electron
   e. Anti-bottom quark.

   See Fig. 1.3.

   a. Tau lepton: τ^-, charge -1, mass $\simeq 1.8$ GeV/c^2.
   b. Anti muon-neutrino: $\bar{\nu}_\mu$, charge 0, mass $\simeq 0$.
   c. Charm quark: c, charge $+2/3$, mass $\simeq 1.3$ GeV/c^2.
   d. Anti-electron: e^+, charge $+1$, mass $\simeq 0.5$ MeV/c^2.
   e. Anti-bottom quark: $\bar{b}$, charge $1/3$, mass $\simeq 4.2$ GeV/c^2.

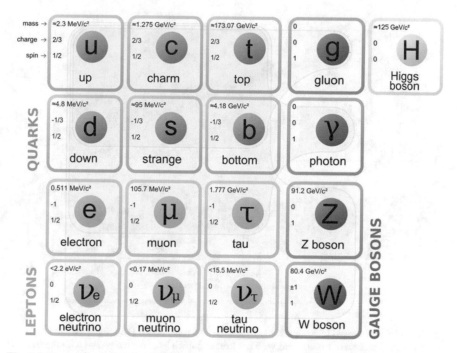

Fig. 1.3 Presently observed elementary particles. Fermions are listed in the first three columns; gauge bosons are listed in the fourth column. Adapted from MissMJ [CC BY 3.0 (http://creativecommons.org/licenses/by/3.0)], via Wikimedia Commons

5. *Strange mesons (5).* How many quark combinations can you make to build a strange neutral meson, using u, d and s quark flavors?

You can build two combinations: $d\bar{s}$ and $s\bar{d}$. At the lowest energy state (with total momentum = 0), the first one is called K^0, the second (it is the antiparticle of the first) is called $\overline{K}^0$.

6. *A look to the Universe (6).* Find a dark place close to where you live, and go there in the night. Try to locate the Milky Way and the Galactic Center (GC). Comment on your success (or failure).

This answer depends on where you live. But if, as two of the authors, you live in Portugal at a latitude of $\sim 40\,°$N, then you can find the Milky Way at a given time of the night just above your head, crossing from south to north. In reality, you are not seeing the whole Galaxy, only one of its *arms*. The center of the Galaxy is visible in the Southern hemisphere. But at these coordinates, you can catch a glimpse of it during the summer near the horizon, just slightly below the *Sagittarius* constellation. While the GC is full of stars, we cannot see it through visible light as there are clouds of dust in between, which completely hide it.

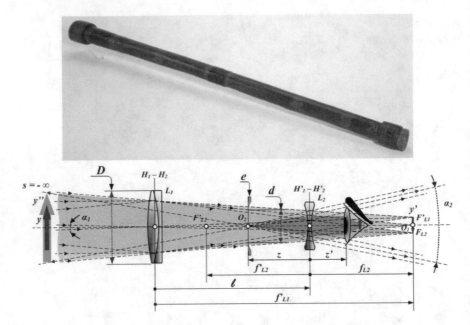

Fig. 1.4 Top: an original telescope made by Gailei. The tube, about 1 m long, is formed by strips of wood joined together. It is covered with red leather (which has become brown with the passage of time) with gold tooling. The instrument's magnification is 21. It was called by Galilei "cannone occhiale"; however, in 1611, Prince Federico Cesi, founder of the Accademia dei Lincei, suggested calling this instrument telescopio, from the Greek tele ("far'") and scopeo ("look")]. Institute and Museum of the History of Science, Firenze. Bottom: sketch of the principle of operation of a Galilean telescope. y - Distant object; y' - Real image from objective; y" - Magnified virtual image from eyepiece; D - Entrance pupil diameter; d - Virtual exit pupil diameter; L1 - Objective lens; L2 - Eyepiece lens; e - Virtual exit pupil. By Tamasflex (Own work) [CC BY-SA 3.0 or GFDL], via Wikimedia Commons

7. *Galilean and Newtonian telescopes: refraction vs. reflection (7).* Make a research on the differences between Newtonian and Galilean telescopes; discuss such a difference.

When in 1609 Galilei heard in Venice about a new optical device, the telescope, which had come to public attention in the Netherlands, he built his own version. He fitted a convex lens in one extremity of a tube and a concave lens in the other one. Some time later, having succeeded in making an improved version, he took it to Venice, where he presented the instrument to the Doge Leonardo Donà, who was sitting in the town's council. The senate, in return, made his position in Padova permanent and doubled his salary.

Galilei devoted his time to improving and perfecting the telescope and soon succeeded in producing instruments of increasing power (Fig. 1.4). He used them as a gadget for high-society in Venezia and for astronomical observations. Galilei's astronomical findings, published in 1610 in the Sidereus Nuncius

Fig. 1.5 Galilei's drawing
of the Moon. From
Wikimedia Commons

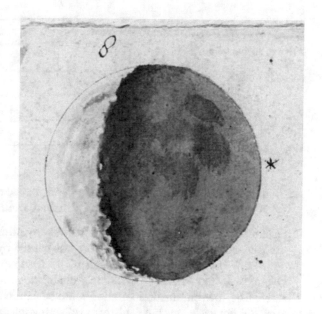

("The starry messenger", or "A message from the stars": Latin is ambiguous), had
important implications (see for example A. De Angelis, *I diciotto anni migliori
della mia vita (Galileo a Padova 1592–1610)*, Castelvecchi 2021), in particular
related to:

- The Moon. According to Aristotle the Moon had to be a perfect sphere. Galilei
  found the "surface of the Moon to be not smooth, even and perfectly spherical
  [...], but on the contrary, to be uneven, rough, and crowded with depressions
  and bulges. And it is like the face of the earth itself, which is marked here
  and there with chains of mountains and depths of valleys." He calculated
  the heights of the mountains by measuring the lengths of their shadows. He
  also detected earthshine on the lunar surface, that is the Moon was lit up by
  reflected light from the Earth just like we receive reflected light from the Moon
  (Fig. 1.5).
- Stars in the Milky Way. Even through a telescope the stars still appeared as
  points-like. Galilei suggested that this was due to the fact that their distance
  from Earth was very large. Observing the Milky Way, he could resolve what
  was appearing as a cloud, and he discovered thousands of yet unknown stars.
- The moons of Jupiter. Observations of the planet Jupiter over successive nights
  (Fig. 1.6) revealed four star-like objects in line with it, moving from night to
  night, sometimes disappearing behind or in front of the planet. Galilei correctly
  inferred that these objects were moons of Jupiter and orbited it just as our Moon
  orbits Earth. Today these four moons are known as the Galilean satellites: Io,
  Europa, Ganymede and Callisto. Galilei studied the orbits of these moons - his

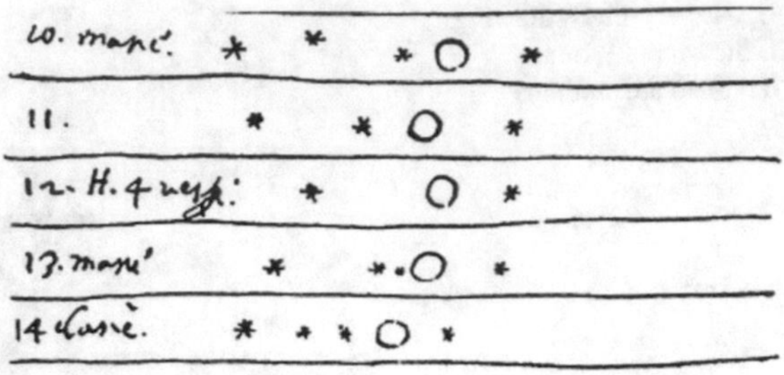

Fig. 1.6 Galilei's drawings
of Jupiter's satellites in
different nights, from
Galilei's logbook. Public
domain

data will later be used by Newton to confirm the law of universal gravitation. For the first time, objects had been observed orbiting another planet.

Color distortion (chromatic aberration) was the primary limit of Galilean telescopes. Newton concluded this defect was caused by the lens of the refracting telescope behaving the same as prisms he was experimenting with, breaking white light into a rainbow of colors around bright astronomical objects. If this were true, then chromatic aberration could be eliminated by building a telescope that did not use a lens but a mirror: a reflecting telescope.

Newton's idea for a reflecting telescope was not a new one. In particular, two Galilei's scholars (Francesco Sagredo and Daniele Antonini) had discussed using a mirror as the image forming objective as an alternative to the refracting telescope, but never built such an instrument. However, Newton was able to rapidly implement the idea. In 1668 he built his first reflecting telescope. He chose a speculum of tin and copper for his objective mirror. He added a secondary diagonally mounted mirror near the primary mirror's focus to reflect the image at 90° to an eyepiece mounted on the side of the telescope. This addition allowed the image to be viewed with minimal obstruction of the objective mirror and relaxed observations (Fig. 1.7). Newton's first version had a primary mirror diameter of 33 cm and a focal ratio of f/5, i.e., a focal length 5 times the reflector diameter. He found that the telescope worked without color distortion and that he could see the four Galilean moons of Jupiter and the crescent phase of the planet Venus with it. Newton demonstrated his invention to king Charles II in 1672.

Newton found it hard to construct an effective reflector since the metal of the speculum was hard to get to a regular curvature. The surface also tarnished rapidly; the consequent low reflectivity of the mirror and also its small size meant that the view through the telescope was very dim compared to contemporary refractors. Because of these difficulties in construction, Newtonian telescopes were initially not widely adopted, while they are today the favorite tool of astronomers.

8. *Number of stars in the Milky Way (8).* Our Galaxy consists of a disk of a radius $r_d \simeq 30$ kpc about $h_d \simeq 300$ pc thick, and a spherical bulge at its center roughly 3 kpc in diameter. The distance between our Sun and our nearest neighboring stars, the Alpha Centauri system, is about 1.3 pc. Estimate the number of stars in our Galaxy.

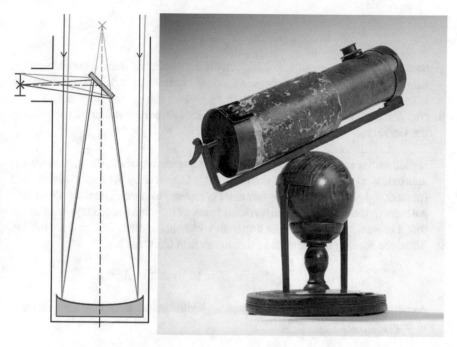

Fig. 1.7 Left: sketch of the principle of operation of a Newtonian telescope. By Krishnavedala (Own work) [CC BY-SA 4.0], via Wikimedia Commons. Right: a Newtonian telescope, probably a refurbished version of Newton's first telescope. From the museum of the Royal Society, London

If we approximate the shape of the disk as a cylinder, the volume of the disk is

$$V_d = \pi r_d^2 h_d = \pi \times (30\,\text{kpc})^2 \times (0.3\,\text{kpc}) = 848\,\text{kpc}^3\,.$$

We also know that the closest star to the Sun is about 1.3 pc away, and that we are in an "ordinary" part of the Galactic disk. Assuming that the average distance between stars throughout the Galaxy is 1.3 pc, there are a total of

$$N_d \simeq \frac{V_d}{V_s} \simeq 9.2 \times 10^{10}\,\text{stars}\,,$$

where V_s is the volume where there is only one star is estimated through

$$V_s = \frac{4}{3}\pi r_s^3\,, \qquad r_s = 1.3\,\text{pc}\,.$$

The volume of the bulge is nearly two orders of magnitude smaller than the disk:

$$V_b = \frac{4}{3}\pi r_b^3 \simeq 14.1 \text{kpc}^3 ,$$

so it would contain about 1.5×10^9 stars if the density of stars in the bulge is the same as the disk - indeed we observe the density to be larger. Still, its contribution is so small that we can approximate to $N_s \sim 10^{11}$ stars in the full Galaxy.

9. *Number of nucleons in the Universe (9)*. Estimate the number of nucleons in the Universe.

If dark matter is, as indicated by observations, not made of nucleons, then we can approximate the total mass of the observable Universe by the total mass of stars (planets and clouds have negligible mass compared to stars). There are about 10^{11} galaxies in the observable Universe, and about 10^{11} stars per galaxy. We assume that a typical star weights like our Sun, which has a mass $M_\odot \simeq 2 \times 10^{30}$ kg. Thus, the mass of all the stars in the observable Universe is

$$M_U \simeq 10^{11} \times 10^{11} \times (2 \times 10^{30}) \simeq 2 \times 10^{52} \text{kg} ,$$

dominated by nucleons (electrons have negligible mass). Being the nucleon mass $m \simeq m_p \simeq 1.67 \times 10^{-27}$ kg,

$$N_{nucleons} \simeq \frac{M_U}{m_p} \simeq 1.2 \times 10^{79} .$$

We expect most of them to be protons, due to the relatively short neutron lifetime.

10. *Hubble's law (10)*. The velocity of a galaxy can be measured using the Doppler effect. The radiation coming from a moving object is shifted in wavelength, the relation being, for $\Delta\lambda/\lambda \ll 1$,

$$z = \frac{\Delta\lambda}{\lambda} \simeq \frac{v}{c} ,$$

where λ is the rest wavelength of the radiation, $\Delta\lambda$ is the observed wavelength minus the rest wavelength, and v is defined as positive when the object parts away from the observer. Notice that in this case (for v small compared to the speed of light) the formula is the same as for the classical Doppler effect.

An absorption line from a galaxy that is found at 500 nm in the laboratory is found at 505 nm when analyzing the spectrum of a particular galaxy. Estimate the distance of the galaxy.

The apparent wavelength of the line observed from the galaxy is larger than the neutral wavelength: the galaxy is thus receding, with speed

$$v \simeq c\frac{505 - 500}{500} = \frac{c}{100} .$$

Using Hubble's law, one has thus

$$v \simeq H_0 d \Longrightarrow d \simeq \frac{v}{H_0} \simeq \frac{c}{100\, H_0} \simeq 140\,\text{Mly}\,.$$

11. *Wavelength and energy of CMB photons.* Evaluate the typical wavelength of a photon from cosmic background radiation.

The CMB has a thermal blackbody spectrum at a temperature T of 2.725 K, corresponding to an energy $k_B T \sim 0.23$ meV, and to a wavelength hc/E of about 5.39 mm. Due to the shape of the blackbody distribution, the spectral radiance $dE_\nu/d\nu$ peaks at about 160 GHz, corresponding to a photon energy of about 0.66 meV, and a wavelength of about 1.87 mm. Alternatively, if spectral radiance is defined as $dE_\lambda/d\lambda$, then the peak wavelength is 1.06 mm (about 282 GHz).

12. *Luminosity and magnitude (11).* Suppose that you burn a car on the Moon, heating it at a temperature of 3000 K. What is the absolute magnitude of the car? What is the apparent magnitude m seen at Earth?

The absolute magnitude (brightness) of a celestial object is defined as its magnitude at a standard distance of 10 parsec.
We can express the total (bolometric) flux of a thermal source with surface S as observed from a distance D by means of the Stefan-Boltzmann's law:

$$F_{bol} = \frac{\sigma T^4 S}{4\pi D^2}, \tag{1.12}$$

where

$$\sigma = 5.67 \times 10^{-8}\,\text{J}\,\text{s}^{-1}\,\text{m}^{-2}\,\text{K}^{-4} \tag{1.13}$$

is the Stefan's constant. If we know the flux of a source, we can calculate its magnitude m by the formula:

$$m = -2.512 \log\left(\frac{F}{\text{J}}\,\text{m}^{-2}\,\text{s}^{-1}\right) + m_0 \tag{1.14}$$

where the symbol log indicates the base 10 logarithm and m_0 gives the normalization of our photometric system with respect to the employed measurement units. m_0 can be obtained from the properties of the Sun, which has a total luminosity:

$$L_{bol,\odot} = 3.828 \times 10^{26}\,\text{J}\,\text{s}^{-1} \tag{1.15}$$

and an absolute magnitude:

$$M_{bol,\odot} = 4.76. \tag{1.16}$$

It is, therefore:

$$m_0 = M_{bol,\odot} + 2.512 \log \left[\frac{L_{bol,\odot}}{4\pi (10\,\text{pc})^2} \right], \tag{1.17}$$

which, knowing that $1\,\text{pc} = 3.086 \times 10^{16}\,\text{m}$, brings to $m_0 = -19.09$.
If we assume that the car is approximately $5 \times 2\,\text{m}^2$ large, according to Eq. (1.12), it has a flux:

$$F_{bol} = 2\frac{\sigma T^4 S}{4\pi D^2} \tag{1.18}$$

where the factor 2 roughly accounts for the fact that the car radiates isotropically (i.e. both upwards and downwards, in our approximation). With $T = 3000\,\text{K}$, the flux at distance D is:

$$F_{bol} = \frac{7.31 \times 10^6}{D^2}\,\text{J s}^{-1}. \tag{1.19}$$

The absolute magnitude M_{bol} is obtained by plugging the above flux in Eq. (1.14), with $D = 10\,\text{pc} = 3.086 \times 10^{17}\,\text{m}$. This gives:

$$M_{bol} = -2.512 \log \left[\frac{7.31 \times 10^6}{(3.086 \times 10^{17})^2} \right] - 19.09 = 51.53. \tag{1.20}$$

If we now consider the fire to be located at the distance of the Moon ($D = 3.85 \times 10^8\,\text{m}$), we get an apparent magnitude of:

$$m_{bol} = -2.512 \log \left[\frac{7.31 \times 10^6}{(3.85 \times 10^8)^2} \right] - 19.09 = 6.80. \tag{1.21}$$

Finally, if we are interested in the apparent visual magnitude, we need to apply a bolometric correction:

$$\Delta m(T) \simeq \frac{29500\,\text{K}}{T} + 10 \log \left(\frac{T}{1\,\text{K}} \right) - 42.62, \tag{1.22}$$

which leads to $\Delta m(3000\,\text{K}) = 1.98$ and ultimately to:

$$m_V = 8.78. \tag{1.23}$$

This is approximately 20 times below the perception limit of the naked human eye.

13. *Hertzsprung-Russell diagram.* Figure 1.8 shows the Hertzsprung-Russell diagram for nearby stars whose parallaxes were accurately measured by the Hipparcos satellite.

a. The color index (B-V) of the star measures its color. What physical property of the star determines its color?

b. What shows that the solar neighborhood contains stars of different ages, including some which are younger than the Sun?

---

a. The temperature. Indeed, if we consider the definition of the color index, as the difference between the magnitude measured in two filters, by means of the logarithmic scale of the magnitude system, a color index is determined by the ratio of the fluxes entering the corresponding band-passes. Now, we know that stars approximately radiate as black bodies. This implies that hotter bodies will be more luminous than cold ones in every band. But, due to Wien's displacement law, the peak emission of a hot body will also fall at a shorter wavelength than that of the cold body. This means that, taking into account any two bandpasses, a hot body will always have a flux ratio between the short and the long wavelength which is higher than the corresponding ratio of a cold body, resulting in a different (more negative) color index.

b. The most important evidence for the presence of young stars is the fact that the Main Sequence extends towards bluer and brighter regions than the spot where the Sun is located. Such positions can only be occupied by high mass stars. Since the lifetime of a star on the Main Sequence is roughly depending on $(M/M_\odot)^{-2.5}$, those stars can only be still there if they were formed more recently than the Sun. Proofs of the existence of older stars, instead, are the Red Giant Branch, which intercepts the Main Sequence approximately at the Sun location, and the existence of some White Dwarfs. White Dwarfs are low mass stellar remnants, with ages typically above 7 billion years, while Sub-Giants of solar luminosity are stars with a lower mass than the Sun that, however, formed earlier and already evolved off the Main Sequence.

As a final consideration, we might point out that the position of a star on the Main Sequence is mainly controlled by its mass, since this will be the driver of the stellar radius and temperature. For what concerns the Main Sequence only, therefore, the color is determined by the stellar mass.

14. *Cosmic ray fluxes and wavelength (12).* The most energetic particles ever observed at Earth are cosmic rays. Make an estimation of the number of such events with an energy between 3×10^{18} eV and 10^{19} eV that may be detected in one year by an experiment with a footprint of $1000\,\mathrm{km}^2$. Evaluate the structure scale that can be probed by such particles.

---

i. Estimation of the number of events.

The flux of cosmic rays follows basically a power law in energy, with the differential spectral index, γ, depending on the energy range. We have approximated here $\gamma \sim 3$ everywhere.

Looking to Fig. 1.9, one has a flux of about 1 particle per square kilometer per year at 10^{10} GeV. One can thus write, close to such an energy,

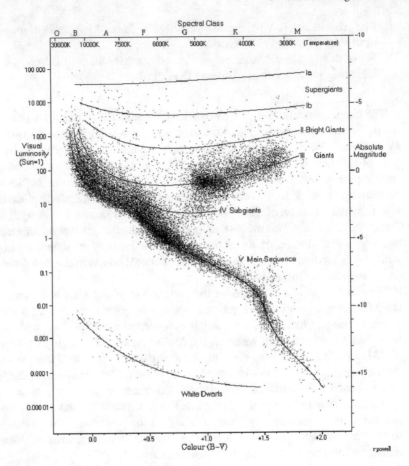

Fig. 1.8 The Hertzsprung-Russell diagram for a representative sample of near stars: 22000 stars from the Hipparcos Catalog and with 1000 low-luminosity stars (red and white dwarfs) from the Gliese Catalog of nearby Stars. From Wikimedia Commons

$$\frac{dN(E)}{dE\,dA\,dt} \simeq k \left(\frac{E}{10^{10}\mathrm{GeV}}\right)^{-3} \mathrm{GeV}^{-1}\,\mathrm{km}^{-2}\,\mathrm{yr}^{-1}, \qquad (1.24)$$

with $k = 1$ particle per square kilometer per year.
Integrating in energy, taking $x \equiv \frac{E}{10^{10}\mathrm{GeV}}$, one has

$$\frac{dN}{dA\,dt} = -\frac{k}{2}\left[x^{-2}\right]_{0.3}^{1} \mathrm{km}^{-2}\,\mathrm{yr}^{-1}. \qquad (1.25)$$

The total number of events that may be detected in one year in an area of $1000\,\mathrm{km}^2$ is thus

Fig. 1.9 Energy spectrum (number of incident particles per unit of energy, per second, per unit area, and per unit of solid angle) of primary cosmic rays. The vertical band on the left indicates the energy region in which the emission from the Sun is supposed to be dominant; the central band the region in which most of the emission is presumably of Galactic origin; the band on the right the region of extragalactic origin. From Wikimedia Commons

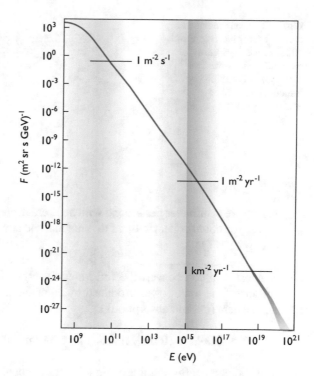

$$N \simeq 5000 . \qquad (1.26)$$

ii. Structure scale.

The wavelength of the most energetic particles in the energy range considered, $E \sim 10^{19}$ eV, is :

$$\lambda = \frac{hc}{E} = \frac{1.24 \, \text{GeV fm}}{E} \simeq 10^{-10} \, \text{fm} . \qquad (1.27)$$

This gives the scale that can be probed by such particles. Notice that this is ten orders of magnitude smaller than the proton radius.

15. *Energy from cosmic rays: Nikola Tesla's "free" energy generator (13).* "This new power for the driving of the world's machinery will be derived from the energy which operates the Universe, the cosmic energy, whose central source for the earth is the sun and which is everywhere present in unlimited quantities." Immediately after the discovery of of natural radioactivity, in 1901, Nikola Tesla patented an engine using the energy involved (and expressed a conjecture about the origin of such radioactivity). In Fig. 1.10 (made by Tesla himself) it is shown his first radiant energy receiver. If an antenna is wired to one side of a capacitor (the other going to ground,) the potential difference will charge the capacitor. Suppose you can intercept all high-energy cosmic radiation (assume 1 particle

Fig. 1.10 Tesla scheme of a radiant energy receiver. From Nikola Tesla, *Apparatus for the Utilization of Radiant Energy*, US-Patent 685,957, issued on November 05,1901, Fig. 1.2

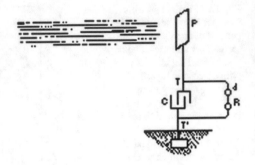

per square centimeter per second with an average energy of 3 GeV); what is the power you could collect with a 1 m^2 antenna, and how does it compare with solar energy?

Suppose you can convert the full kinetic energy of cosmic rays into an useful form. The total power produced by an antenna of 1 m^2 placed outside the atmosphere (to limit absorption) is

$$P \simeq 3\,\text{GeV} \times 10^4\,\text{m}^2/\text{cm}^2 \times 1\,\text{particle}\,\text{cm}^{-2}\text{s}^{-1} \simeq 5 \times 10^{-6}\,\text{W/m}^2.$$

The average energy coming from sunlight is about 1400 W/m^2: photoelectric cells are by far more convenient than the Tesla engine!

16. *Galactic and extragalactic emitters of gamma rays (14).* A plot in Galactic coordinate shows that more than half of the emitters of high-energy photons lie in the Galactic plane (the equatorial line). Guess why.

Two effects can play a role: (a) present detectors do not have a resolution large enough to observe the structure of galaxies out of the Milky Way, and (b) the signal from distant galaxies is attenuated—in first approximation, as $1/d^2$, where d is the distance. Notice that the size of the Milky Way is about 100 kly, while the distance of the nearest large galaxy, Andromeda, is about 2.5 Mly.

17. *Systems of coordinates.* Although astroparticle physicists love the Galactic coordinate system, many astronomers like to talk about right ascension and declination (equatorial or celestial coordinate system). This system is analogous to the latitude-longitude system.

In the celestial coordinate system the North and South Celestial Poles are determined by projecting the rotation axis of the Earth to intersect the celestial sphere, which in turn defines a celestial equator. The celestial equivalent of latitude is called declination δ and is measured in degrees North (positive numbers) or South (negative numbers) of the celestial equator. The celestial equivalent of longitude is called right ascension RA, and for historical reasons it is measured in time (hours, minutes, seconds). The sky apparently turns 360° in 24 h (15° every hour); thus, 1 h of right ascension is equivalent to 15° of sky rotation.

The North Galactic Pole is at Right Ascension 12 h 49 m, declination +27°24'. What is the tilt of the Galactic plane to the celestial equator?

The distance from the Northern Celestial Pole P to the Northern Galactic Pole $\hat{G}$ is just (90° - declination of $\hat{G}$) $\simeq$ 62.6°. This is also the tilt of the Galactic plane to the equator.

Chapter 2
Basics of Particle Physics

1. *Rutherford formula (1).* Consider the Rutherford formula.

   a. Determine the distance of closest approach of an α particle with an energy of 7.7 MeV to a gold target.
   b. Determine the de Broglie wavelength of that α particle.
   c. Explain why the classical Rutherford formula survived the revolution of quantum mechanics.

   You can find the numerical values of particle data and fundamental constants in your Particle Data Book(let).

---

   a. The α particle will be deflected as it approaches the nucleus, describing a curved trajectory. Both the energy, E, and the angular momentum, L, will be conserved.
   Taking b as the impact parameter, the angular momentum can be written as

   $$L = mvb = \sqrt{2Em}\,b \qquad (2.1)$$

   where v and m are the velocity and the mass of the α particle, respectively. The relation between momentum and energy $p = mv = \sqrt{2Em}$ was used. The system energy is given by

   $$E = \frac{1}{2}mv^2 + \frac{e^2 Z_1 Z_2}{4\pi\varepsilon_0 d} \qquad (2.2)$$

   e being the absolute value of the electron charge, d the distance between the α particle and the nucleus, and Z_1 and Z_2 are the atomic number of the α and of the nucleus, respectively.

© Springer Nature Switzerland AG 2021
A. De Angelis et al., *Particle and Astroparticle Physics*, Undergraduate Lecture Notes in Physics,
https://doi.org/10.1007/978-3-030-73116-8_2

It can be easily proved that the situation in which the α particle will be at the closest distance from the nucleus corresponds to $b = 0$ and $v = 0$. Then using equation Eq. 2.2 one obtains,

$$d_{min} = \frac{e^2 Z_1 Z_2}{4\pi \varepsilon_0 E} \simeq 2.95 \times 10^{-14}\, m = 29.5\, fm \qquad (2.3)$$

taking $E = 7.7\, MeV$, $Z_1 = 2$, $Z_2 = 79$ and $b = 0\, m$.

b. Using the de Broglie relation,

$$\lambda = \frac{h}{p} \qquad (2.4)$$

the momentum p of the α particle can be obtained from the equation

$$E^2 = p^2 c^2 + m^2 c^4 . \qquad (2.5)$$

From Eqs. 2.4 and 2.5 one gets that the α particle's wavelength is

$$\lambda = \frac{h}{\sqrt{\left(\frac{E}{c}\right)^2 - m^2 c^2}} \simeq 5.17 \times 10^{-15}\, m = 5.17\, fm \qquad (2.6)$$

with $m = 4 \times 938\, MeV/c^2$ and noting that the energy of the α particle is now its kinetic energy plus a mass term, $E = T + mc^2 = 7.7\, MeV + mc^2$.

c. Since $\lambda \ll d$ the particle's wave can be assumed as point like and so there are no interference phenomena. Hence, the classical approximation is valid.

2. *A simplified diagram for beta decay?* Why was the process $n \rightarrow pe^-$ thought to be impossible even without invoking the principle of conservation of the lepton number?

Because during the decay of the neutron it was seen that the electron had an energy spectrum distribution and not a fixed energy which is the expected behavior in case of a two-body decay.

3. *Uncertainty relations (5).* Starting from the relation

$$\langle (\Delta A)^2 \rangle \langle (\Delta B)^2 \rangle \geq \frac{\langle C^2 \rangle}{4}$$

with $[\hat{A}, \hat{B}] = iC$, demonstrate the uncertainty relation for position and momentum.

The commutator between $\hat{p}_x$ and $\hat{x}$ can be computed as follows:

$$[\hat{p}_x, \hat{x}]\Psi = (\hat{p}_x \hat{x} - \hat{x} \hat{p}_x)\Psi = -i\hbar \frac{\partial}{\partial x}(x\Psi) + ix\hbar \frac{\partial}{\partial x}(\Psi) = -i\hbar\Psi - ix\hbar \frac{\partial \Psi}{\partial x} + ix\hbar \frac{\partial \Psi}{\partial x}$$

and thus

$$[\hat{p}_x, \hat{x}] = -i\hbar.$$

One has thus

$$\delta x \, \delta p_x \geq \frac{\hbar}{2},$$

where we called $\delta A = \sqrt{\langle \Delta A^2 \rangle}$.

4. *Uncertainty principle and ground state of the harmonic oscillator.* Consider a particle of mass m subject to the harmonic potential

$$V(x) = \frac{1}{2}kx^2.$$

The energy of the ground state is nonzero:

$$E_0 = \frac{1}{2}\hbar\omega$$

with $\omega = \sqrt{k/m}$.

Does this satisfy the uncertainty principle? Comment.

Let us start by considering the energy of the quantum harmonic oscillator, which must be at least,

$$E = \frac{(\Delta p)^2}{2m} + \frac{1}{2}m\omega^2(\Delta x)^2 \tag{2.7}$$

where Δp and Δx is the momentum and the position uncertainty, respectively, and ω the frequency of the harmonic oscillator.

Using the uncertainty principle, $\Delta x \, \Delta p \sim \frac{\hbar}{2}$, one gets

$$E = \frac{(\hbar)^2}{8\,m(\Delta x)^2} + \frac{1}{2}m\omega^2(\Delta x)^2. \tag{2.8}$$

By finding the minimum of the above equation for the position uncertainty, i.e., making the derivative to Δx and setting it equal to zero, we discovery the solution,

$$\Delta x = \sqrt{\frac{\hbar}{2\,m\omega}}. \tag{2.9}$$

Substituting this solution in Eq. 2.8 we finally get,

$$E_0 = \frac{\hbar\omega}{2}. \tag{2.10}$$

This means that the minimum energy that the harmonic oscillator can have is different from zero. There are many physical systems that can be described by an harmonic oscillator, for instance atoms in a solid lattice. This result indicates that even at the absolute zero temperature, one will always find a vibrational state.

5. *Lifetime and width of the π^0 (18).* The lifetime of the π^0 meson is $\tau_\pi \simeq 0.085$ fs. What is the width of the π^0 (absolute, and relative to its mass)?

One has
$$\Gamma_\pi \sim \frac{\hbar}{\tau_\pi} \sim \frac{6.6 \times 10^{-7} \text{MeV fs}}{0.085 \text{ fs}} \sim 8 \text{ eV}/c^2$$

(this is a typical width of electromagnetic decaying particles).
The width relative the mass is

$$\Gamma_\pi / m_\pi \sim 5 \times 10^{-8} \,.$$

6. *Width and lifetime of $\rho(770)$ (19).* The width of the $\rho(770)$ meson is $\Gamma_\rho \simeq 150$ MeV/c^2. What is its lifetime?

$\Gamma_\rho \simeq 150$ MeV is the typical width of hadronic decays. One has

$$\tau_\rho \sim \frac{\hbar}{\Gamma_\rho} \sim \frac{6.6 \times 10^{-7} \text{MeV fs}}{150 \text{ MeV}} \sim 10^{-24} \text{ s}$$

($10^{-24} - 10^{-23}$ s is the typical lifetime of hadronically decaying particles).

7. *Decay chains.* A nucleus A decays a into a nucleus B with an average lifetime $\tau_A \simeq 2$ min. B decays into C with average life $\tau_B \simeq 5\,000$ s. At the beginning there are $N_{0A} = 30$ million nuclei of type A. Calculate the activity (number of disintegrations per second) f_B, in units of the initial sample of A nuclei, of B after 1.2 s and after 5 000 s. Describe the approximate behavior of the number of B nuclei in the limits $\tau_B \gg \tau_A$ and $\tau_A \gg \tau_B$, respectively.

Let us call $\lambda_A = 1/\tau_A$, $\lambda_B = 1/\tau_B$. For the number of A nuclei one has

$$\frac{dN_A}{dt} = -\lambda_A \implies N_A(t) = N_{0A} e^{-\lambda_A t} \,.$$

The number of B nuclei is instead given by the equation

$$\frac{dN_B}{dt} = -\lambda_B N_B(t) - \frac{dN_A}{dt} = -\lambda_B N_B(t) + N_{0A} e^{-\lambda_A t}$$

(the number of B nuclei increases due to the decay of A nuclei, and decreases due to its own decay). The last equation, solved, provides

$$N_B(t) = N_{0A} \frac{\lambda_A}{\lambda_B - \lambda_A} (e^{-\lambda_A t} - e^{-\lambda_B t}).$$

Inserting the numerical values one has

$$f_B(1.2\,\mathrm{s}) \simeq 0.63$$

$$f_B(5000\,\mathrm{s}) \simeq 0.37.$$

Note that, if $\tau_B \gg \tau_A \Longrightarrow \lambda_A \gg \lambda_B$, one has, for large t,

$$N_B(t) \simeq N_{0A} e^{-\lambda_B t}$$

(i.e., the B population is approximately just an exponential regulated by the time decay constant of B.)
If instead $\tau_A \gg \tau_B \Longrightarrow \lambda_B \gg \lambda_A$, one has, for large t,

$$N_B(t) \simeq N_{0A} \frac{\lambda_A}{\lambda_B} e^{-\lambda_A t}$$

(i.e., the B population is approximately proportional to the A population).

8. *Unstable nucleons and carbon dating.* The half-life of the carbon 14 isotope, ^{14}C, is 5730 years. The concentration of this isotope in the atmosphere is 10^{-12} of the stable isotope ^{12}C. What is the ratio of the two isotopes in an artefact made of wood in Athens, V Century B.C.?

Cosmic rays hit the atmosphere and produce showers of hadrons. Some of them are neutrons; they can induce the transition

$$n + {}^{14}N \rightarrow p + {}^{14}C.$$

Due to the almost stationary cosmic ray flux on Earth, the production of ^{14}C is constantly occurring with a fixed rate since a very long time, so the rate of ^{14}C to the stable isotope ^{12}C in the atmosphere is constant. From the moment of vegetal or animal death, the ^{14}C fraction in organic material decreases due to its radioactive decay.
From $t_{1/2}$ compute the lifetime τ and then $e^{-\tau/2500\,\mathrm{yr}}$.

9. *Cross-section at fixed target (2).* Consider a fixed target experiment with a monochromatic proton beam with a energy of 20 GeV and a 2 m length liquid hydrogen (H_2) target ($\rho = 60\,\mathrm{kg/m^3}$). In the detector placed just behind the target beam fluxes of 7×10^6 protons/s and 10^7 protons/s were measured respectively with the target full and empty. Determine the proton-proton total cross-section at this energy and its statistical error:

a. Without taking into account the attenuation of the beam inside the target;

b. Taking into account the attenuation of the beam inside the target.

The proton density in the target is:

$$\rho_p = \rho/m_p = 3.6 \times 10^{28} \, \text{m}^{-3} \, .$$

Due to interaction with the target protons, beam is lost. The loss rate is given by the relation:

$$d\phi/dx = -\phi\rho_p\sigma \, .$$

If we neglect the beam attenuation within the target, we have:

$$\Delta\phi = \phi_f - \phi_e = -\phi_e\rho_p\sigma L$$

$$\sigma = -\frac{\Delta\phi}{\phi_e} \frac{1}{\rho_p L} = 41.6 \, \text{mb} \, .$$

By properly integrating the flux loss rate we obtain instead

$$\frac{d\phi}{\phi} = -\rho_p\sigma dx \longrightarrow \ln(\phi_f/\phi_e) = -\rho_p\sigma L$$

$$\sigma = -\ln(\phi_f/\phi_e)\frac{1}{\rho_p L} = 49.5 \, \text{mb} \, .$$

10. *LHC collisions (3)*. The LHC running parameters in 2012 were, for a c.m. energy $\sqrt{s} = 8$ TeV: number of bunches = 1400; time interval between bunches = 50 ns; number protons per bunch = 1.1×10^{11}; beam width at the crossing point = 16 μm.

a. Determine the maximum instantaneous luminosity of the LHC in 2012.

b. Determine the number of interactions per collision ($\sigma_{pp} \sim 100$ mb).

c. As you probably heard, LHC found a particle called Higgs boson, that Leon Lederman called the "God particle" (a name the news like very much). If Higgs bosons are produced with a cross-section $\sigma_H \sim 21$ pb, determine the number of Higgs bosons decaying into 2 photons ($BR(H \rightarrow \gamma\gamma) \simeq 2.28 \times 10^{-3}$) which might have been produced in 2012 in the LHC, knowing that the integrated luminosity of the LHC (luminosity integrated over the time) during 2012 was around 20 fb^{-1}. Compare it to the real number of detected Higgs in this particular decay mode reported by the LHC collaborations (about 400). Discuss the difference.

a. The instantaneous luminosity can be computed using the following equation,

$$\mathscr{L} = \frac{f_{col} N_1 N_2}{A} \tag{2.11}$$

where f_{col} is the frequency at which the bunch cross (not the revolutions) and N_1 and N_2 are the number of particles in each bunch. In this problem $N_p \equiv N_1 = N_2 = 1.1 \times 10^{11}$ protons and $f_{col} = (50 \times 10^{-9})^{-1} \, \mathrm{s}^{-1} = 20 \, \mathrm{MHz}$. The quantity A is the transverse area of each bunch. Using a 2D-gaussian approximation it can be computed as $A = 4\pi\sigma^2 = 32 \times 10^{-6} \, \mathrm{cm}^2$, where σ is the width of the beam at the crossing point. Hence, the maximum instantaneous luminosity achieved by such an experiment is $\mathscr{L} = 7.3 \times 10^{33} \, \mathrm{cm}^{-2}\mathrm{s}^{-1} = 7.3 \, \mathrm{nb}^{-1}\mathrm{s}^{-1}$.

b. The number of collisions per second can be computed through $N_{col}/s = \mathscr{L}\sigma_{pp}$. As the bunches cross at a frequency f_{col} then the number of interactions per collisions is

$$N_{col} = \frac{\mathscr{L}\sigma_{pp}}{f_{col}} \simeq 37, \tag{2.12}$$

noting that $\sigma_{pp} = 10^{-25} \, \mathrm{cm}^2$.

c. The number of Higgs produced in 2012 can be computed via,

$$N_{\mathrm{Higgs}} = \underbrace{\mathscr{L} \Delta t}_{\mathscr{L}_{int}} \sigma_H B R_H = 958, \tag{2.13}$$

where it was used $\mathscr{L}_{int} = 20 \times 10^{39} \, \mathrm{cm}^{-2}$, $\sigma_H = 2.1 \times 10^{-36} \, \mathrm{cm}^2$ and $B R_H = 2.28 \times 10^{-3}$.

Comparing the value obtained in Eq. 2.13 with the one quoted in the problem ($N_{\mathrm{Higgs}} = 400$) one can see and only about 42% of the predicted Higgs were effectively measured. This difference is expected and is connected to acceptance and detector efficiencies, DAQ operation and even possible analysis cuts to mitigate the experimental background.

11. *Experimental determination of cross sections (4).* A thin (1.4 mg/cm^2) target made of ^{22}Na is bombarded with a beam of α particles of intensity $I = 5$ nA. A detector with area $A = 16$ cm^2 is placed at 1 m from the target perpendicular to the line between the detector and the target. The detector records $w = 45$ protons/s, independent of the angular position (θ, ϕ) of the detector. Find the cross section in mb for the ^{22}Na $+\alpha \rightarrow p + X$ reaction – also written as ^{22}Na(α, p).

The total reaction rate is given by

$$W = \sigma J N$$

where J is the beam flux (number of incoming particles per unit area per unit time), N the total number of targets and σ the reaction cross section.

We start by calculating W: we have a detector of area $A = 16$ cm^2 at a distance of $R = 100$ cm. The solid angle subtended by this detector is $\Delta\Omega = a/R^2 = 16/10^4 = 1.6 \times 10^{-3}$ steradians.

The relation holds

$$\frac{w}{W} = \frac{\Delta\Omega}{4\pi} \implies W = \frac{4\pi w}{\Delta\Omega} = 3.53 \times 10^5 \text{s}^{-1}.$$

Since $J = i/A$ and $N = nAD$, where n is the number density of the target Na atoms, D is the depth of the target, and i is the particle current (number of incident projectiles per unit time $i = I/(2 \times 1.6 \times 10^{-19})$ C/particle), one has

$$J N = (i/A)(nAD) = i(nD).$$

The mass density ρ of the target is given by $\rho D = mn D = 1.4$ mg/cm^2; the mass m of each target atom is about (each mol of ^{22}Na has a weight of 22 g) $m \simeq 22$ g$/6.02 \times 10^{23}$.

Finally

$$\sigma = \frac{W}{JN} = \frac{W}{inD} \simeq 5.91 \times 10^{-29} \text{m}^2 = 591 \text{ mb}.$$

12. *Classical electromagnetism is not a consistent theory (6)*. Consider two electrons at rest, and let r be the distance between them. The (repulsive) force between the two electrons is the electrostatic force

$$F = \frac{1}{4\pi\varepsilon_0} \frac{e^2}{r^2},$$

where e is the charge of the electron; it is directed along the line joining the two charges. But an observer is moving with a velocity v perpendicular to the line joining the two charges will measure also a magnetic force (still directed as F)

$$F' = \frac{1}{4\pi\varepsilon_0} \frac{e^2}{r^2} - \frac{\mu_0}{2\pi r} v^2 e^2 \neq F.$$

The expression of the force is thus different in the two frames of reference. But masses, charges and accelerations are invariant. Comment.

Electrodynamics is a Lorentz invariant theory, so that all quantities have to be transformed according to Lorentz transformations when we move from a given inertial system to another one. To show that in this frame no contradiction arises we will proceed as follows:

a. Starting from the rest frame of the two charges, transform first the electromagnetic field into the moving frame, and then calculate the Lorentz force acting on one of the charges;

b. *Correctly* transform the 3-force from the rest system of the two charges into the moving frame, by considering the transformation law for the associated 4-force.

We will see that the obtained results are consistent.

To simplify the expressions, we will call $E = e/(4\pi\varepsilon_0 r^2)$, so that $F = eE$, and we will set the speed of light $c \equiv 1$. Let the frame K be such that the charges are at rest in K, the x-axis contains the segment connecting the charges, the y-axis has the positive direction of the velocity v, and the z-axis is chosen to make a right-handed spatial frame. The frame K' is the one in which the axes are directed as in K, but the system is translating uniformly with respect to K in the $y = y'$ direction with velocity v. Quantities in K' will be identified by a prime, all other quantities are intended with respect to K, or do not change in the transformation from K to K'.

Part 1. In K, with the chosen coordinates, the electromagnetic field is described by the electromagnetic tensor

$$\mathscr{F}_{\mu\nu} = \begin{pmatrix} 0 & E & 0 & 0 \\ -E & 0 & 0 & 0 \\ 0 & 0 & 0 & 0 \\ 0 & 0 & 0 & 0 \end{pmatrix}.$$

The matrix associated to the Lorentz transformation from K into K' is given by

$$\Lambda^\alpha{}_\beta = \begin{pmatrix} \gamma & 0 & -v\gamma & 0 \\ 0 & 1 & 0 & 0 \\ -v\gamma & 0 & \gamma & 0 \\ 0 & 0 & 0 & 1 \end{pmatrix},$$

where $\gamma^2(1 - v^2) = 1$, $\gamma > 0$. Since $\mathscr{F}' = \Lambda\,\mathscr{F}\,\Lambda^T$ (where Λ^T is the transposed matrix of Λ), we obtain

$$\mathscr{F}'_{\mu\nu} = \begin{pmatrix} 0 & \gamma E & 0 & 0 \\ -\gamma E & 0 & v\gamma E & 0 \\ 0 & -v\gamma E & 0 & 0 \\ 0 & 0 & 0 & 0 \end{pmatrix},$$

that corresponds to an electric field $E'_{x'} = \gamma E$ and a magnetic field $B'_{z'} = \gamma v E$. According to the Lorentz force equation, we then have

$$\mathbf{F}' = e\left(\mathbf{E}' + \mathbf{v}' \times \mathbf{B}'\right)$$

or[1]

$$(F', 0, 0)' = e \left((\gamma E, 0, 0) + \begin{vmatrix} \hat{x}' & \hat{y}' & \hat{z}' \\ 0 & -v & 0 \\ 0 & 0 & \gamma v E \end{vmatrix} \right)$$

$$= e(\gamma E - \gamma v^2 E, 0, 0)' = (F/\gamma, 0, 0)'.$$

The magnitude of the 3-force in K' calculated in this way is then F/γ.

Part 2. We now first calculate the force in K and then transform the result in K'. For completeness we will work with the 4-force g^μ. In K the 4-force if given by

$$g^\mu = (\gamma \mathbf{F} \cdot \mathbf{v}, \gamma \mathbf{F}),$$

where $\mathbf{v} = \mathbf{0}$ and

$$\mathbf{F} = e (\mathbf{E} + \mathbf{v} \times \mathbf{B}) = (F, 0, 0).$$

Then $g^\mu = (0, F, 0, 0)$, as $\mathbf{v} = \mathbf{0}$ and $\gamma = 1$. We observe, incidentally, that this expression is consistent with

$$e \eta^{\mu\alpha} F_{\alpha\beta} u^\beta = \begin{pmatrix} 1 & 0 & 0 & 0 \\ 0 & -1 & 0 & 0 \\ 0 & 0 & -1 & 0 \\ 0 & 0 & 0 & -1 \end{pmatrix} \begin{pmatrix} 0 & E & 0 & 0 \\ -E & 0 & 0 & 0 \\ 0 & 0 & 0 & 0 \\ 0 & 0 & 0 & 0 \end{pmatrix} \begin{pmatrix} 1 \\ 0 \\ 0 \\ 0 \end{pmatrix}$$

$$= (0, qE, 0, 0) = (0, F, 0, 0)'.$$

We can now transform the components of the four force g^μ to find the components g'^μ in K'. We have $g' = \Lambda g$, or

$$g'^\mu = \Lambda^\mu{}_\alpha g^\alpha = \begin{pmatrix} \gamma & 0 & -v\gamma & 0 \\ 0 & 1 & 0 & 0 \\ -v\gamma & 0 & \gamma & 0 \\ 0 & 0 & 0 & 1 \end{pmatrix} \begin{pmatrix} 0 \\ F \\ 0 \\ 0 \end{pmatrix} = (0, \gamma F/\gamma, 0, 0)'.$$

We thus see that the three force in K' is $(F/\gamma, 0, 0)'$. Naturally, if we would have calculated the 3-force acting on the charge in K' by using $(i = x, y, z)$

---

[1]The quantities $\hat{x}'$, $\hat{y}'$, $\hat{z}'$ represent unit vectors in the direction of the spatial axis of K', and expressions like $(-, -, -)'$ represent spatial 3-vectors in K'.

$$g^{\prime i}/\gamma = \frac{e}{\gamma'} \, \eta^{iv} F'_{v\alpha} u'^{\alpha}$$

$$= \frac{e}{\gamma} \begin{pmatrix} 0 & -1 & 0 & 0 \\ 0 & 0 & -1 & 0 \\ 0 & 0 & 0 & -1 \end{pmatrix} \begin{pmatrix} 0 & \gamma E & 0 & 0 \\ -\gamma E & 0 & v\gamma E & 0 \\ 0 & -v\gamma E & 0 & 0 \\ 0 & 0 & 0 & 0 \end{pmatrix} \begin{pmatrix} \gamma \\ 0 \\ \gamma v \\ 0 \end{pmatrix}$$

$$= \begin{pmatrix} F/\gamma \\ 0 \\ 0 \end{pmatrix}, \tag{2.14}$$

we would have obtained the same result.

As the results in Part 1 and Part 2 are the same, we see that there is no inconsistency when the appropriate transformation laws are used to transform the various quantities between the two reference systems.

13. *Classical momentum is not conserved in special relativity (7).* Consider the completely inelastic collision of two particles, each of mass m, in their c.m. system (the two particles become one particle at rest after the collision). Now observe the same collision in the reference frame of one particle. What happens if you assume that the classical definition of momentum holds in relativity as well?

Let us call v the velocity of particle A in the c.m. reference frame; then the velocity of particle B is $v_B = -v$.

The total momentum p in the c.m. is, by definition (and by symmetry considerations), zero, both according to the classical and to the relativistic definitions of momentum, before and after the collision.

The reference frame in which particle B is at rest moves with velocity $-v$ with respect to the c.m. Thus, before the collision, the total classical momentum p' is

$$p'_{\text{classical, before}} = mv'_A + 0 = m\frac{2v}{1 + v^2/c^2} = \frac{2mv}{1 + v^2/c^2}.$$

After the collision one has

$$p'_{\text{classical, after}} = 2mv \neq p'_{\text{classical, before}}.$$

It is thus clear that in special relativity momentum cannot be equal to the product of the mass by the velocity.

14. *Energy is equivalent to mass (8).* How much more does a hot potato weigh than a cold one (in kg)?

Let us assume that the mass of a potato is $m \simeq 0.2\,\mathrm{kg}$ at room temperature ($20°$ C), and we heat it to the boiling point of water ($100°$ C). Approximating the specific heat of a potato with the specific heat of water ($c_s = 1\,\mathrm{cal\,kg^{-1}\,K^{-1}} \simeq 4200\,\mathrm{J\,kg^{-1}\,K^{-1}}$), the energy absorbed by the potato is

$$Q = mc_s \Delta T \simeq (4200 \times 0.2 \times 80)\, \text{J} \simeq 8 \times 10^4 \text{J}.$$

The increase in mass is thus

$$\Delta m = Q/c^2 \simeq 10^{-12} \text{kg}.$$

15. *Mandelstam variables (9)*. Demonstrate that, in the $1 + 2 \to 3 + 4$ scattering,

$$s + t + u = m_1^2 + m_2^2 + m_3^2 + m_4^2.$$

Let us start by noting that the square of the 4-momentum of a particle is simply the square of its mass $p_i^2 = E_i^2 - P_i^2 = m_i^2$, where E_i and P_i is the particle's energy and momentum, respectively, in any reference frame. Moreover, the energy-momentum conservation implies that,

$$p_1 + p_2 = p_3 + p_4 \tag{2.15}$$

So, knowing that

$$s = (p_1 + p_2)^2 = p_1^2 + p_2^2 + 2p_1 \cdot p_2$$
$$t = (p_1 - p_3)^2 = p_1^2 + p_3^2 - 2p_1 \cdot p_3$$
$$u = (p_1 - p_4)^2 = p_1^2 + p_4^2 - 2p_1 \cdot p_4$$

(the symbol $\cdot$ between two 4-vectors a and b stands for $a^\mu b_\mu$), and adding the three Lorentz invariant quantities we obtain

$$s + t + u = m_1^2 + m_2^2 + m_3^2 + m_4^2 + 2p_1^2 + 2p_1 \cdot p_2 - 2p_1 \cdot p_3 - 2p_1 \cdot p_4 \tag{2.16}$$

Finally, noting that the last four terms add up to zero using the four-momentum conservation, we get

$$2p_1^2 + 2p_1 \cdot p_2 - 2p_1 \cdot p_3 - 2p_1 \cdot p_4 = 2p_1 \cdot (p_1 + p_2 - p_3 - p_4) = 0 \tag{2.17}$$

and

$$s + t + u = m_1^2 + m_2^2 + m_3^2 + m_4^2. \tag{2.18}$$

16. *Photon conversion into $e^+ e^-$ (12)*. Consider the conversion of a photon in an electron-positron pair. Determine the minimal energy that the photon must have in order that this conversion would be possible if the photon is in presence of:

   a. A proton;
   b. An electron;

c. When no charged particle is around.

In the presence of a charged particle of mass M, considered to be initially at rest, the minimal energy that the photon must have is the energy required to produce a final state where all particles are at rest in the center-of-mass frame. The total 4-momenta in the initial and final state, p_μ^i and p_μ^f, expressed respectively in the laboratory and in the center-of-mass frames are:

$$p_\mu^{i,Lab} = (E_\gamma + M, \mathbf{P}_\gamma) \tag{2.19}$$

$$p_\mu^{f,CM} = (M + 2\, m_e, \mathbf{0}). \tag{2.20}$$

Since $p_\mu^i\, p^{i,\mu}$ and $p_\mu^f\, p^{f,\mu}$ are Lorentz invariants and 4-momentum is conserved, one has:

$$\left(p^{i,Lab}\right)^2 = \left(p^{i,CM}\right)^2 = \left(p^{f,CM}\right)^2 \tag{2.21}$$

yielding the relation:

$$(E_\gamma + M)^2 - P_\gamma^2 = (M + 2\, m_e)^2 \tag{2.22}$$

which leads to

$$E_\gamma^2 + M^2 + 2\, E_\gamma\, M - P_\gamma^2 = M^2 + 4\, m_e^2 + 4\, M\, m_e \tag{2.23}$$

and finally:

$$E_\gamma = 2\, m_e \left(1 + \frac{m_e}{M}\right). \tag{2.24}$$

a. For a spectator particle with mass $M \gg m_e$ one has:

$$E_\gamma \simeq 2\, m_e. \tag{2.25}$$

In particular, for the case of the conversion in presence of a proton, the minimal energy that the photon must have is just a fraction of about 5×10^{-4} above the mass of the electron-positron pair. In fact, for a fixed momentum transferred to the spectator particle, its gain in kinetic energy decreases as the mass increases. In the limit of a very large mass, where one can assume that the velocity is small:

$$T \simeq \frac{P^2}{2\, M}, \tag{2.26}$$

the spectator recoils in order to conserve the momentum, but carries only a small fraction of the energy of the converted photon and most of the available energy is converted into the masses of the electron and of the positron. Hence, as M increases the minimal energy that the photon must have approaches the limit $2\,m_e$.

b. In the case of the photon conversion in the presence of an electron, $M = m_e$, and one has:

$$E_\gamma = 4\,m_e \, . \tag{2.27}$$

c. Considering now the photon conversion in vacuum, the minimum energy of the photon is obtained by letting $M \to 0$. In this case, according to Eq. 2.24 we have $E_\gamma \to \infty$. A photon cannot convert in an electron-positron pair in the absence of a spectator charged particle, otherwise the total momentum would not be conserved. This can be visualised by placing ourselves in the center-of-mass frame of the electron-positron. Here we would see an initial photon converting into a final state with total momentum $\mathbf{P} = \mathbf{0}$, which would violate momentum conservation.

17. π^- *decay (13)*. Consider the decay of a flying π^- into $\mu^-\bar{\nu}_\mu$ and suppose that the μ^- was emitted along the flight line of flight of the π^-. Determine:

a. The energy and momentum of the μ^- and of the $\bar{\nu}_\mu$ in the π^- frame.
b. The energy and momentum of the μ^- and of the $\bar{\nu}_\mu$ in the laboratory frame, if the momentum $P_\pi^- = 100$ GeV/c).
c. Same as the previous question but considering now that was the $\bar{\nu}_\mu$ that was emitted along the flight line of the π^- .

---

a. Energy-momentum conservation in the π^- frame leads to:

$$m_\pi = E_\mu + E_\nu \tag{2.28}$$
$$0 = \mathbf{P}_\mu + \mathbf{P}_\nu \, . \tag{2.29}$$

Using the first equality and the relation between energy and momentum gives:

$$E_\mu^2 = (m_\pi - E_\nu)^2 = m_\pi^2 + E_\nu^2 - 2\,m_\pi\,E_\nu \tag{2.30}$$
$$E_\mu^2 = P_\mu^2 + m_\mu^2 \tag{2.31}$$

and, since

$$E_\nu = |\mathbf{P}_\nu| = |\mathbf{P}_\mu| \tag{2.32}$$

we get:

$$E_\nu = \frac{m_\pi^2 - m_\mu^2}{2\,m_\pi}\,. \tag{2.33}$$

Inserting E_ν in Eq. 2.28 we obtain:

$$E_\mu = \frac{m_\pi^2 + m_\mu^2}{2\,m_\pi}\,. \tag{2.34}$$

Taking the particle masses $m_\pi = 139.58\,\text{MeV}$ and $m_\mu = 105.66\,\text{MeV}$, gives:

$$E_\mu \simeq 109.78\,\text{MeV} \quad ; \quad E_\nu \simeq 29.80\,\text{MeV}\,. \tag{2.35}$$

Finally, the momenta are computed through Eq. 2.32.

b. Defining the z axis along the direction of flight of the π^-, the momentum vectors of the final state particles in the π^- frame are:

$$\mathbf{P}_\mu \equiv (0, 0, P_\mu) \tag{2.36}$$
$$\mathbf{P}_\nu = -\mathbf{P}_\mu \tag{2.37}$$

and the Lorentz transformation to the laboratory frame gives

$$E_\mu^{lab} = \gamma\left(E_\mu + \beta P_\mu\right) \quad ; \quad E_\nu^{lab} = \gamma\left(E_\nu - \beta P_\nu\right) \tag{2.38}$$
$$P_\mu^{lab} = \gamma\left(\beta E_\mu + P_\mu\right) \quad ; \quad P_\nu^{lab} = \gamma\left(\beta E_\nu - P_\nu\right) \tag{2.39}$$

where γ and β are the Lorentz boost and the velocity of the π^-:

$$\gamma = \frac{E_\pi}{m_\pi} \quad ; \quad \beta = \frac{P_\pi}{E_\pi} \tag{2.40}$$

or

$$\beta\gamma = \frac{P_\pi}{m_\pi} = 719 \quad ; \quad \gamma = \sqrt{(\beta\gamma)^2 + 1} \simeq \beta\gamma\,. \tag{2.41}$$

Note that from the above relations one gets $\beta \sim 1$ ($\beta \simeq 0.99999902$). Using the results for the energy and momentum of the μ^- and of the $\bar{\nu}_\mu$ in the π^- frame, Eqs. 2.32, 2.33 and 2.34, their energy and momentum in the laboratory frame can be written as:

$$E_\mu^{lab} = \frac{E_\pi}{2} \left[(1 + \beta) + \frac{m_\mu^2}{m_\pi^2} (1 - \beta) \right] \; ; \; E_\nu^{lab} = \gamma \, E_\nu \, (1 - \beta) \quad (2.42)$$

$$P_\mu^{lab} = \frac{E_\pi}{2} \left[(1 + \beta) + \frac{m_\mu^2}{m_\pi^2} (\beta - 1) \right] \; ; \; P_\nu^{lab} = -E_\nu^{lab} . \quad (2.43)$$

For $\beta \to 1$ the energy of the neutrino goes to zero and the muon energy approaches E_π. In particular this is the case for $P_\pi^- = 100$ GeV/c, see Eq. 2.41.

c. In this case we have

$$\mathbf{P}_\mu \equiv (0, 0, -P_\mu) \quad (2.44)$$
$$\mathbf{P}_\nu = -\mathbf{P}_\mu \quad (2.45)$$

and the Lorentz transformation to the laboratory frame yields,

$$E_\mu^{lab} = \gamma \left(E_\mu - \beta P_\mu \right) \; ; \; E_\nu^{lab} = \gamma \left(E_\nu + \beta P_\nu \right) \quad (2.46)$$
$$P_\mu^{lab} = \gamma \left(\beta E_\mu - P_\mu \right) \; ; \; P_\nu^{lab} = \gamma \left(\beta E_\nu + P_\nu \right) \quad (2.47)$$

or, using again the results from Eqs. 2.32, 2.33 and 2.34,

$$E_\mu^{lab} = \frac{E_\pi}{2} \left[(1 - \beta) + \frac{m_\mu^2}{m_\pi^2} (1 + \beta) \right] \; ; \; E_\nu^{lab} = \gamma \, E_\nu \, (1 + \beta) \quad (2.48)$$

$$P_\mu^{lab} = \frac{E_\pi}{2} \left[(\beta - 1) + \frac{m_\mu^2}{m_\pi^2} (1 + \beta) \right] \; ; \; P_\nu^{lab} = E_\nu^{lab} . \quad (2.49)$$

In the limit $\beta \to 1$ we have

$$E_\mu^{lab} \simeq E_\pi \frac{m_\mu^2}{m_\pi^2} \; ; \; E_\nu^{lab} \simeq 2 \, \gamma \, E_\nu \quad (2.50)$$

$$P_\mu^{lab} \simeq E_\mu^{lab} \; ; \; P_\nu^{lab} = E_\nu^{lab} . \quad (2.51)$$

It should be noted that the boost due to the large momentum of the π^- results in that the μ^- reversed its direction of motion (in the π^- frame it was emitted opposite to the flight line of the π^-). Also, in this case $P_\mu^{lab} \sim E_\mu^{lab}$, i.e. the muon mass is negligible in comparison with its energy.

18. π^0 *decay (14)*. Consider the decay of a π^0 into $\gamma\gamma$ (with pion momentum of 100 GeV/c). Determine:

   a. The minimal and the maximal angles that the two photons may have in the laboratory frame.

b. The probability of having one of the photons with an energy smaller than an arbitrary value E_0 in the laboratory frame.
c. Same as (a) but considering now that the decay of the π^0 is into e^+e^-.
d. The maximum momentum that the π^0 may have in order that the maximal angle in its decay into $\gamma\gamma$ and in e^+e^- would be the same.

---

a. Let us start by evaluating the decay in the CM reference frame (see Fig. 2.1). In this reference frame the π^0 is at rest and the two photons, that arise from the π^0 decay, have opposite directions such that the relation between their momentum is $\mathbf{p}_1 = -\mathbf{p}_2$. Since for photons, in natural units (NU), momentum is equal to energy, then

$$E^*_{\gamma 1} = E^*_{\gamma 2} = \frac{M_{\pi^0}}{2}. \tag{2.52}$$

In order to obtain the energy of the photons one needs to apply the Lorentz transformations

$$\begin{pmatrix} E \\ P \end{pmatrix} = \begin{pmatrix} \gamma & \gamma\beta \\ \gamma\beta & \gamma \end{pmatrix} \begin{pmatrix} E^* \\ P^* \end{pmatrix} \tag{2.53}$$

where (E, P) are the energy and the momentum of the particle in the laboratory reference frame while (E^*, P^*) are the same quantities in the CM frame.

Using Eq. 2.53 one can now transform the energy of the photons from the center-of-mass (CM) frame to the laboratory (LAB) frame,

$$E_{\gamma 1} = \gamma(E^*_{\gamma 1} + \beta E^*_{\gamma 1} \cos\theta^*)$$
$$E_{\gamma 2} = \gamma(E^*_{\gamma 2} - \beta E^*_{\gamma 2} \cos\theta^*).$$

The γ and β that relate reference frames can be obtained from

$$\gamma = \frac{E_\pi}{M_\pi} \quad ; \quad \beta = \frac{P_\pi}{E_\pi} \tag{2.54}$$

leading to

$$E_{\gamma 1} = \frac{E_\pi}{2}\left(1 + \frac{P_\pi}{E_\pi}\cos\theta^*\right)$$
$$E_{\gamma 2} = \frac{E_\pi}{2}\left(1 - \frac{P_\pi}{E_\pi}\cos\theta^*\right).$$

In order to determine the minimum and maximum angle between the photons in the laboratory frame one can start by identifying two limiting situations:

the photons directions are aligned with the boost direction; both photons are emitted transversely to the boost direction.

- Maximum angle: in the CM reference frame the angle between the photons is always $\alpha^* = 180°$. As the boost cannot reverse the momentum of the photon travelling against the boost direction, the angle that the photons will do in the laboratory (LAB) frame will continue to be $\alpha = 180°$. Furthermore, for $\theta^* = 0°$,

$$E_{\gamma 1} = \frac{E_\pi}{2} + \frac{P_\pi}{2} \qquad (2.55)$$

$$E_{\gamma 2} = \frac{E_\pi}{2} - \frac{P_\pi}{2} . \qquad (2.56)$$

- Minimum angle: this occurs when both photons are emitted transversely to the boost direction, i.e., $\cos \theta^* = 0$. To evaluate the minimum angle let us start by noting that the Lorentz transformation only alter physical quantities that have a non-zero projection along the boost axis. Therefore, the transverse momentum of γ_1 in the LAB is equal to its transverse momentum in the CM frame, leading to $P_{\gamma 1}^T = E_{\gamma 1}^*$. The longitudinal momentum (in the boost direction) can be obtained through the Lorentz transformations (Eq. 2.53),

$$P_{\gamma 1}^{\parallel} = \gamma \beta E_{\gamma 1}^* + \underbrace{\gamma P_{\gamma 1}^* \cos \theta^*}_{=0} \qquad (2.57)$$

Therefore, the minimum angle between the two photons is given by

$$\alpha = 2 \arctan \left(\frac{E_{\gamma 1}}{\gamma \beta E_{\gamma 1}} \right) \simeq 2 \frac{M_\pi}{P_\pi} \qquad (2.58)$$

where the approximation for small angles $\tan \theta \sim \theta$ was used; the factor of 2 comes from the symmetry of the system, i.e., both photons will have the same angle with the boost direction.

b. As seen in the previous problem

$$E_{\gamma 1} \in \left[\frac{E_\pi}{2} ; \frac{E_\pi}{2} + \frac{P_\pi}{2} \right]$$

$$E_{\gamma 2} \in \left[\frac{E_\pi}{2} - \frac{P_\pi}{2} ; \frac{E_\pi}{2} \right] .$$

Therefore, the photon that fulfils the problem condition is the one moving in the opposite direction to the boost, i.e., photon 2.

The photon energy in the LAB has to be smaller than E_0, and so

$$E_{\gamma 2} = \frac{E_\pi}{2}\left(1 - \frac{P_\pi}{E_\pi}\cos\theta^*\right) \leq E_0 \tag{2.59}$$

leading to the condition

$$\delta \equiv \cos\theta^* \geq \frac{E_\pi}{P_\pi} - \frac{2E_0}{P_\pi}. \tag{2.60}$$

The emission of the photons in the CM frame is isotropic. Thus, the probability of finding a photon with a given solid angle is uniform. As a consequence, we can write

$$\text{Prob} = \int_\delta^1 d\cos\theta^* = 1 - \delta. \tag{2.61}$$

Notice that the cosine of an angle is bounded between -1 and $+1$, which justifies the upper bound of the integral.

Substituting Eq. 2.60 in Eq. 2.61 and noticing that $E_\pi/P_\pi \simeq 1$ we finally obtain

$$\text{Prob} = \frac{2E_0}{P_\pi}. \tag{2.62}$$

c. In this problem we are in a situation similar to the one depicted in Fig. 2.1 but now photons 1 and 2 are substituted by the e^- and the e^+, respectively (this is an arbitrary choice). Like before we have two limiting situations when $\cos\theta^* = \pm 1$ and $\cos\theta^* = 0$. However, as now the particles produced during the decay are massive, their momentum might be reversed by the boost, even if they are produced backwards.

The condition for the particle to be emitted in the LAB frame, always in the direction of the boost, even if the particle is emitted in the opposite

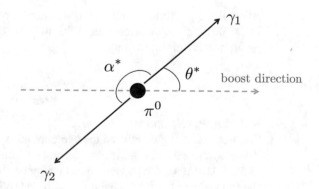

Fig. 2.1 Scheme of the π^0 decay into two photons in the CM reference frame

direction of the boost ($\cos\theta^* = 180°$), can be easily obtained from the Lorentz transformation

$$P_e = \gamma\beta E_e^* - \gamma P_e^* > 0. \tag{2.63}$$

This means that if the particle is always emitted forward in the LAB frame then the following condition has to be fulfilled:

$$\frac{P_e^*}{E_e^*} = \beta_e < \beta_\pi = \frac{P_\pi}{E_\pi}. \tag{2.64}$$

As the electron and the positron have the same mass, then in the CM reference frame, they will share the same amount of energy coming from the decay of the pion and so

$$E_e^* \equiv E_{e^-}^* = E_{e^+}^* = \frac{M_\pi}{2}. \tag{2.65}$$

Hence, using the relation $P = \sqrt{E^2 - m^2}$, it is possible evaluate numerically Eq. 2.64,

$$0.999973 < 0.99999(9). \tag{2.66}$$

The above result confirms that, in the LAB frame, all particles will be produced in the direction of the boost, independent of their production angle. Thus, in the LAB, the minimum angle is zero and will occur when $\cos\theta^* = \pm 1$.

Naturally, the maximum angle between the electron and the positron, α, will be in the opposite limiting situation, i.e. $\cos\theta^* = 0$. The procedure to compute this angle is the same used to evaluate the minimum angle in (a),

$$\alpha = 2\arctan\left(\frac{P_e^T}{P_e^\parallel}\right) \simeq 2\frac{\sqrt{(E_e^*)^2 - m_e^2}}{\gamma\beta E_e^*} \tag{2.67}$$

where the approximation $\tan\theta \sim \theta$ was used for small angles.

By neglecting the electron mass with respect to the pion mass, one gets finally for the minimum angle

$$\alpha \simeq \frac{1}{\beta\gamma} \tag{2.68}$$

with $\beta = P_\pi/E_\pi$ and $\gamma = E_\pi/m_\pi$.

d. The maximum angle between the two photons is $\alpha = 180°$, independent of the pion momentum. For the electron and positron to make an angle of $180°$ in the LAB frame, the particle must be aligned with the boost direction and the condition in Eq. 2.64 must not be fulfilled (otherwise the momentum of the particle emitted backwards would be reversed). Therefore, the condition is

$$\beta_e^* > \beta_{CM} \tag{2.69}$$

and thus

$$P_\pi < \sqrt{\frac{m_\pi^2 \left(\frac{P_\pi^2}{P_\pi^2 + m_\pi^2}\right)^2}{1 - \left(\frac{P_\pi^2}{P_\pi^2 + m_\pi^2}\right)^2}} \simeq 19\,\text{GeV}/c. \tag{2.70}$$

19. *Three-body decay (15)*. Consider the decay $K^+ \to \pi^+\pi^+\pi^-$. Determine the maximum value of the π^- momentum in the K^+ rest system. Do the same for the electron in the decay $n \to pe\bar{\nu}_e$.

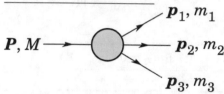

Defining $p_{ij} = p_i + p_j$ and $m_{ij}^2 = p_{ij}^2$, then $m_{12}^2 + m_{23}^2 + m_{13}^2 = M^2 + m_1^2 + m_2^2 + m_3^2$ and $m_{12}^2 = (P - p_3)^2 = M^2 + m_3^2 - 2ME_3$, where E_3 is the energy of particle 3 in the c.m. system (the rest frame of the M particle). In that frame, the momenta of the three decay particles lie in a plane.

The momenta in the laboratory system can then be recovered by giving three Euler angles defining the orientation of the final system relative to the initial particle.

The number of degrees of freedom for the process in the rest frame of the M particle is 2: 9 (number of independent variables in three four-momenta) minus 4 (conservation of 4-momentum) minus three (Euler angles which describe the orientation of a plane with a preferred direction, say the 3-momentum of particle 1, in space). These two degrees of freedom can be chosen, for example, to be the quantities (relativistically invariant) m_{23}^2 and m_{12}^2.

The plot of m_{23}^2 versus m_{12}^2 is called *Dalitz plot*, from the Australian physicist Richard Dalitz (1925–2006) who proposed it.

Due to kinematical constraints, not all points in the Dalitz plot, an example of which is given in Fig. 2.2, are allowed. In addition, the density of the points in a Dalitz plot can reveal some aspects of the dynamics of a decay process (for example an accumulation of points around a fixed value of m_{12}^2 can reveal that particles 1 and 2 are produced through a resonance).

The 3-momenta of particles 1 and 3 in the rest frame are given by

Fig. 2.2 Dalitz plot for the decay $D^+s \to \pi^+\pi^+\pi^-$ into two photons in the c.m. reference frame. The accumulation of points around $m(\pi^+\pi^-) \simeq 0.96\ \mathrm{GeV^2}$ evidences that many $\pi^+\pi^-$ pairs come from the resonance $f_0(980)$, which often mediates the D_s^+ decay. From The BaBar Collaboration, Phys. Rev. D 79 (2009) 032003

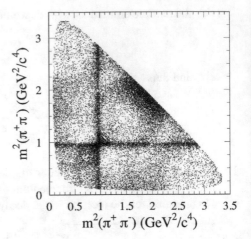

$$|\mathbf{p}_1|^* = \frac{\sqrt{(m_{12}^2 - (m_1 + m_2)^2)(m_{12}^2 - (m_1 - m_2)^2)}}{2m_{12}} \qquad (2.71)$$

$$|\mathbf{p}_3|^* = \frac{\sqrt{(M^2 - (m_{12} + m_3)^2)(M^2 - (m_{12} - m_3)^2)}}{2M}. \qquad (2.72)$$

Of interest are the maximum values of the three-momenta of the daughter particles in the rest frame of the mother particle. The maximum of, say, $|\mathbf{p}_3|^*$, is achieved when $m_{12} = m_1 + m_2$, i.e., when particles 1 and 2 have the same vector velocity in the rest frame of the decaying particle. A straightforward calculation then leads to

$$|\mathbf{p}_3|^*_{\mathrm{max}} = \frac{\sqrt{(M^2 - (m_1 + m_2 + m_3)^2)(M^2 - (m_1 + m_2 - m_3)^2)}}{2M}. \qquad (2.73)$$

In the case of the decay $K^+ \to \pi^+\pi^+\pi^-$ one has:

$$|\mathbf{p}_\pi|^*_{\mathrm{max}} = \frac{\sqrt{(M_K^2 - 9m_\pi^2)(M_K^2 - m_\pi^2)}}{2M_K} \simeq 125\ \mathrm{MeV}/c.$$

In the case of the decay $n \to pe^-\bar{\nu}_e$ one has:

$$|\mathbf{p}_e|^*_{\mathrm{max}} = \frac{\sqrt{(M_n^2 - (m_p + m_{\bar{\nu}} + m_e)^2)(M_n^2 - (m_p + m_{\bar{\nu}} - m_e)^2)}}{2M_n} \simeq 1.515\ \mathrm{MeV}/c.$$

(notice also that for a nucleus at rest the c.m. system corresponds to the laboratory system). The maximum value depends on the presently unknown mass of the (almost) invisible neutrino.

In addition, the distribution of m_{12} values possesses an end-point or maximum value at $m_{12} = M - m_3$. This can be used to constrain the mass difference of a parent particle and an invisible decay product.

20. *A classical model for the electron (16).* Suppose we interpret the electron as a classical solid sphere of radius R and mass m, spinning with angular momentum $\hbar/2$. What is the speed, v_p of a point on its "equator"? Experimentally, it is known that R is less than $R_{exp} = 10^{-18}$ m. What is the corresponding equatorial speed? What do you conclude from this result?

At the end of the XIX century, scientists built a classical model for the electron, and computed the electron radius under this assumption.

A classical electron radius length can be motivated by considering the energy necessary to assemble an amount of charge e into a sphere of a radius R. The electrostatic potential at a distance r from a charge q is

$$V(r) = \frac{1}{4\pi\varepsilon_0}\frac{q}{r}.$$

To bring an additional amount of charge dq from infinity necessitates putting energy into the system

$$dU = V(r).$$

If the sphere is assumed to have constant charge density, ρ, then

$$q = \rho\frac{4}{3}\, dq = \rho 4\pi r^2 dr.$$

Performing the integration from $r = 0$ up to a final radius R to assemble the electron's charge e leads to the expression for the total energy

$$U = \frac{1}{4\pi\varepsilon_0}\frac{3}{5}\frac{e^2}{R}.$$

This is called the electrostatic self-energy of the object. If U is set equal to the relativistic mass-energy of the electron, $m_e c^2$, and the numerical factor 3/5 is ignored as being specific to the special case of a uniform charge density, one obtains for the classical radius of the electron

$$R = \frac{1}{4\pi\varepsilon_0}\frac{e^2}{m_e c^2} \simeq 2.82 \times 10^{-15} \text{m}.$$

The angular momentum of the rotating classical electron is

$$I\omega = \frac{\hbar}{2} \implies \omega = \frac{\hbar}{2I} \simeq \frac{\hbar}{2m_e R^2}$$

where $I \simeq m_e R^2$ is the momentum of inertia (again we omit form-specific factors), and ω is the angular velocity.

The speed v_p of a point of the equator of the "classical" electron would be:

$$v_p = \omega R \simeq \frac{\hbar}{2m_e R} \simeq 1.3 \times 10^{11} \text{m/s} \simeq 430c \,.$$

From the above expression, if we use instead of the radius computed from the electrostatic energy the experimental limit R_{exp}, we obtain

$$v_p = \omega R_{\text{exp}} \simeq \frac{\hbar}{2m_e R_{\text{exp}}} > 1.2 \times 10^6 c \,.$$

The classical model of the electron is thus inconsistent with the fact that superluminal velocities are not allowed.

21. *The Ω^- baryon.* The Ω^- baryon was discovered in the reaction $K^- p \rightarrow K^+ K^0$.

   a. Knowing that the proton was in the LAB reference frame, determine the minimum energy of the K^- so that the above reaction is possible. Present the answer as a function of the masses of all involved particles.
   b. Consider now that the K^0 is produced along the beam line with a velocity of $\beta = 0.8$ and that it decays after traveling a distance L in two neutral pions.
      i. What is the distribution of L in the LAB frame for 100 decays.
      ii. Determine the maximum angle that the pions can do with the beam line direction, in the LAB reference frame.

---

   a. In the LAB frame we have before the interaction,

   $$p_\mu^{i,LAB} = (E_{K^-} + m_p, \mathbf{P}_{K^-}). \tag{2.74}$$

   The minimum energy of the K^- occurs when all the particle are produced at rest. As such, one can write in the CM reference frame,

   $$p_\mu^{f,CM} = (m_{\Omega^-} + m_{K^-} + m_{K^0}, \mathbf{0}). \tag{2.75}$$

   Computing the Lorentz invariant quantity $s = p_\mu p^\mu$, we can relate the two 4-vectors,

   $$(E_{K^-} - m_p)^2 - (\mathbf{P}_{K^-})^2 = (m_{\Omega^-} + m_{K^-} + m_{K^0})^2 \tag{2.76}$$

   from which one can extract the minimum energy of the kaon,

$$E_{K^-} = \frac{(m_{\Omega^-} + m_{K^-} + m_{K^0})^2 - m_{K^-}^2 - m_p^2}{2m_p} \simeq 3183 \, \text{MeV}. \quad (2.77)$$

b. i. Since we are dealing with a decay we have that

$$\frac{dN}{dt} = -\frac{N}{t_m} \implies N(t) = N_0 e^{-t/t_m} \quad (2.78)$$

where t_m is the mean lifetime of the decaying particle in the LAB frame. Using the Lorentz transformations we can easily relate it with the lifetime in its proper reference frame, τ, with t_m, obtaining $t_m = \gamma \tau$.

The K^0 will travel a distance L with a velocity βc and so the time that it will take to cover a distance x is $t = x/\beta c$. Hence, Eq. 2.78 becomes

$$N(L) = N_0 e^{-\frac{L}{\gamma \beta c \tau}} \quad (2.79)$$

with $\beta = 0.8$ and $\gamma = \left(\sqrt{1 - \beta^2}\right)^{-1}$. To find τ one should note that the meson K^0 is a combined state of K_S^0 and K_L^0, which have decay times substantially different. However, the decay channel $K^0 \to \pi^0 \pi^0$ is dominated by K_S^0 and consequently $\tau(K^0 \to \pi^0 \pi^0) = \tau(K_S^0) = 0.89 \times 10^{-10}$ s. From Eq. 2.79 one should expect to measure, after detecting 100 K^0 decays, an exponential distribution with slope $(\gamma \beta c \tau)^{-1}$.

ii. First let us start by noting that we are in the presence of a 2 body decay in which the decay products have the same mass. Consequently, in the CM reference frame the energy of the emerging pions will be $E_\pi^* = m_{K^0}$. For simplicity lets address the pions as (1) and (2). In the CM pion (1) can be produced in any direction provided that pion (2) is emitted in the opposite direction (carrying the same absolute value in momentum). To find the maximum angle between the pions in the LAB frame one should first test if the boost applied is sufficiently high to reserve the pion momentum when emitted backward with respect with the K^0 travelling line. From the Lorentz transformations for the particle momentum it can be easily shown that the particle momentum gets reversed only if the particle velocity in the CM, β_π^* is smaller than the CM frame velocity with respect to the LAB frame, β_{CM}, i.e. the velocity of the K^0 in the LAB. From the problem we know that $\beta_{CM} = 0.8$ while β_π^* can be computed as

$$\beta_\pi^* = \frac{P_\pi^*}{E_\pi^*} = \frac{1}{m_K}\sqrt{m_K^2 - 4m_\pi^2} \simeq 0.84. \quad (2.80)$$

Since $\beta_\pi^* > \beta_{CM}$ one can conclude that a particle emitted backward in the CM reference frame continues to travel backward in the LAB and consequently the maximum angle that the two pions can do in the LAB is 180°.

22. *Antiproton capture.* Antiprotons can be captured at rest in deuterium via the reaction $pd \rightarrow n\pi^0$. Determine:

- The deuterium binding energy;
- The total energy of the emitted pion.

The masses of the involved particles are: $m_p \simeq 938.3$ MeV, $m_p \simeq 939.6$ MeV, $m_d \simeq 1875.6$ MeV, $m_{\pi^0} \simeq 140$ MeV, The binding energy is thus $BE \simeq (m_p + m_n) - m_d \simeq 2.3$ MeV.

$$E - \pi = \frac{s - m_n^2 - m_\pi^2}{2\sqrt{s}} \simeq 1253 \text{ MeV}.$$

23. *Invariant flux (17).* In a collision between two particles a and b the incident flux is given by $F = 4|\mathbf{v_a} - \mathbf{v_b}| E_a E_b$ where $\mathbf{v_a}$, $\mathbf{v_b}$, E_a and E_b are respectively the vectorial speeds and the energies of particles a and b.

   a. Verify that the above formula is equivalent to: $F = 4\sqrt{(P_a P_b)^2 - (m_a m_b)^2}$ where P_a and P_b are respectively the four-vectors of particles a and b, and m_a and m_b their masses.
   b. Obtain expression of the flux in the center-of-mass and in the laboratory reference frames.

---

   a. We start from the definition of the four-momentum $p^\mu = (E/c, \mathbf{p})$, or $p_\mu = (E/c, -\mathbf{p})$, and calculate the product $p_a p_b = p_a^\mu p_{\mu b} = (E_a E_b)/c^2 - \mathbf{p_a p_b}$. Squaring this term, and using the relativistic relations $p = \gamma m\mathbf{v}$ and $E = \gamma mc^2$, where γ is the Lorentz factor, we recover the desired expression. F is a Lorentz's invariant because it is a product of two four-vectors.
   b. In the CM frame, $\mathbf{p_b} = -\mathbf{p_a}$, from what follows, after some algebra, that $F = 4c^2 \mathbf{p_a}(E_{a,CM} + E_{b,CM})$.
   Introducing the Mandelstam variable $s = (p_a + p_b)^2$, which is a Lorentz's invariant, and rewriting, we get: $F = 4|\mathbf{p_a}|\sqrt{s}$, making $c = 1$.
   In the laboratory frame, let us start from the expression: $F = (4/c^3) E_a E_b |\mathbf{v_a} - \mathbf{v_b}|$, where $\mathbf{v_a} - \mathbf{v_b} = (p_a/\gamma_a m_a - p_b/\gamma_b m_b)\mathbf{i}$. Thus, $F = 4/c(p_a E_b - p_b E_a)$.

24. *Classical Schwarzschild radius for a Black Hole (20).* Compute the radius of a spherical planet of mass M for which the escape velocity is equal to c, both in SI and solar mass units.

Let us start by imagine that we have an object at the surface of the planet which will be through with a velocity high enough to escape the planet's gravitational attraction. Using energy conservation it can be easily seen that it can only escape if its kinematical energy is greater than the gravitic potential. Hence, in classical Newtonian physics, we can right the following limiting condition:

$$\frac{GMm}{r} = \frac{1}{2}mv^2, \tag{2.81}$$

where M is the mass of the planet, m of the object, r the radius of the planet and v the initial velocity of the object.

Inverting 2.81 to get r and putting $v = c$ one obtains finally that the radius for which light cannot escape the planet gravity would be in S.I.,

$$r_S = \frac{2GM}{c^2} \text{ m.} \tag{2.82}$$

and in solar masses $M/M_\odot$,

$$r_S = \frac{2GM_\odot}{c^2}\left(\frac{M}{M_\odot}\right) \simeq 3\,\text{km}\left(\frac{M}{M_\odot}\right). \tag{2.83}$$

25. *Natural Units and everyday life quantities (21).* Determine in natural units:

    a. Your own dimensions (height, weight, mass, age).
    b. The mean lifetime of the muon ($\tau_\mu = 2.2\,\mu s$).

In NU the speed of light c and the reduced Planck constant are set to be equal to 1. Of the quantities with different dimensions that can be formed with time, length, and mass, the two conditions $c \equiv 1$ and $\hbar \equiv 1$ allow for one more quantity only to be chosen: in NU this is energy, which is expressed in multiples of the electron Volt (eV). As

$$[\text{Energy}] = [M] \times [L]^2 \times [T]^{-2}$$
$$[c] = [L] \times [T]^{-1}$$
$$[\hbar] = [M] \times [L]^2 \times [T]^{-1},$$

we have

$$[T] = [\hbar] \times [\text{Energy}]^{-1}$$
$$[L] = [\hbar] \times [c] \times [\text{Energy}]^{-1}$$
$$[M] = [c]^{-2} \times [\text{Energy}].$$

Remembering $\text{Energy}_J = 0.62 \times 10^{19} \times \text{Energy}_{eV}$, the equations above provide the conversion factors for the time, length and mass units from the SI system to the NU system. Indeed

$$1s = (0.62 \times 10^{19})^{-1} \times (1.05 \times 10^{-34})^{-1}eV^{-1} = 0.15 \times 10^{16}eV^{-1}$$
$$1m = (0.62 \times 10^{19})^{-1} \times (1.05 \times 10^{-34})^{-1} \times (3 \times 10^{8})^{-1}eV^{-1} = 0.5 \times 10^{7}eV^{-1}$$
$$1kg = 0.62 \times 10^{19} \times (3 \times 10^{8})^{2}eV = 0.56 \times 10^{36}eV.$$

Then, an age of $44yr = 44 \times 365 \times 24 \times 60 \times 60\,s = 1\,387\,584\,000$ s corresponds to about $2.1 \times 10^{24}eV^{-1}$. A mass of $80\,kg$ corresponds to about $4.5 \times 10^{37}eV$. An height of $1.8\,m$ corresponds to about $0.9 \times 10^{7}eV^{-1}$.
By using the same approach as above, in natural units we also have

$$[\text{Force}] = [\text{Energy}]^{2} \times [c]^{-1} \times [\hbar]^{-1}.$$

The force will then be measured in eV^{2}, and

$$1N = (0.62 \times 10^{19})^{2} \times (1.05 \times 10^{-34}) \times (3 \times 10^{8})$$
$$= 1.2 \times 10^{12}eV^{2}.$$

Then, a weight of 780 N corresponds to about 9.4×10^{14} eV^{2}.

26. *Natural Units and coupling constants (22).* In NU the expression of the muon lifetime is given by:

$$\tau_\mu = \frac{192\pi^{3}}{G_F^{2}\, m_\mu^{5}}$$

where G_F is the Fermi constant.

a. Is the Fermi constant dimensionless? If not compute its dimension in NU and in SI.
b. Obtain the conversion factor for transforming G_F from SI to NU.

In natural units the Fermi constant has dimensions of eV^{-2}. This is because from the expression above

$$eV^{-1} = [G_F]^{-2} \times eV^{-5} \quad \Rightarrow \quad [G_F]^{2} = eV^{-4} \quad \Rightarrow \quad [G_F] = eV^{-2}.$$

In SI units, the relation between G_F, m_μ, τ_μ and the fundamental constants c and $\hbar$ can be written as

$$\frac{G_F}{(\hbar c)^3} = \sqrt{\frac{\hbar}{\tau_\mu} \frac{192\pi^3}{(m_\mu c^2)^5}}.$$

As $[\hbar/\tau_\mu] = [m_\mu c^2] = [\text{Energy}]$, we see that $[G_F/(\hbar c)] = [\text{Energy}]^{-2} = [\text{J}]^{-2}$ in SI units. By replacing the values of the constants we obtain

$$\frac{G_F}{(\hbar c)^3} = 4.5 \times 10^{14} \text{J}^{-2} \quad \text{and} \quad G_F = 1.4 \times 10^{-62} \text{J m}^3.$$

In natural units we have $G_F = 1.16 \times 10^{-23} \text{eV}^{-2}$. The conversion constant from SI to NU is then about $2.6 \times 10^{-38} \text{J}^2\text{eV}^{-2}$.

Chapter 3
Cosmic Rays and the Development of Particle Physics

1. *The measurement by Hess (1).* Discuss why radioactivity decreases with elevation up to some 1000 m, and then increases. Can you make a model? This was the subject of the thesis by Schrödinger in Wien in the beginning of XX century.

Ionizing radiation must carry more than 10 eV in energy in order to be able to ionize atoms and molecules. The level of radioactivity decreases with elevation up to about 1000 m because in these altitudes the sources of ionising radiation are mostly on ground. This can be modelled as $e^{-h/\lambda}$. There is an inflection on the profile density of ionising particles in the atmosphere, when radiation from the soil attains a minimum. After this height, radiation begins to increase with altitude because of the contribution of cosmic rays to ionisation. Therefore, the particle flux can be written as

$$F = Ae^{-h/\lambda} + Bh + c$$

where h is the height measured from ground level, λ a parameter related to atmospheric attenuation, the other terms being model constants.

2. *Klein-Gordon equation (2).* Show that in the non-relativistic limit $E \simeq mc^2$ the positive energy solutions Ψ of the Klein-Gordon equation can be written in the form

$$\Psi(\mathbf{r}, t) \simeq \Phi(\mathbf{r}, t)e^{-\frac{mc^2}{\hbar}t},$$

where Φ satisfies the Schrödinger equation.

Let us define $\mu = mc/\hbar$ and $x^\mu = (x^0, \mathbf{x})$, with $x^0 = ct$ and $\mathbf{x} = (x^1, x^2, x^3)$, and,

$$\Psi(\mathbf{x}, t) = \Phi(\mathbf{x}, t)e^{-i\mu x^0}.$$

© Springer Nature Switzerland AG 2021

A. De Angelis et al., *Particle and Astroparticle Physics*, Undergraduate Lecture Notes in Physics, https://doi.org/10.1007/978-3-030-73116-8_3

We start by calculating the first partial derivatives of Ψ, expressed in the form above, with respect to all its arguments.

$$\partial_0 \Psi = (\partial_0 \Phi)e^{-i\mu x^0} - i\mu \Phi e^{-i\mu x^0} = (\partial_0 \Phi)e^{-i\mu x^0} - i\mu \Psi$$
$$\partial_i \Psi = (\partial_i \Phi)e^{-i\mu x^0}.$$

We now move to the second, non mixed, partial derivatives

$$\partial_0^2 = (\partial_0^2 \Phi)e^{-i\mu x^0} - i\mu(\partial_0 \Phi)e^{-i\mu x^0} - i\mu \partial_0 \Psi$$
$$= (\partial_0^2 \Phi)e^{-i\mu x^0} - 2i\mu(\partial_0 \Phi)e^{-i\mu x^0} - \mu^2 \Phi e^{-i\mu x^0}$$
$$\Delta\Psi = (\Delta\Phi)e^{-i\mu x^0}.$$

We can now use the above results inside the Klein-Gordon equation,

$$(\partial_0^2 - \Delta + \mu^2)\Psi = 0,$$

to obtain, after simplification of the common factor $\exp(-i\mu x^0)$,

$$\partial_0^2 \Phi - 2i\mu\partial_0 \Phi - \Delta\Phi = 0.$$

By rewriting everything in terms of the non rescaled quantities t and m, we obtain

$$\partial_t^2 \Phi - \frac{2imc^2}{\hbar}\partial_t \Phi - c^2 \Delta\Phi = 0. \tag{3.1}$$

Let us now assume that the first term can be neglected with respect to the other two. The above equation then becomes

$$-\frac{\hbar^2}{2m}\Delta\Phi \simeq i\hbar\partial_t \Phi,$$

which is nothing but Schrödinger equation for $\Phi(\mathbf{x}, t)$.

We now show that for the positive energy solutions under the condition $E \simeq mc^2$ it is reasonable to drop the first term in (3.1). To this end let us consider an *ansatz* for the solution of the Klein-Gordon equation in the form

$$\Psi(\mathbf{x}, t) \sim e^{-ip_\mu x^\mu}.$$

By substitution into the Klein-Gordon equation we obtain that the above *ansatz* is a solution if $-p_\nu p^\nu + \mu^2 = 0$, i.e. if

$$p_0^2 = \mu^2 + \mathbf{p}^2.$$

Positive energy solutions are those for which

$$p_0 = (\mu^2 + \mathbf{p}^2)^{1/2} = \mu \left(1 + \frac{\mathbf{p}^2}{\mu^2}\right)^{1/2};$$

in the limit in which $\mathbf{p}^2 \ll \mu^2$, we can write

$$p_0 \simeq \mu \left(1 + \frac{\mathbf{p}^2}{2\mu^2}\right) = \mu + \frac{\mathbf{p}^2}{2\mu}.$$

Our ansatz for Ψ above becomes then

$$\Psi(\mathbf{x}, t) \sim e^{-i\mu x^0} e^{-i[\mathbf{p}^2 x^0/(2\mu) + \mathbf{p}\cdot\mathbf{x}]},$$

where, with reference to the first part of the proof, it is now possible to identify

$$\Phi(\mathbf{x}, t) \sim e^{-i[\mathbf{p}^2 x^0/(2\mu) + \mathbf{p}\cdot\mathbf{x}]}.$$

It is now just a matter to calculate

$$\partial_0^2 \Phi(\mathbf{x}, t) \sim - \left(\frac{\mathbf{p}^2}{2\mu}\right)^2 \Phi(\mathbf{x}, t),$$
$$-2i\mu\partial_0 \Phi(\mathbf{x}, t) \sim -\mathbf{p}^2 \Phi(\mathbf{x}, t),$$
$$\Delta\Phi(\mathbf{x}, t) \sim -\mathbf{p}^2 \Phi(\mathbf{x}, t),$$

to see that under the approximation that we are considering the first term is suppressed with respect to the other two. The approximation that we did dropping the second order time derivatives in (3.1) is then justified under the same assumptions.

3. *Production of antiprotons at accelerators (3), (2.11), (5.1).* The total number of nucleons minus the total number of antinucleons is constant in a reaction— you can create nucleon-antinucleon pairs. What is the minimum energy (i.e., the kinematic threshold) of a proton hitting a proton at rest to generate an antiproton?

The reaction involving the minimal number of new particles while assuring charge and baryon number conservation is:

$$pp \to ppp\bar{p}.$$

The minimum energy is obtained when the antiproton and the three protons are produced at rest in the center-of-mass frame, i.e.:

$$\sqrt{s} = 4m_p, \quad s = 16m_p^2$$

(proton and antiproton have the same mass).

In the laboratory frame, the value of s before the reaction is

$$s = (p_1 + p_2)^2 = 2\,m_p^2 + 2E_p m_p \, .$$

As s is conserved in the reaction and is also Lorentz-invariant, we can equate the two expressions, obtaining:

$$16 m_p^2 = 2\,m_p^2 + 2E_p m_p \, ;$$

$$E_p = 7 m_p \simeq 6.57 \, \text{GeV} \, .$$

Anti-protons were first produced in laboratory in 1955, in proton-proton fixed target collisions at an accelerator called the Bevatron (it was named for its ability to impart energies of billions of eV, i.e., Billions of eV Synchrotron), located at Lawrence Berkeley National Laboratory, US. The discovery resulted in the 1959 Nobel Prize in physics for Emilio Segrè and Owen Chamberlain.

4. *Fermi maximum accelerator (4).* According to Enrico Fermi, the ultimate human accelerator, the "Globatron", would be built around 1994 encircling the entire Earth and attaining energy of around 5000 TeV (with an estimated cost of 170 million US1954 dollars...). Discuss the parameters of such an accelerator.

For the sake of reference, the circumference of the LHC is 27 km, using 8.3 T magnets. The corresponding proton-proton center-of-mass energy achieved is 14 TeV.

The CM energy that a proton-proton collider can achieve is related to its perimeter and the magnetic field strength of its bending magnets. Fermi assumed a magnetic field of 2 T, which would allow obtaining a maximum energy of 5 PeV, comparable to that of CR at the "knee" of the spectrum. This "Globatron", going around Earth's equator would be some 40 000 km long. Using the same magnetic field as in the LHC, the Globatron would achieve energies of about 20 PeV.

5. *Cosmic pions and muons (5).* Pions and muons are produced in the high atmosphere, at a height of some 10 km above sea level, as a result of hadronic interactions from the collisions of cosmic rays with atmospheric nuclei. Compute the energy at which charged pions and muons respectively must be produced to reach in average the Earth's surface.

You can find the masses of the lifetimes of pions and muons in your Particle Data Booklet.

An unstable particle produced in the high atmosphere can only reach the Earth's surface if the time of flight, t, is smaller than the average lifetime in the Earth reference frame, τ. This condition imposes that

$$\frac{L}{c} = \beta \tau = \beta \gamma \tau_0 \tag{3.2}$$

with τ_0 the average lifetime in the rest frame. Given that

$$\beta \gamma = \sqrt{\gamma^2 - 1}, \tag{3.3}$$

one has

$$\gamma = \sqrt{\left(\frac{L}{c \tau_0}\right)^2 + 1}. \tag{3.4}$$

Taking

$$m_\pi = 139.57 \, \text{MeV}/c^2 \;\; ; \;\; c \tau_\pi = 7.8 \, \text{m} \tag{3.5}$$
$$m_\mu = 105.66 \, \text{MeV}/c^2 \;\; ; \;\; c \tau_\mu = 658.6 \, \text{m} \tag{3.6}$$

the minimum energy at which charged pions and muons must be produced at a height of 10 km above sea level to reach in average the Earth's surface is:

$$E_\pi \simeq 180 \, \text{GeV} \tag{3.7}$$
$$E_\mu \simeq 1.6 \, \text{GeV}. \tag{3.8}$$

Given that the pion and muon masses are of the same order, they have identical boost for the same production energy. Hence, in this case it is the difference in the lifetimes that dictates the distance they can travel before decaying. In fact the minimum energies computed above differ by a factor of ~ 100, which is approximately the factor between the pion and muon lifetimes.

6. *Geomagnetic cutoff.* The Earth's magnetic field shields us from cosmic rays below a given value of momentum, times the charge. The geomagnetic cutoff varies with geomagnetic latitude. Compute an expression for the minimum energy of cosmic rays able to penetrate the Earth's magnetic field and to reach sea level as a function of the momentum M of Earth's magnetic dipole, of the Earth radius, and of latitude. Comment on the East-West effect, and why it is more visible close to the equator.

To check if a particle can cross the Earth's magnetic field and reach its atmosphere one can start by asking if there is a trajectory starting from the point considered on Earth that could reach $r = \infty$. Naturally, for very small energies this limit is not validity as the particle trajectory can be deflected back to Earth.

Let us start, for simplicity, to consider that the Earth magnetic field is generated by a dipole which intensity decrease with the distance to the center of the Earth. In these circumstances, a particle that has a momentum large enough to describe a circular uniform orbit around the planet, in the equatorial plane, will be able to escape if its momentum directly is tilted to the out space.

In these conditions one can write for a particle with charge q

$$qv \times B = m\frac{v^2}{r} \tag{3.9}$$

where the Lorentz force is said to be equal to the centripetal force. The Earth magnetic field is induced by the Earth magnetic momentum $\mathcal{M}$,

$$B = \frac{\mu_0}{4\pi}\frac{\mathcal{M}}{r^3}. \tag{3.10}$$

Taking $r = R_\oplus = 6.38 \times 10^6$ m and knowing that the Earth magnetic field intensity at the equator is roughly $B = 0.307 \times 10^{-4}$ T, from Eq. 3.10 we get $\mathcal{M} = 7.94 \times 10^{22}$ Am^2. Combining Eqs. 3.9 and 3.10 we obtain the following equation for the radius of the orbit,

$$r = \sqrt{\frac{\mu_0}{4\pi}\frac{q\mathcal{M}}{p}}, \tag{3.11}$$

where $p = mv$ is the momentum of the particle. Hence, from the previous expression, we can finally write a formula for the minimum rigidity of a particle for it to reach the Earth atmosphere,

$$\frac{p}{q} = \frac{\mu_0}{4\pi}\frac{\mathcal{M}}{r_\oplus^2}. \tag{3.12}$$

Using the previously stated value for $\mathcal{M}$ and considering that the particle has an orbit of $r = R_\oplus$, the minimum allowed closed orbit, we obtain a minimum electromagnetic rigidity of,

$$R \sim 5.96 \times 10^7 \text{ V} = 59.6 \text{ GV}. \tag{3.13}$$

The Earth dipole decreases is the increase of latitude and so, according to Eq. 3.12, also the charge particles' minimum allowed rigidity.

Finally, as demonstrated in Fig. 3.1, the curvature induced to charged cosmic rays cause by the geomagnetic field combined with the Earth's *shadow* can lead to the suppression of a set of trajectories from the East direction leading to the observed East-West cosmic ray asymmetry. This effect is only possible because there is a dominance of positive charged particles (protons) at the ground.

7. *Very-high-energy cosmic rays (6).* Justify the sentence "About once per minute, a single subatomic particle enters the Earth's atmosphere with an energy larger than 10 J" in Chapter 1.

The differential CR spectrum can be written as $dF/dE = AE^{-\alpha}$, so that the flux above a certain energy E_1 is given for $\alpha = -3$ by

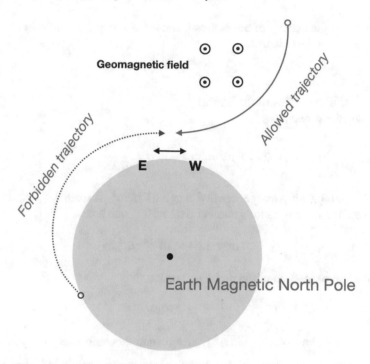

Fig. 3.1 Scheme on the East-West cosmic ray asymmetry

$$F(E > E_1) = \int_{E_1}^{\infty} \frac{dF}{dE} dE = \frac{1}{2} \frac{A}{E_1^2}, \tag{3.14}$$

Here we use $A = 3.375 \times 10^{22}$ eV^{2}m^{-2}s^{-1}sr^{-1}, derived from the fact that, at $E \simeq 1.5 \times 10^9$ eV, the differential flux is of $\simeq 1 \times 10^{-5}$ eV^{2}m^{-2}s^{-1}sr^{-1}. So, $F(E > E_1) = 4.33 \times 10^{-18}$ m^{-2}s^{-1}sr^{-1}, and finally, the number of particles reaching the Earth per minute is, $N = 2\pi A F(E > E_1)$, where A is the surface area of the Earth and 2π sr counts the entire solid angle of arrival of particles. From which we obtain the value of one sub atomic particle per minute.

8. *Very-high-energy neutrinos (7).* The IceCube experiment in the South pole can detect neutrinos crossing the Earth from the North pole. If the cross section for neutrino interaction on a nucleon is $(6.7 \times 10^{-39} E)$ cm^2 with E expressed in GeV (note the linear increase with the neutrino energy E), what is the energy at which half of the neutrinos interact before reaching the detector? Comment the result.

The mean free path of the neutrino can be calculated from the cross section:

$$\lambda \simeq \frac{1}{\mathcal{N} \rho \sigma}$$

where ρ is the density of the material, and $\mathcal{N}$ is Avogadro's number.
One has thus the equation

$$e^{-2R/\lambda} = \frac{1}{2},$$

where R is the radius of the Earth.
Solving for σ one has

$$2R\mathcal{N}\rho\sigma = \ln 2 \Longrightarrow \sigma = \frac{\ln 2}{2R\mathcal{N}\rho}$$

and assuming an average density $\rho \simeq 5.51\,\text{g/cm}^3$, a diameter $2R \simeq 12.7 \times 10^8\,\text{cm}$, and the Avogadro's number 6.02×10^{23}, one has

$$\sigma \simeq 1.64 \times 10^{-34}\,\text{cm}^2\,.$$

In the approximation of the problem, one has thus

$$E \simeq 25\,\text{TeV}\,.$$

9. *Pions in air showers (8).* If a π^0 from a cosmic shower has an energy of 2 GeV:

   a. Assuming the two γ rays coming from its decay are emitted in the direction of the pion's velocity, how much energy does each have?
   b. What are their wavelengths and frequencies?
   c. How far will the average neutral pion travel, in the laboratory frame, from its creation to its decay? Comment on the difficulty to measure the pion lifetime.

   ---

   a. In the π^0 proper frame, each *gamma* will have an energy of $E^* = m_\pi/2$ and will be emitted with an angle of $180°$ between them. According to the problem, in the CM frame, one of the photons would have an angle $\theta^* = 0^*$ with the pion's velocity and the other an angle of $\theta^* = 180°$. Applying the Lorentz transformations to pass from the CM to the LAB framework we get

$$E = \gamma \left(E^* + \beta P^* \cos\theta^* \right) \tag{3.15}$$

$$E = \frac{E_\pi}{m_\pi} \left(E^* + \frac{P_\pi}{m_\pi} E^* \cos\theta^* \right) \tag{3.16}$$

   where it was used that for the photon $E = P$ and the relation between the reference frame can be obtain computing $\gamma = \frac{E_\pi}{m_\pi}$ and $\beta = \frac{P_\pi}{m_\pi}$.

From Eq. 3.16 we can compute the energy of the photons in the LAB frame as being $E_1 \simeq 2\,\mathrm{GeV}$, for the photon emitted along the boost direction (velocity of the pion) and $E_2 \simeq 2.5\,\mathrm{MeV}$ for the photon emitted backwards. Notice that, since the photon is massless, the momentum of the photon emitted backwards cannot be reversed, contrary to what might happen to particle if the boost is high enough.

b. The energy of the photons can be computed using $E_\gamma = h\nu$ and the wavelength through $c = \lambda\nu$, where $h = 4.14 \times 10^{-15}\,\mathrm{eV\,s}$ and $c = 3 \times 10^8\,\mathrm{m\,s}^{-1}$.

Therefore, the frequency and wavelength for the photon travelling forward is $\nu_1 \simeq 4.8 \times 10^{23}\,\mathrm{s}^{-1}$ and $\lambda_1 \simeq 6.2 \times 10^{-16}\,\mathrm{m}$, respectively, while for the photon emitted backward is $\nu_2 \simeq 5.9 \times 10^{20}\,\mathrm{s}^{-1}$ and $\lambda_2 \simeq 5.1 \times 10^{-13}\,\mathrm{m}$.

c. The boost of the pion is given by the equation:

$$E = \gamma m c^2 \implies \gamma \simeq 7.4. \tag{3.17}$$

In the laboratory frame, the muon will decay after a time of $t = \gamma\tau$, where τ is the decay time of the π^0 at rest ($\tau = 8.52 \times 10^{-17}\,\mathrm{m}$). So the pion will travel in average

$$x = \beta c t = \frac{P}{E}\frac{E}{m}c\tau = \frac{\sqrt{E^2 - m_\pi^2}}{m_\pi}c\tau \simeq 3.8 \times 10^{-7}\,\mathrm{m} = 0.38\,\mu\mathrm{m}. \tag{3.18}$$

On one hand, the neutral pion does not leave traces in a tracker that measures the tracks of charged particles. On the other hand, it lives very little time to be identified as the origin of the two emitted photons. This short lifetime comes from the mechanism by which this particle decays (electromagnetic decay) instead of weak decay (longer lifetime).

10. *Discovery of the positron.* Carl D. Anderson received the 1936 Nobel Prize in physics for the discovery of the positron, the antiparticle of the electron. In 1932, during an experiment designed to observe cosmic rays, he observed a track of a positive particle with a mass about equal to that of the electron in a cloud chamber.

   a. One striking feature in Fig. 3.2 is that although produced by cosmic rays, the positron seems to be coming from below. Show that a vertical muon, going down, cannot produce a positron with an energy of 63 MeV in the opposite direction? Why? Remember that $\mu^+ \to \bar\nu_\mu e^+ \nu_e$.
   b. Discuss a possible origin for the muon that gave origin to this positron.

   _____

   a. The most favourable scenario for a e^+ to be produced upwards is if the μ^+ is at rest, i.e., no boost. We want the positron to the the maximum energy,

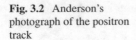

Fig. 3.2 Anderson's photograph of the positron track

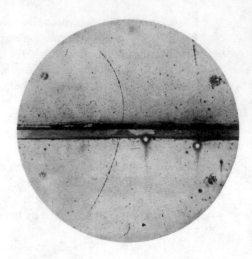

so the best configurations of the decay is when both the neutrinos travel in the opposite direction of the e^+ (see Fig. 3.3).

For this configuration the 4-momentum of the muon is $P_\mu = (m_\mu, \mathbf{0})$, for the electron $P_e = (E_e, \mathbf{P})$ and for the neutrinos $P_\nu = (E_\nu, -\mathbf{P})$, where E_ν is the sum of the energy carried by both neutrinos.

From the energy-momentum conservation one gets $P_\mu = P_e + P_\nu$. Squaring this equation and noting that the mass of the of the neutrinos is negligible ($P_\nu = E_\nu$) one gets

$$E_\nu = \frac{m_\mu^2 - m_e^2}{2\, m_\mu} \tag{3.19}$$

And therefore the maximum energy that the positron would be $E_e = \sqrt{E_\nu^2 + m_e^2} = 52.8\,\text{MeV}$ which is less than the measured invalidating the possibility that this positron was produced by a vertical muon coming from above.

b. Previously we have demonstrated that the positive muon needs to be traveling upward to give origin to a positron with an energy of 63 MeV. The most likely possibility would be the production of the muon by a secondary cosmic ray shower charged pion with $\beta \simeq 0$. As such, there is a chance that the muon gets produced in the opposite direction of the flight direction of the pion and the pion *boost* does not reverse its momentum.

Fig. 3.3 Scheme of the
kinematics of the muon
decay

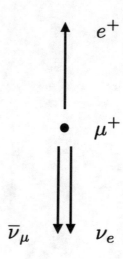

Chapter 4
Particle Detection

1. *Muon energy loss (1).* A muon of 100 GeV crosses a layer of 1m of iron. Determine the energy loss and the expected scattering angle.

$E_\mu = 100$ GeV
$m_\mu \simeq 106$ GeV/c^2
For iron: Z = 26, A = 56, $\rho = 7.9$ g/cm^3, $X_0 = 1.76$ cm
The particle is relativistic and $E_\mu \simeq p_\mu$. We can derive γ:

$$\gamma = \frac{E_\mu}{m_\mu c^2} = \frac{10^2 \times 10^3 \, \text{MeV}}{106 \, \text{MeV}/c^2} \simeq 1000. \tag{4.1}$$

We are still dominated by ionization-excitation losses (rather than radiative ones) and we can write the simplified expression, because $\beta \simeq 1$:

$$-\frac{dE}{dX} \simeq D \frac{Z}{A} z^2 \ln\left(\frac{m_e c^2 \gamma^2}{I}\right) \tag{4.2}$$

where X is expressed in g/cm^2. We derive:

$$-\frac{dE}{dX} \simeq 0.307 \, \text{MeV} \frac{\text{cm}^2}{\text{g}} \frac{26}{56} \ln\left(\frac{0.5 \frac{\text{MeV}}{c^2} c^2 (10^3)^2}{13.6 \times 26 \times 10^{-6} \, \text{MeV}}\right) \simeq 2.02 \, \text{MeV} \frac{\text{cm}^2}{\text{g}} \tag{4.3}$$

Note that Z/A is about 0.5 for all elements except for hydrogen and heaviest elements.

X is ρx, with x thickness of the traversed material. Given the density of iron we have:

$$-\frac{dE}{dx} = 7.9 \frac{\text{g}}{\text{cm}^3} \times 2.02 \, \text{MeV} \frac{\text{cm}^2}{\text{g}} \simeq 16 \frac{\text{MeV}}{\text{cm}}. \tag{4.4}$$

© Springer Nature Switzerland AG 2021
A. De Angelis et al., *Particle and Astroparticle Physics*, Undergraduate Lecture Notes in Physics,
https://doi.org/10.1007/978-3-030-73116-8_4

Since the muon traversed 1 m of iron we can evaluate the total energy loss:

$$\Delta E = 16 \frac{\text{MeV}}{\text{cm}} \cdot 10^2 \, \text{cm} = 1600 \, \text{MeV} = 1.6 \, \text{GeV}. \tag{4.5}$$

Notice that some approximations are made to reach the mean energy loss by ionization. From Fig. 4.1 it can be seen that a slightly higher value is expected for $dE/dX \simeq 2.19 \, \text{MeV} \, \text{g}^{-1} \, \text{cm}^2$. This would lead to a change in the total energy loss of less than 9%.

The scattering angle is:

$$\theta_0 \simeq \frac{13.6 \, \text{MeV}}{E} \sqrt{\frac{x}{X_0}} \simeq 0.77 \, \text{mrad}. \tag{4.6}$$

2. *Energy loss in a water Cherenkov detector (2).* In the Pierre Auger Observatory, the surface detectors are composed of water Cherenkov tanks 1.2 m height, each containing 12 tons of water. These detectors are able to measure the light produced by charged particles crossing them. Consider one tank crossed by a single vertical muon with an energy of 5 GeV. The refraction index of water is $n \simeq 1.33$ and can be in good approximation considered constant for all the relevant photon wavelengths. Determine the energy lost by ionization and compare it with the energy lost by Cherenkov emission. Assume that the muon's ionization energy loss in the water is similar to the one lost while crossing carbon or helium gas.

Let us start by evaluating the energy lost by ionization. Using the dE/dX plot, shown in Fig. 4.1, one gets that a 5 GeV muon in a material with characteristic between carbon and helium gas loses about

$$\frac{dE}{dX} \simeq 2.1 \, \text{MeV} \, \text{cm}^2 \, \text{g}^{-1}. \tag{4.7}$$

The amount of matter traversed by the muon is

$$\Delta X = l \, \rho_{water} \simeq 120 \, \text{g} \, \text{cm}^{-2} \tag{4.8}$$

where $l = 120$ cm is the height of the tank and ρ_{water} the water density. Thus, the muon energy lost by ionization is

$$E_{loss}^{ionization} = \Delta X \frac{dE}{dX} \simeq 252 \, \text{MeV}. \tag{4.9}$$

Let us now compute the energy lost due to Cherenkov radiation. Assuming that in average each photon carried about 3.5 eV, then

$$E_{loss}^{Cherenkov} = \langle E_\gamma \rangle N_\gamma \tag{4.10}$$

Fig. 4.1 Mean energy loss rate in liquid (bubble chamber) hydrogen, gaseous helium, carbon, aluminum, iron, tin, and lead. Radiative effects, relevant for muons and pions, are not included. These become significant for muons in iron for $\beta\gamma \geq 1000$, and at lower momenta for muons in higher-Z absorbers. Taken from https://pdg.lbl.gov/2019/reviews/rpp2018-rev-passage-particles-matter.pdf

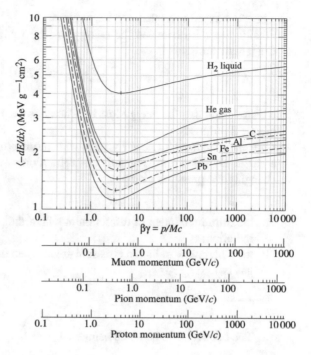

with N_γ being the number of produced Cherenkov photons. This last quantity can be obtained through

$$\frac{d^2N}{dE\,dx} = 370 \sin^2(\theta_c)\ \text{photons eV}^{-1}\,\text{cm}^{-1}. \tag{4.11}$$

Knowing that $E_\mu = 5\,\text{GeV}$ and that the boost is $\gamma = E_\mu/m_\mu$ one gets that the velocity of the particle is

$$\beta = \sqrt{1 - \frac{1}{\gamma^2}} = \sqrt{1 - \left(\frac{m_\mu}{E_\mu}\right)^2} \simeq 0.9998 \simeq 1 \tag{4.12}$$

and the Cherenkov emission angle is

$$\theta_c = \arccos\left(\frac{1}{\beta\,n}\right) = 41.25°. \tag{4.13}$$

We can now obtain the number of produced Cherenkov photons by integrating Eq. 4.11 in energy and path:

$$N_\gamma \simeq 370 \sin^2(\theta_c)\,\Delta E\,\Delta X = 19\,300\ \text{photons}, \tag{4.14}$$

where it was assumed that $n(E)$ is constant.

Finally, the ratio between the energy lost by ionization and the one lost by Cherenkov emission is

$$\frac{E_{loss}^{ionization}}{E_{loss}^{Cherenkov}} \simeq 3\,730 \tag{4.15}$$

which means that the energy lost by Cherenkov emission can be neglected for practical purposes.

3. *Cherenkov radiation (3).* A proton with momentum $1\,\text{GeV}/c$ passes through a gas at high pressure. The index of refraction of the gas can be changed by changing the pressure. Compute:

  a. the minimum index of refraction at which the proton will emit Cherenkov radiation;
  b. the Cherenkov radiation emission angle when the index of refraction of the gas is 1.6.

---

  a. The boost can be obtained using

$$\gamma = \frac{E}{m} = \frac{\sqrt{p^2 + m^2}}{m} = 1.46 \tag{4.16}$$

where m is the mass of the proton and p its momentum. The velocity, β, can obtained inverting the boost formula:

$$\gamma = \frac{1}{\sqrt{1 - \beta^2}}. \tag{4.17}$$

The condition for Cherenkov light emission is

$$\cos\theta = \frac{1}{n\beta} \leq 1 \tag{4.18}$$

where n is the medium refraction index.

Hence, using Eqs. 4.18 and 4.17, one finds that the refraction index necessary to produce Cherenkov light is

$$n \geq \frac{1}{\sqrt{1 - \frac{1}{\gamma^2}}} \simeq 1.37. \tag{4.19}$$

  b. Again, using Eq. 4.18, and taking now $n = 1.6$, and β from Eq. 4.17, one gets

$$\cos \theta = \frac{1}{n\beta} = \frac{1}{1.6 \times 0.729} \simeq 0.86 \qquad (4.20)$$

which means that $\theta \simeq 31°$.

4. *Range (8).* Compare approximately the ranges of two particles of equal velocity and different mass and charge traveling through the same medium.

Accordingly with the problem we have $m_1 \neq m_2$, $z_1 \neq z_2$ and $v \equiv v_1 = v_2$. The ionization energy loss can be expressed as,

$$-\frac{dE}{dX} = z^2 N Z f(v) \qquad (4.21)$$

where N and Z are respectively the number density of atoms and the atomic number of the traversed medium. As a first approximation:

$$f(v) \propto v^{-2} \ln v^2. \qquad (4.22)$$

Since $v_1 = v_2$, this function is the same for both particles and it is equivalent to a $f(\gamma)$. The *range* is defined as:

$$R = \int_{E_0}^{mc^2} dX = \int_{E_0}^{mc^2} \frac{dE}{dE} dX = \int_{mc^2}^{E_0} \frac{dE}{-dE/dX} \qquad (4.23)$$

where X, and consequently R, can be expressed in cm (or g cm^{-2}, if $X = \rho x$ with x a length). E_0 is the initial total energy of the particle.
We want to express the integral in Eq. (4.23) as a function of γ.

$$dE = d(\gamma mc^2) = mc^2 d\gamma .$$

We have therefore:

$$R = \frac{1}{z^2 N Z} \int_1^{\gamma_0} \frac{mc^2}{f(\gamma)} d\gamma . \qquad (4.24)$$

This integral is the same for both particles and the traversed medium is the same, so we finally derive the following relationship between the ranges of the two particles R_1, R_2:

$$\frac{R_1}{R_2} = \frac{m_1}{m_2} \frac{z_2^2}{z_1^2} . \qquad (4.25)$$

5. *Pair production and multiple scattering (4).* What is the optimal thickness (in radiation lengths) of a layer of silicon in a gamma-ray telescope with hodoscopic structure in order that the multiple scattering does not deteriorate the information from the opening angle of the electron-positron pair in a photon conversion?

The angle of emission of an electron-positron pair can be approximated by:

$$\theta_{pair} = \frac{0.8\,\text{MeV}}{E}.$$ (4.26)

The scattering angle is also approximated by

$$\theta_0 = \frac{13.6\,\text{MeV}}{E}\sqrt{\frac{x}{X_0}}$$ (4.27)

where $X_0 = 21.82\,\text{g/cm}^2$ is the radiation length of silicon.

If we compare both equations, we obtain that the optimal thickness is given when the angle for the electron-positron pair equals that of the multiple scattering:

$$\frac{0.8\,\text{MeV}}{E} = \frac{13.6\,\text{MeV}}{E}\sqrt{\frac{x}{X_0}}$$ (4.28)

and finally $x = 0.0755\,\text{g/cm}^2$. Since silicon density is $\rho = 2.33\,\text{g/cm}^3$, the optimal width of the silicon layer would be $w = 0.03\,\text{cm}$.

6. *Compton scattering (5)*. A photon of wavelength λ is scattered off a free electron initially at rest. Let λ' be the wavelength of the photon scattered in the direction θ. Compute:

   a. λ' as a function of λ, θ and universal parameters;
   b. the kinetic energy of the recoiling electron.

---

a. We will apply to this scattering problem the conservation of four-momentum in the frame in which the electron is initially at rest: we will choose the x-asis along the line containing the photon and the electron before the scattering. As in the text, we will indicate with primes the quantities after the scattering, e.g. λ', and without primes those before the scattering, e.g. λ.

Let the four momenta (before and after the scattering) with q_μ and q'_μ for the photon, and with p_μ and p'_μ for the electron. In the chosen frame we have

$$q_\mu = (h\nu/c, h\nu/c, 0, 0) \qquad q'_\mu = (h\nu'/c, h\nu'\cos\theta/c, h\nu'\sin\theta/c, 0),$$
$$p_\mu = (m_e c, \mathbf{0}) \qquad p'_\mu = (\gamma' m_e c, \mathbf{p}'),$$

where $\gamma' = (1 - v'^2/c^2)^{-1/2}$. From the time component of energy momentum conservation,

$$q_\mu + p_\mu = q'_\mu + p'_\mu,$$

we obtain

$$\frac{h\nu}{c} + m_e c = \frac{h\nu'}{c} + m_e\gamma'c \quad \Rightarrow \quad h\nu - h\nu' + m_e c^2 = m_e\gamma'c.$$ (4.29)

Conservation of three-momentum can be expressed using the angle θ between the incident and scattered photon. Since from the spatial part of momentum conservation the total three momentum before the scattering (and only related to the photon in the chosen frame) must equal the sum of the three momenta after scattering, the three momenta $\mathbf{q}$, $\mathbf{q}'$ and $\mathbf{p}'$ form a triangle, with an angle θ at the vertex formed by $\mathbf{q}$ and $\mathbf{q}'$. Using Carnot rule on this triangle, we get

$$m_e^2 \gamma'^2 v'^2 = \frac{h^2 v'^2}{c^2} + \frac{h^2 v'^2}{c^2} - 2\frac{h^2 vv'}{c^2} \cos\theta. \tag{4.30}$$

Equation 4.29 can be first rewritten as

$$(hv - hv' + m_e c^2)^2 = m_e^2 \gamma'^2 c^2, \tag{4.31}$$

and, solving for v'^2 by remembering that $\gamma'^2 c^2 = c^2 - v'^2$, results in

$$v'^2 = c^2 - \frac{m_e^2 c^6}{(hv - hv' + m_e c^2)^2}.$$

From the above and (4.31), with a little algebra, we then obtain

$$m_e^2 \gamma'^2 v'^2 = \frac{(hv - hv' + m_e c^2)^2}{c^2} - m_e^2 c^2,$$

and, by comparison with (4.30),

$$\frac{(hv - hv' + m_e c^2)^2}{c^2} - m_e^2 c^2 = \frac{h^2 v^2}{c^2} + \frac{h^2 v'^2}{c^2} - 2\frac{h^2 vv'}{c^2} \cos\theta.$$

The above result can be now simplified into

$$-\frac{2h^2 vv'}{c^2} + 2hm_e(v - v') = -\frac{2h^2 vv'}{c^2} \cos\theta \quad \Rightarrow \quad \frac{h}{m_e c^2}(1 - \cos\theta) = \frac{1}{v'} - \frac{1}{v}.$$

The same expression in terms of the photon wavelength before and after the scattering, $\lambda = c/v$ and $\lambda' = c/v'$, respectively, is

$$\lambda' - \lambda = \frac{h}{m_e c}(1 - \cos\theta),$$

and gives us the relationship between λ, λ', the photon scattering angle θ, and the universal parameters c, h, and m_e.

b. The kinetic energy of the scattered electron, T_e', can then be calculated as

$$T'_e = m_e \gamma' c^2 - m_e c^2$$

$$= \frac{hc}{\lambda} - \frac{hc}{\lambda'}$$

$$= \frac{hc}{\lambda} \left(1 - \frac{1}{1 + h(1 - \cos\theta)/(\lambda m_e c)} \right)$$

$$= \frac{hc}{\lambda} \left(\frac{(1 - \cos\theta)}{\lambda m_e c/h + (1 - \cos\theta)} \right), \qquad (4.32)$$

again in terms of c, h, m_e, λ, and the scattering angle of the photon θ.

7. *Nuclear reactions (7)*. The mean free path of fast neutrons in lead is of the order of 5 cm. What is the total fast neutron cross section in lead?

The relation between mean free path λ and cross section σ is

$$\lambda = \frac{1}{n\sigma} \implies \sigma = \frac{1}{n\lambda}$$

where n is the number density, i.e., the density of nuclei per unit of volume. For lead, the density ρ is ~ 11.35 g/cm^3. One has thus, calling m the mass of a lead atom,

$$n = \frac{\rho}{m} = \frac{\rho}{(207.2\,\text{g/mol})/N_A} \simeq 3.3 \times 10^{22}\,\text{cm}^{-3}$$

where N_A is Avogadro's number, and finally

$$\sigma = \frac{1}{n\lambda} \simeq 6.1 \times 10^{24}\,\text{cm}^2 = 6.1\,\text{barn}.$$

8. *Hadron therapy (9)*. The use of proton and carbon ion beams for cancer therapy can reduce the complications on the healthy tissue compared to the irradiation with MeV gamma rays. Discuss why.

Protons and carbon ions are more advantageous in cancer radiation therapy with respect to X- and gamma-rays mainly because of two reasons.

- The release of energy along their path inside the patient's body is characterized by a large deposit localized in the last few millimetres at the end of their range, in the so called Bragg peak region, where they produce severe damage to the cells while sparing both traversed and deeper located healthy tissues. The situation is different for photons, for which the absorption rate is more or less constant (and thus the superficial damage is maximal).
- Hadrons penetrate the patient with minimal diffusion and, using their electric charge, few millimeter wide pencil beams of variable penetration depth can be precisely guided towards any part of the tumor.

A third reason for a higher effectiveness of hadron therapy applies to C ions and pertains to radiation biology. Since, for the same range, carbon ions deposit about a factor 24 more energy in the Bragg peak region with respect to protons, the produced ionization column is so dense to be able to induce direct multiple strand brakes in the DNA, thus leading to nonrepairable damage.

9. *Neutrino interaction in matter (10).* The typical energies of neutrinos produced in nuclear reactors are $E_\nu \sim 1$ MeV.

    a. What is the probability to interact in a water detector with the thickness of one meter?

    b. What is the probability to interact inside the Earth traveling along a trajectory that passes through its center?

    c. Redo the previous questions for a neutrino with an energy of 1 PeV.

---

    a. The cross-section of a neutrino of energy between 1 MeV and 10 TeV with ordinary matter is given by,

$$\sigma = (6.7\,E)\,\text{fb} \tag{4.33}$$

with E given in GeV. For a neutrino of $E = 1$ MeV, we have a cross-section of:

$$\sigma_{\nu N}|_{E_\nu=1\,\text{MeV}} = 6.7 \times 10^{-3}\,\text{fb} = 6.7\,\text{ab}. \tag{4.34}$$

The mean free path of these neutrinos in water is given by:

$$\langle \mu \rangle = \frac{m_N}{\rho \sigma} \tag{4.35}$$

where $m_N = 1.67 \times 10^{-24}$ g is the mass of a nucleon and ρ the density of the material. Here we are assuming that the neutrinos will interact essentially with the nucleons present in the atoms in the way.

From Eq. 4.35 and taking the density of the water as $\rho_{\text{water}} = 1\,\text{g cm}^{-3}$ we obtain finally that the neutrino would travel in average

$$\langle \mu \rangle_{\text{water}} = 2.5 \times 10^{17}\,\text{cm} \tag{4.36}$$

before interacting.

The probability of a neutrino to interact with a nucleon can be calculated through

$$\frac{dN}{dx} = -\frac{N}{\langle \mu \rangle} \implies P_{\text{int}} = 1 - e^{-\frac{x}{\langle \mu \rangle}}, \tag{4.37}$$

giving an interaction probability in 100 cm of water of $P_{\text{int}} \simeq 4.4 \times 10^{-16}$.

Table 4.1 Some parameters resulting from the so-called Rossi approximation B

	Incident electron	Incident photon
Peak of shower t_{max}	$1.0 \times (\ln y - 1)$	$1.0 \times (\ln y - 0.5)$
Center of gravity t_{med}	$t_{max} + 1.4$	$t_{max} + 1.7$
Number of e^+ and e^- at peak	$0.3y/\sqrt{\ln y - 0.37}$	$0.3y/\sqrt{\ln y - 0.31}$
Total track length	y	y

b. The Earth average density can be computed from

$$\rho_{Earth} \simeq \frac{M_\oplus}{\frac{4}{3}\pi R_\oplus^3} \simeq 5.5\,\mathrm{g\,cm}^{-3}, \tag{4.38}$$

with $M_\oplus = 5.972 \times 10^{24}\,\mathrm{kg}$ and $R_\oplus = 6371\,\mathrm{km}$. As such, from Eq. 4.35 we have that the mean path travelled by the neutrino before interacting with the Earth's atoms is $\langle\mu\rangle_{Earth} = 4.5 \times 10^{16}\,\mathrm{cm}$. Considering the crossing of the Earth $x = 2R_\oplus$, from Eq. 4.37 we get a probability of interaction of $P_{int} \simeq 2.8 \times 10^{-8}$. Notice, that in reality, if the neutrino crosses the center of the Earth then the average density that it would encounter would be of roughly $\rho \sim 8\,\mathrm{g\,cm}^{-3}$. However, even taking into account the high density core of the Earth it would be still very unlikely that this neutrinos would interact.

c. At $E = 1\,\mathrm{PeV}$, the cross-section is given by the equation:

$$\sigma = \left(6.7\sqrt{\frac{E}{10\mathrm{TeV}}}\right)10^{-34}\mathrm{cm}^2 \tag{4.39}$$

with E in TeV. The cross-section for 1 PeV neutrinos with ordinary matter is then

$$\sigma_{1\,PeV} = 6.7\,\mathrm{nb}.$$

Repeating the calculations performed before with this new value for the cross-section we obtain,

$$P_{int}(\mathrm{water}) \simeq 4 \times 10^{-7}$$
$$P_{int}(\mathrm{Earth}) \simeq 1.$$

So these neutrinos do not interact with 1 m of water but cannot cross the Earth without interacting with it.

10. *Electromagnetic showers (11).* How does an electromagnetic shower evolve as a function of the penetration depth in a homogeneous calorimeter? What is the difference between an incoming photon and an incoming electron/positron?

A common parameterization of the longitudinal profile for a shower of initial energy E_0 as a function of the penetration depth t in radiation length units is

$$\frac{dE}{dt} = E_0 \frac{\beta}{\Gamma(\alpha)} (\beta t)^{\alpha-1} e^{-\beta t}, \tag{4.40}$$

where Γ is Euler's Gamma function $\Gamma(z) = \int_0^{+\infty} t^{z-1} e^{-t} dt$. In the above approximation, $t_{max} = (\alpha - 1)/\beta$, which should be thus equal to $\ln(E_0/E_c) - C$ with $C = 1$ for an electron and $C = 0.5$ for a photon (i.e., a photon-initiated shower is displaced forward by 0.5 radiation lengths with respect to a shower initiated by an electron of the same energy).

In the so-called Rossi approximation B, one has, calling y the energy in units of the critical energy:

11. *Hadronic showers (12).* Let us approximate the effective cross section for protons on nucleons in air with a value of 40 mb per nucleon. Calculate the interaction length of a proton (in g/cm^2, and in meters at NTP). What is the average height above the sea level where this interaction takes place? In hadronic showers we find also an electromagnetic component, and muons. Where do these come from?

For the interaction length λ one has

$$\lambda = \frac{1}{n\sigma}. \tag{4.41}$$

Let us assume for simplicity that the atmosphere is composed only by N_2. In this case the average number density of nucleons in the atmosphere is

$$n = \frac{\rho N_A A}{M_a} \tag{4.42}$$

where $\rho \simeq 1.2 \, \text{kg m}^{-3} = 1.2 \times 10^{-3} \, \text{g cm}^{-3}$ is the density of air at sea level, N_A is the Avogadro number, $M_a \simeq 28 \, \text{g mol}^{-1}$ is the molar mass of N_2 (notice that the dry air molar mass is around 29 g mol^{-1}). Combining the two previous equation, and using $\sigma = 4 \times 10^{-26} \, \text{cm}^2$ one has thus $\lambda = 345$ m or multiplying by the air density at NTP conditions, $\lambda \simeq 41.5 \, \text{g cm}^{-2}$.

Using a simple model for the atmosphere density evolution, in particular the isothermal approximation, we can transform depth into height using the following equation,

$$X = X_0 e^{-\frac{h}{H}} \tag{4.43}$$

where $H \simeq 6.5$ km and $X_0 \simeq 1030 \, \text{g cm}^{-2}$. Considering $X = 41.5 \, \text{g cm}^{-2}$ have then that the height above sea level of the first interaction should occur in average at

$$h = H \ln\left(\frac{X_0}{X}\right) = 20.9 \, \text{km.} \tag{4.44}$$

The γ component in the cascade comes mostly from decays of neutral pions generated in the development of the shower. The muon component arises mostly from the decay of charged pions.

12. *Tracking detectors (13)*. Could you build a tracking detector for photons? And for neutrinos?

A tracker and a photon detector have conflicting prescriptions. The photon needs a lot of converting material to materialize and create a cascade that will measure its energy, but this fact changes its direction. However, the pair production and Compton interaction generate secondary products which keep memory of the direction of the primary photon. In the first case, the outcoming electron and positron can be tracked and eventually their energy can be measured in a calorimeter. This is the principle of operation of hodoscopes like the *Fermi* Large Area Tracker, in which planes of position-sensitive detectors (for example Si strips of pixels) are alternated to heavy converters (Pb or W). In the second case, one needs to detect the outoming photon (for example with a calorimeter), the point of interaction and, if possible, the outcoming electron. The secondary photon might, in turn, interact, and its kinematical parameters be measured.

The same problems apply to a neutrino tracking detector, with the additional difficulty that, due to charge and lepton number conservation, the interaction often generates another outcoming neutrino (which makes it impossible to determine completely the kinematics) and that, being the cross section much smaller, you need much more material. Tracking calorimeters such as the MINOS detectors at FNAL use alternating planes of absorber material and detector material (the active detector is often liquid or plastic scintillator, read out with photomultiplier tubes, although various kinds of ionization chambers have also been used; the converter is generally steel). The NOνA detector eliminates the absorber planes using a very large active detector volume. At high-energy (above several GeV) neutral current interactions of neutrinos appear as a shower of hadronic debris and charged current interactions are identified by the presence of the charged lepton's track (and possibly some hadronic debris as well).

13. *Momentum measurement in magnetic field*. In a drift chamber there is a magnetic field of 0.8 T, a photon converts to an e^+e^- pair and two tracks with a curvature radius of $R = 20$ cm are observed, initially parallel. Calculate the photon energy.

The momentum of the electrons expressed in GeV/c is:

$$p_e = 0.3\,R\,B = 0.3 \times 20 \times 10^{-2} \times 0.8 \simeq 0.048\,\text{GeV}/c$$

where R is expressed in m and B in T.
Since $m_e \ll p_e$, $E_e \simeq 48$ MeV. The energy of the photon is

$$E_\gamma = 2E_e = 96\,\text{MeV}.$$

14. *Time of flight.* A proton and an electron, both of total energy 2 GeV, pass through two scintillators placed 30 m apart. Which are the flight times of the two particles?

$$m_p \simeq 0.938 \, \text{GeV}/c^2$$
$$m_e \simeq 0.511 \, \text{MeV}/c^2.$$

Since $E = \gamma mc^2$, we can evaluate the Lorentz factor for both particles,

$$\gamma_p = \frac{2}{0.94} \simeq 2.13 \tag{4.45}$$

$$\gamma_e = \frac{2}{0.5 \times 10^{-3}} \simeq 3914. \tag{4.46}$$

The relative velocity of the particles can be computed through:

$$\beta = \sqrt{\frac{\gamma^2 - 1}{\gamma^2}} \tag{4.47}$$

leading to, $\beta_p = 0.88$ and $\beta_e \simeq 1$.
The time of flight therefore is:

$$t_p = \frac{l}{\beta_p c} \simeq 0.11 \, \mu s$$

$$t_e = \frac{l}{\beta_e c} \simeq 0.10 \, \mu s$$

where $l = 30$ m and c is the speed of light in vacuum.

15. *Pion production.* A 100 GeV proton beam passes through a 0.1 cm thick iron plate. The proton flow corresponds to a current of 0.016 nA. The inelastic interaction length of iron is 17 cm. Estimate the number of charged pions and neutral pions per unit time produced the plate. Consider that in this interaction the average multiplicity is roughly $\langle m \rangle \sim 10$.

The proton current is giving us the number of protons entering the iron plate:

$$N_0 = \frac{I}{q} = \frac{1.6 \times 10^{-11} \text{C}/s}{1.6 \times 10^{-19} \text{C}} = 10^8 \text{protons}/s. \tag{4.48}$$

The number of protons exiting produced in the plate will be:

$$\frac{dN}{N} = -\frac{dX}{X_n} \tag{4.49}$$

where X_n is the nuclear interaction length. The final number of particles is:

$$N_f = N_0 \exp\left(-\frac{d}{X_n}\right) \tag{4.50}$$

where d is the width of the plate. The number of particles that interacted will then be:

$$N_i = N_0 - N_f = N_0(1 - \exp(-d/X_n)) = 5.87 \times 10^5 \text{ protons}. \tag{4.51}$$

Since the energy of the beam is 100 GeV, we can make the approximation that each proton that interacted produced in equal quantities π^+, π^- and π^0. Therefore the number of charged pions is:

$$N(\pi^\pm) = \frac{2}{3}N_i(\langle m \rangle - 2) = 3.13 \times 10^6. \tag{4.52}$$

Notice that due to baryon number conservation, the outcome of the interaction needs to have at least two baryons. Since the average multiplicity is three, then only one pion per interaction can be in average produced.

Following the same line of thought, we obtain for neutral pions

$$N(\pi^0) = \frac{N_i}{3}(\langle m \rangle - 2) = 1.57 \times 10^6. \tag{4.53}$$

16. *Space resolution of a multiwire proportional chamber.* For a multiwire proportional chamber with anode spacing d, estimate the RMS error of the localization of a track in the plane of the chamber.

A cell with signal is insensitive to the particle crossing position within the cell dimension (d). This means that the detection probability can be taken as an uniform distribution. It can be shown that the variance of an uniform distribution is given by the square of the range of the distribution divided by 12. Hence, as $\sigma \simeq \sqrt{Var}$, we have,

$$\sigma = d/\sqrt{12}.$$

17. *Detection of muons in a transparent medium.* A 400 GeV/c muon vertically enters the sea. Through which physical process can it be revealed? Estimate the depth at which it arrives and decays.

A vertical muon that enters the sea is mainly loosing its energy via collisions with the atomic electrons (ionization losses). From Fig. 4.1 it can be seen that the energy loss rate depends on the crossed material properties but for all the particles and material there is a minimum around $\beta\gamma \simeq 3.5$. For this minimum, known as minimum ionising particle (MIP)

$$\frac{dE}{dX} \sim 2\,\mathrm{MeVg^{-1}cm^2}.$$

Assuming that the energy lost by the muon does not depend on its momentum (actually it has a logarithmic grow with the particle momentum) we can simply write,

$$E_{\mathrm{loss}} = \int \frac{dE}{dX}dX \sim \frac{dE}{dX}\Delta X \Leftrightarrow \Delta X = \frac{E_{\mathrm{loss}}}{dE/dX}, \qquad (4.54)$$

where E_{loss} is the energy lost by the muon until it stops and ΔX the amount of water traversed by the muon. According with Fig. 4.1, for a momentum of $P_\mu \simeq 200\,\mathrm{MeV}/c$ the ionization losses becomes very high and the muon should stopped after travelling very short distances. As such, taking into account that the mass of the muon is much smaller than it initial momentum, we can say that $E_{\mathrm{loss}} \simeq 400\,\mathrm{GeV}$. Thus, using Eq. 4.54, the muon would travel,

$$\Delta X = \frac{400 \times 10^3}{2} = 2 \times 10^5\,\mathrm{g\,cm^{-2}}. \qquad (4.55)$$

Dividing ΔX by the water density, $\rho_{\mathrm{water}} = 1\,g\,cm^{-3}$ we would get that a vertical muon would reach a depth of 2 km.

We have used the MIP to perform this calculation. Again, from Fig. 4.1 it can be seen that the relation between the minimum dE/dX (MIP) and the dE/dX for a 400 GeV/c muon is roughly 50% (depends on the material). This means that the mean depth that a vertical muon could reach would be smaller than 3 km. When the muon enters into the water, since it is a relativistic particle, it also emits Cherenkov light. The Cherenkov light is produced mostly in the ultra-violet and visible energy range and therefore it can be measured by a photo-detector.

18. *Photodetectors (14)*. What gain would be required from a photomultiplier in order to resolve the signal produced by three photoelectrons from that due to two or four photoelectrons? Assume that the fluctuations in the signal are described by Poisson statistics, and consider that two peaks can be resolved when their centers are separated by more than the sum of their standard deviations.

The number of electrons, N, produced by $N^{(pe)}$ photoelectrons is

$$N = G N^{(pe)} \qquad (4.56)$$

where G is the photomultiplier (PMT) gain.

From the problem one has that the condition to have resolved peaks is

$$\mu_1 - \mu_2 \geq \sigma_1 + \sigma_2 \tag{4.57}$$

where μ is the distribution mean and σ its standard deviation. If the distributions follow Poisson statistics then

$$\mu_i = N_i \quad \text{and} \quad \sigma_i = \sqrt{N_i}. \tag{4.58}$$

Therefore, Eq. 4.57 becomes

$$N_{i+1} - N_i \geq \sqrt{N_{i+1}} + \sqrt{N_i}. \tag{4.59}$$

Using Eq. 4.56 the above equation provides

$$G(N_{i+1}^{(pe)} - N_i^{(pe)}) \geq \sqrt{G}\left(\sqrt{N_{i+1}^{(pe)}} + \sqrt{N_i^{(pe)}}\right) \tag{4.60}$$

and so the gain must be

$$G \geq \left(\frac{\sqrt{N_{i+1}^{(pe)}} + \sqrt{N_i^{(pe)}}}{N_{i+1}^{(pe)} + N_i^{(pe)}}\right)^2. \tag{4.61}$$

From the above condition one has that for $N_i^{(pe)} = 2$ and $N_{i+1}^{(pe)} = 3$

$$G \gtrsim 10 \tag{4.62}$$

and for $N_i^{(pe)} = 3$ and $N_{i+1}^{(pe)} = 4$,

$$G \gtrsim 14. \tag{4.63}$$

19. *Silicon PMTs versus conventional PMTs.* Which SiPM properties make them better than conventional PMTs for use as photodetectors in a Cherenkov telescope?

The main advantages of silicon PMTs (SiPMs) over conventional PMTs in Cherenkov telescopes are their lower power consumption and the low Voltage at which they are operated. This allows them to be used during moonlight observations where the background light level is very high and PMTs might be damaged or worsen their performance due to aging effects. Moreover, the mean quantum efficiency of a SiPM is higher than that of the PMT.

20. *Cherenkov counters (15).* Estimate the minimum length of a gas Cherenkov counter used in the threshold mode to be able to distinguish between pions and kaons with momentum 20 GeV. Assume that 200 photons need to be radiated to ensure a high probability of detection and that radiation covers the whole visible spectrum (neglect the variation with wavelength of the refractive index of the gas).

The number of Cherenkov photons emitted per unit of track length and per unit of the photon energy interval is:

$$\frac{d^2 N_\gamma}{dE\,dx} \simeq 370 \sin^2 \theta_c \, \text{eV}^{-1} \, \text{cm}^{-1}. \tag{4.64}$$

In the whole visible spectrum ($\lambda \in [300, 700]$ nm) the Cherenkov light yield in a length L (in cm) is:

$$N_\gamma = \int_0^L \int_{E(700\,\text{nm})}^{E(300\,\text{nm})} \frac{d^2 N_\gamma}{dE\,dx} dE\,dx \simeq 875 \sin^2 \theta_c \, L \tag{4.65}$$

where

$$\sin^2 \theta_c = 1 - \cos^2 \theta_c = 1 - \frac{1}{n^2 \beta^2} \tag{4.66}$$

and the variation with wavelength of the refractive index was neglected.
From these relations it is clear that for a fixed number of radiated photons, the length of the Cherenkov counter can be minimised by increasing the refractive index of the gas. However, in order to be able to distinguish between pions and kaons, by operating the detector in threshold mode, the maximum n that can be used is the value matching the kaon threshold velocity, $\beta = 1/n$. Writing the kaon threshold velocity in terms of its momentum, p_K, gives for the refractive index,

$$n^2 = \frac{1}{\beta^2} = \frac{m_K^2 + p_K^2}{p_K^2} \simeq 1 + 6 \times 10^{-4} \tag{4.67}$$

and, inserting this relation in Eq. 4.66, the angle of the Cherenkov photons emitted by the pion is then

$$\sin^2 \theta_c = 1 - \frac{1}{n^2 \beta_\pi^2} = 1 - \frac{m_\pi^2 + p_\pi^2}{m_K^2 + p_K^2} = \frac{m_K^2 - m_\pi^2}{m_K^2 + p_K^2} \tag{4.68}$$

where use was made of the assumption $p_K = p_\pi$. From Eqs. 4.65 and 4.68, the length of the Cherenkov counter is then

$$L = \frac{N_\gamma}{875} \frac{m_K^2 + p_K^2}{m_K^2 - m_\pi^2} \, \text{cm} = \frac{N_\gamma}{875} \frac{1 + \left(\frac{p_K}{m_K}\right)^2}{1 - \left(\frac{m_\pi}{m_K}\right)^2} \, \text{cm} \simeq \frac{N_\gamma}{875} \left(\frac{p_K}{m_K}\right)^2 \, \text{cm}, \quad (4.69)$$

using the approximations $p_K^2 \gg m_K^2$ and $m_\pi^2 \ll m_K^2$.
Taking $N_\gamma = 200$, $m_\pi = 139.58 \, \text{MeV}$ and $m_K = 493.68 \, \text{MeV}$ the minimum length of the gas Cherenkov counter is

$$L \simeq 3.6 \, \text{m}. \qquad (4.70)$$

21. *Energy resolution of a scintillator.* A scintillator emits 10^4 photons/MeV. Calculate the resolution (FWHM) obtainable for 4 MeV particles assuming a light collection efficiency of 1.

If we have a scintillator that produces 10^4 photons per MeV and a particle of 4 MeV, the number of photons produced is:

$$N_\gamma = 4 \cdot 10^4 \, \text{photons}. \qquad (4.71)$$

The resolution will be given by the error on the measurement. Considering that we have Poissonian fluctuations, the error and therefore the resolution is given by:

$$\sigma = \sqrt{N_\gamma} = 63. \qquad (4.72)$$

22. *Electromagnetic calorimeters (16).* Electromagnetic calorimeters have usually 20 radiation lengths of material. Calculate the thickness (in cm) for calorimeters made of BGO, $PbWO_4$ (as in the CMS experiment at LHC), uranium, iron, tungsten, and lead. Take the radiation lengths from Appendix B or from the Particle Data Book.

Let us compute the thickness, and compare it to the weight per unit area per radiation length:

$$L_{BGO} = 22.3 \, \text{cm} \; ; \quad \rho X_0 = 7.97 \, \text{g/cm}^2$$
$$L_{PBWO_4} = 17.8 \, \text{cm} \; ; \quad \rho X_0 = 7.39 \, \text{g/cm}^2$$
$$L_U = 6.4 \, \text{cm} \; ; \quad \rho X_0 = 6.00 \, \text{g/cm}^2$$
$$L_{Fe} = 35.2 \, \text{cm} \; ; \quad \rho X_0 = 13.84 \, \text{g/cm}^2$$
$$L_W = 7.0 \, \text{cm} \; ; \quad \rho X_0 = 6.76 \, \text{g/cm}^2$$
$$L_{Pb} = 11.2 \, \text{cm} \; ; \quad \rho X_0 = 6.37 \, \text{g/cm}^2 \, .$$

Besides uranium, a material quite expensive and complicated to treat, if your main problem is space a good material to build a converter for an electromagnetic calorimeter is lead, or, even better, tungsten—but tungsten will be a bit heavier and quite more expensive. Iron is cheap and performs reasonably well. BGO and lead tungstate ($PbWO_4$) occupy a space a larger than Pb or W, but they are active materials, and thus you can use them for a very performant homogeneous calorimeter (you do not need to add a sensitive detector).

23. *Particle accelerators.* What is the function of quadrupoles in a particle accelerator?

One uses in accelerators a sequence of two types of quadrupoles, called "F" (horizontally focusing but vertically defocusing) and "D" (vertically focusing but horizontally defocusing) - Maxwell's equations show that it is impossible for a quadrupole to focus in both planes at the same time. Figure 4.2 shows an example of a quadrupole focusing in the vertical direction a beam of positively charged particles going into the image plane. If an F quadrupole and a D quadrupole are placed next, and the distance has been correctly chosen, the overall effect is focusing in both horizontal and vertical planes. A lattice can then be built up enabling the transport of the beam over long distances-for example an entire acceleration ring. A common layout is the so-called FODO lattice consisting of a basis of a focusing quadrupole, "nothing" or a bending magnet, a defocusing quadrupole and another length of "nothing".

24. *The HERA collider (17).* The HERA accelerator collided protons at energy $E_p \simeq 920$ GeV with electrons at $E_e \simeq 27.5$ GeV. Which value of $\tilde{E}_e$ would be needed to obtain the same center-of-mass energy at an ep fixed target experiment?

The center-of-mass energy E_{cm} of HERA can be computed as

$$E_{cm}^2 \simeq |(E_p, E_p, 0, 0) + (E_e, -E_e, 0, 0)|^2 = (E_p + E_e)^2 - (E_p - E_e)^2$$
$$= 4E_p E_e \simeq (318\,\text{GeV})^2$$

assuming the beam axis to be the x axis, and taking the proton direction to be positive. Note that masses are negligible with respect to momenta.

In the hypothesis that electrons collide with protons at rest, one would have

$$E_{cm}^2 \simeq |(m_p, 0, 0, 0) + (E_e, -\tilde{E}_e, 0, 0)|^2 = (m_p + \tilde{E}_e)^2 - (m_p - \tilde{E}_e)^2$$
$$= 4m_p \tilde{E}_e \simeq (318\,\text{GeV})^2 \implies \tilde{E}_e \simeq 27\,\text{TeV}.$$

Consider the fact that, due to energy losses due to bremsstrahlung, the record energy for the acceleration of an electron is of about 110 GeV (obtained at the CERN LEP collider in phase II).

Fig. 4.2 Magnetic field lines of a quadrupole field in the plane transverse to the beam direction. The arrows marked as "B" (red in color) show the direction of the magnetic field while the "F" (blue) arrows indicate the direction of the Lorentz force on a positive particle going into the image plane. Source: Wikimedia Commons

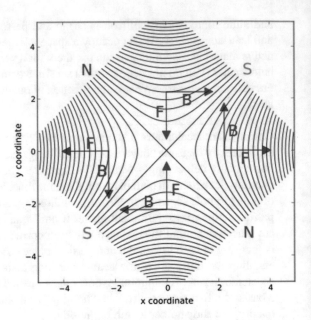

25. *The LHC collider (18).* What is the maximum energy for a tunnel 27 km long with a maximum magnetic field in the vacuum tube of 8.36 T?

Since the perimeter of the LHC is is 27 km, its radius is $R = 4140$ m. Having the maximum magnetic field $B = 8.36$ T, the maximum momentum is given by:

$$p = 0.3\, BR \simeq 10^4 \text{GeV}/c. \tag{4.73}$$

The maximum energy would therefore be:

$$E_{\max} = \sqrt{p^2 c^2 + m^2 c^4} \simeq 10\,\text{TeV}. \tag{4.74}$$

26. *Focusing in the LHC (19).* The diameter of the vacuum tube in LHC is 18 mm. How many turns and for how long can a proton beam stay vertically in the tube if you do not focus it?

If the diameter of the tube is 18 mm, let us assume that the proton falls just but its own gravity, to fall half of the diameter of the tube it would take:

$$x = \frac{gt^2}{2}.$$

If $x = 9$ mm and $g = 9.8$ m/s^2:

$$t \simeq 4.5 \times 10^{-2}\,\text{s}.$$

Assuming a frequency of $\sim 10^4$ Hz for the LHC, the particle could survive for 450 revolutions without hitting the walls.

27. *Collisions in the LHC (20).* The nominal parameters for the LHC are 2835 bunches each with 10^{11} protons in each ring which collide which each other once in each detector. The frequency of revolution is $f \sim 10^4$ Hz, the beam size is $\sigma_x = \sigma_y = 16\,\mu$m the cross section for protons is $\sigma_{pp} \sim 100$ mb. How many collisions of protons are there in one second?

The transverse area of a beam can be approximated to $A = 4\pi\sigma_x\sigma_y$. The instantaneous luminosity is given by:

$$\mathscr{L} = \frac{fn_1n_2N_b}{A} \simeq 8.8 \times 10^{34}\mathrm{cm}^{-2}\mathrm{s}^{-1} \tag{4.75}$$

with $n_1 = n_2 = 10^{11}$ and $N_b = 2835$.
The total number of collision in $\Delta t = 1$ s, can be computed through

$$\mathscr{L}_{int} = \mathscr{L}\,\Delta t\,\sigma_{pp} \simeq 2.25 \times 10^9, \tag{4.76}$$

where σ_{pp} was converted from 100 mb to 10^{-25} cm^2.

28. *Luminosity (21).* How much integrated luminosity does an experiment need to collect in order to measure at better than 1% the rate of a process with cross section of 1 pb?

The integrated luminosity is given by the equation:

$$\mathscr{L}_{int} = \int \mathscr{L}\,dt$$

with $\mathscr{L}$ the differential luminosity. Taking into account that to get a single event from an event with a cross section $\sigma = 1$ pb we need an integrated luminosity of $\mathscr{L} = 1$ pb^{-1}, if we want to have an uncertainty of 1%, the total number of events would be given by

$$\frac{\sqrt{N}}{N} = 0.01$$

since the distribution is Poissonian. This leads to $N = 10^4$ collisions, corresponding to an integrated luminosity of:

$$\mathscr{L}_{int} = 10\,000\,\mathrm{pb}^{-1} = 10\,\mathrm{fb}^{-1}.$$

29. *Luminosity measurement at the LEP collider (22).* The luminosity at the Large Electron–Positron Collider (LEP) was determined by measuring the elastic e^+e^- scattering (Bhabha scattering) as its cross section at low angles is well known

from QED. In fact, assuming small polar angles, the Bhabha scattering cross section integrated between a polar angle θ_{min} and θ_{max} is given at first order by

$$\sigma \simeq \frac{1040\,\text{nb}}{s\,/\,\text{GeV}^2} \left(\frac{1}{\theta_{max}^2 - \theta_{min}^2} \right).$$

Determine the luminosity of a run of LEP knowing that this run lasted 4 h, and the number of identified Bhabha scattering events was 1200 in the polar range of $\theta \in [29; 185]$ mrad. Take into account a detection efficiency of 95 % and a background of 10 % at $\sqrt{s} = m_Z$.

The luminosity can be obtained through the following formula:

$$\mathscr{L} = \frac{N_B - N_{bkg}}{\varepsilon \sigma_B \Delta t} \tag{4.77}$$

with N_B the number of detected events, N_{bkg} the number of background events, ε the detection efficiency, σ_B the Bhabha scattering cross section and Δt the time of run.
Knowing that:

- $\varepsilon = 0.95$;
- $N_B = 1200$;
- $N_{bkg} = 0.1\,N_B$;
- $\Delta = 3600 \times 4 = 14400\,\text{s}$;
- $\sigma_B = 3.75$, where it was used $\theta_{min} = 29\,\text{mrad}$, $\theta_{max} = 185\,\text{mrad}$ and $s = (M_Z)^2 = (91.2)^2\,\text{GeV}^2$,

then the luminosity of the run is

$$\mathscr{L} \simeq 0.021\,\text{nb}^{-1}\,\text{s}^{-1} = 2.1 \times 10^{31}\,\text{cm}^{-2}\,\text{s}^{-1} \tag{4.78}$$

where we used the relation $1\,\text{b} = 10^{-24}\,\text{cm}^2$ to perform the conversion.

30. *Luminosity and cross section (23).* The cross section of a reaction to produce the Z boson at the LEP e^+e^- collides is 32 nb at the beam energy 91 GeV. How long did LEP have to wait for the first event if the luminosity was 23×10^{30} $\text{cm}^{-2}\text{s}^{-1}$?

The luminosity of the beams is given by:

$$\int \mathscr{L}\,dt = 23 \times 10^{30}t.$$

To get one event, the luminosity has to be equal to the cross section:

$$23 \times 10^{30} t = 32 \times 10^{33},$$

therefore the time has to be:

$$t \simeq 1390\,\text{s}.$$

31. *Synchrotron radiation (24).* Consider a circular synchrotron of radius R_0 which is capable of accelerating charged particles up to an energy E_0. Compare the radiation emitted by a proton and an electron and discuss the difficulties to accelerate these particles with this technology.

Electrons lose energy due to bremsstrahlung at a rate $\sim 10^{13}$ times higher than the protons, the reason being that the radiative power scales with $1/m^4$, where m is the mass of the charged particle. Furthermore, radiation losses in the presence of magnetic field also scale with B^2. Therefore, a circular collider, which needs to apply all the way intense magnetic fields to bend and confine the particle's trajectories, cannot be used to accelerate electrons efficiently at very-high energies due to prohibitive, intense radiative losses, a factor which is less critical for protons.

32. *Initial state radiation (25).* The effective energy of the elastic e^+e^- scattering can be changed by the radiation of a photon by the particles of the beam (initial radiation), which is peaked at very small angles. Supposing that a measured e^+e^- pair has the following transverse momenta: $p_1^t = p_2^t = 5\,\text{GeV}$, and the radiated photon is collinear with the beam and has an energy of $10\,\text{GeV}$, determine the effective energy of the interaction of the electron and positron in the center-of-mass, $\sqrt{s_{e^+e^-}}$. Consider that the beam was tuned for $\sqrt{s} = m_Z$.

Let us consider that the electron has a 4-momentum P_1 and the positron a momentum P_2. Assuming that the electron radiates a photon with energy E_γ just before interacting with the positron we have that

$$P_1 = (E - E_\gamma, 0, 0, E - E_\gamma) \;\; ; \quad P_2 = (E, 0, 0, -E) \qquad (4.79)$$

where the masses of the leptons were neglected.
Thus the total 4-momentum immediately before the collision can be written as

$$P_\mu^i = P_1 + P_2 = (2E - E_\gamma, 0, 0, -E_\gamma). \qquad (4.80)$$

As the beam was tuned so that $\sqrt{s} = m_Z$, then $E \equiv E_{\text{beam}} = m_Z/2$. Therefore the effective center-of-mass energy of the interaction can be computed as:

$$\sqrt{s'_{e^-e^+}} = \sqrt{P^i_\mu P^{\mu,i}} = 2E\sqrt{1 - \frac{E}{E_\gamma}} \simeq 80.6\,\text{GeV}. \tag{4.81}$$

33. *Cyclotrons vs. synchrotrons.* What are the essential differences between a cyclotron and a synchrotron?

The main difference is that a cyclotron accelerates the particles in a spiral since the magnetic field is constant, whereas the synchrotron adjusts the magnetic field to keep the particles in a circular orbit.

34. *Maximum energy of LEP.* The LEP dipoles allowed a maximum magnetic field of about 0.135 T and covered 2/3 of the 27 km accumulation ring. What was the maximum energy that can be reached by the electrons?

The maximum momentum reached by the electrons is given by

$$p = 0.3BR. \tag{4.82}$$

Taking into account that the diameter of the system is 27 km, its radius is $R = 4140$ m. Since only 2/3 of the ring was covered by magnets, the momentum is given by:

$$p = 0.3 \times 2/3 \times 0.135 \times 4140 = 112\,\text{GeV}/c. \tag{4.83}$$

Taking into account that the rest energy of the electron is much smaller than its kinetic energy, we can safely say that the maximum energy is:

$$E \simeq 112\,\text{GeV}. \tag{4.84}$$

35. *Muon collider.* Which prospects the construction of a high-energy $\mu^+\mu^-$ collider would open and which technological difficulties must be solved to achieve this goal?

Lepton collisions are relatively clean with respect to hadron collisions; however, electron mass is very small and a lot of energy in electron acceleration is wasted due to bremsstrahlung and synchrotron radiation. Muons have the advantage of being $\sim$200 times heavier than electrons. However, muons have a short lifetime of about 2 μs. Muon-muon collisions are in principle achievable but require being able to produce and manipulate beams of muons.

36. *Bending radius of cosmic rays from the Sun (26).* What is the bending radius of a solar proton, 1 MeV kinetic energy, in the Earth's magnetic field (0.5 G), for vertical incidence with respect to the field?

If the kinetic energy of a proton is $E_k = 1$ MeV and the rest energy is $E_R \simeq 1$ GeV, the γ factor can be calculated as:

$$E = E_k + E_R = \gamma mc^2$$

and $\gamma = 1.001$. The momentum can be calculated from the 4-vector formula:

$$E^2 - p^2 c^2 = m^2 c^4$$

and the total momentum is finally given by:

$$pc = mc^2 \sqrt{\gamma^2 - 1} = 45 \, \text{MeV}.$$

The bending radius of a proton in a magnetic field is given by:

$$p = 0.3 BR$$

with R in meter, B in tesla and p in GeV/c. Substituting, we finally have:

$$R = \frac{45 \times 10^{-3}}{0.3 \times 0.5 \times 10{-4}} = 3000 \, \text{m}.$$

37. *Low Earth Orbit (27).* Low-Earth Orbits (LEOs) are orbits between 300 km and 2000 km from the ground; the altitude is optimal in order to protect them from cosmic rays, thanks to the Van Allen radiation belts. Due to atmospheric drag, satellites do not usually orbit below 350 km. What is the velocity of an Earth satellite in a LEO and how does it compare to the escape velocity from Earth? How many revolutions per day does it make? Apply to the cases of the International Space Station (ISS), orbiting at an average quota of about 410 km, and of the *Fermi* satellite, flying at an average quota of 550 km.

The speed of the satellite is coming from the equality of the centripetal force and the gravitational one:

$$\frac{mv^2}{R} = \frac{GMm}{R^2}.$$

Knowing that $GM/R_E^2 \simeq 9.8$ m/s^2, where $R_E \simeq 6370$ km is the radius of the Earth, the speed of a satellite orbiting at 410 km from the surface of the Earth is given by:

$$v = \sqrt{\frac{GM}{R_E^2}\frac{R_E^2}{R^2}} = 7.9 \times 10^3 \text{ m/s}.$$

The escape velocity v_e of a satellite is given by considering that the satellite escapes from the Earth and ends up with 0 velocity and potential:

$$-\frac{GMm}{R_E} + \frac{1}{2}mv_e^2 = 0$$

and the escape velocity

$$v_e = \sqrt{2\frac{GM}{R_E}} = \sqrt{2\frac{GM}{R_E^2}R_E} = 1.1 \times 10^4 \text{m/s}.$$

The escape velocity and the velocity of a satellite at a distance R from the Earth compare as:

$$\frac{v}{v_e} = \sqrt{\frac{R_E}{2R}}.$$

To calculate how much time it takes to complete an orbit, we first calculate the longitude traveled by the satellite in an orbit:

$$L = 2\pi R = 4.44 \times 10^7 \text{m}$$

and the period is then, at an altitude of 410 km:

$$T_{410} = \frac{L}{v} = \frac{4.44 \times 10^7}{7.9 \times 10^3} = 5.6 \times 10^3 \text{s} \simeq 1.5 \text{h}.$$

One can easily rescale to the Fermi orbit, at 550 km, using the third Kepler's law:

$$\left(\frac{T_{550}}{T_{410}}\right)^2 = \left(\frac{6370 + 550}{6370 + 410}\right)^3 \implies T_{550} = T_{410}\left(\frac{6370 + 550}{6370 + 410}\right)^{3/2} \simeq 1.03\, T_{410}$$

(i.e., the orbit of Fermi is just 3% slower than the orbit of the ISS).
A satellite rolling fast, like the *Fermi* satellite, can "see" the full sky every period (apart from the occultation by the Earth). A satellite with a slower rolling can "see" the sky every $4\pi/(2\pi/5) \simeq 10$ periods, i.e., every 15 h.

38. *Visual scan of an event in a collider detector.* The event display below shows the cross section of a collision in a particle physics detector at a collider (ATLAS at the CERN LHC). Discuss the experimental signatures expect in this experiment for electrons, photons muons, tau-leptons, neutrinos, hadrons and quarks.

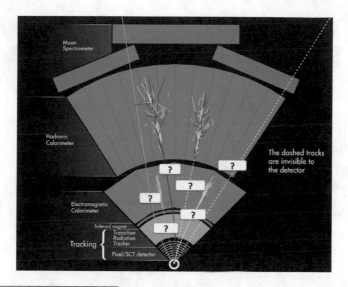

Different particles leave different signals in the various particle detector components allowing almost unambiguous identification:

a. $e^{\pm}$: a track + energy in electromagnetic calorimeters (with absorption);
b. γ: no track + energy in electromagnetic calorimeters (with absorption);
c. $\mu^{\pm}$: a track, small calorimetric energy deposits, penetrating;
d. $\iota^{\pm}$: decay products;
e. ν: not detected;
f. hadrons: track (if charged) + calorimetric energy deposits (in particular in hadronic calorimeters);
g. quarks: seen as jets of hadrons.

These criteria are schematized in the figure below, together with the solution.

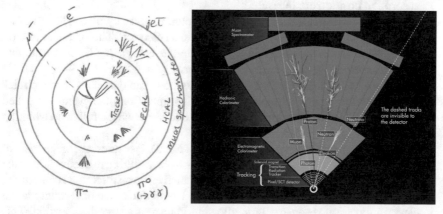

This problem has been taken from the lectures of Dr. Tina Potter at the University of Cambridge.

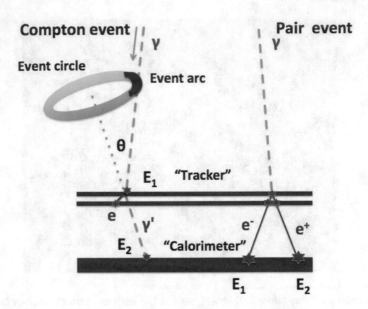

Fig. 4.3 Representative event topologies for a Compton event (left) and for a pair event (right). Photon tracks are dashed (pale blue), and electron and/or positron tracks solid (red). Courtesy of Alex Moiseev

39. *Reconstruction of a Compton interaction event in a gamma-ray telescope (6).*
    Detecting gamma rays by Compton scattering in a gamma-ray telescope with hodoscopic structure (Fig. 4.3) is more complicated than for pair production. The Compton scattering of the incident photon occurs in one of the tracker planes, creating an electron and a scattered photon. The tracker measures the interaction location, the electron energy, and in some cases the electron direction. The scattered photon can be absorbed in the calorimeter where its energy and absorption position are measured.

    Suppose that an incident gamma-ray Compton scatters by an angle Θ in one layer of the tracker, transferring energy E_1 to an electron. The scattered photon is then absorbed in the calorimeter, depositing its energy E_2, Demonstrate that the scattering angle is given by $\cos \Theta = m_e c^2/E_2 + m_e c^2/(E_1 + E_2)$, where m_e is the electron mass. With this information, one can derive an "event circle" from which the original photon arrived—this sort of Compton events are called "untracked". Multiple photons from the same source enable a full deconvolution of the image, using probabilistic techniques.

    For energetic incident gamma rays (above ∼1 MeV), measurement of the track of the scattered electron might in addition be possible, resulting in a reduction of the event circle to a definite direction. If the scattered electron direction is measured, the event circle reduces to an event arc with length due to the uncertainty in the electron direction reconstruction, allowing improved source localization. This event is called "tracked", and its direction reconstruction is somewhat similar to that for pair event—the primary photon direction is reconstructed from the direction and energy of two secondary particles: scattered electron and photon. Comment.

Tracked events constrain completely the kinematics, and are somehow equivalent to pair production events. They are, however, rare, since the process of ejection of electrons is not trivial, especially at the lowest (hundreds keV) energies.

Redundant kinematic information from multiple interactions (i.e., when the secondary photon in turn undergoes a Compton scattering) are also of help in localizing the sources of emission. Also in this case the occurrence of the process is rare at the lowest energies.

At the lowest energies one must thus make use of untracked events for statistically reconstructing the direction of emission. The uncertainty in the event circle reconstruction of untracked events is reflected in its width and is due to the uncertainties in direction reconstruction of the scattered photon and the energy measurements of the scattered electron (E_1) and the scattered photon (E_2). Multiple photons from the same source enable a full deconvolution of the image, using probabilistic techniques. If it exists, the simultaneous information from tracked events is of great help.

40. *Electromagnetic showers in the atmosphere (28).* In the isothermal approximation, the depth x of the atmosphere at a height h (i.e., the amount of atmosphere above h) can be approximated as

$$X_{\max} \simeq X e^{-\frac{h}{7\,\mathrm{km}}}, \tag{4.85}$$

with $X \simeq 1030$ g/cm^2. If a shower is generated by a gamma ray of $E = 1$ TeV penetrating the atmosphere vertically, considering that the radiation length X_0 of air is approximately 36.6 g/cm^2 (440 m) and its critical energy E_c is about 88 MeV, calculate the height h_M of the maximum of the shower in the Heitler model and in the Rossi's approximation B (Table 4.1).

In the Heitler model,

$$X_{\max} = X_0 \left(1 + \frac{\ln(E/E_c)}{\ln 2} \right) \simeq 530 \, \mathrm{g\,cm}^{-2} \tag{4.86}$$

where $E = 1$ TeV, $E_c = 88$ MeV and $X_0 = 36.6 \, \mathrm{g\,cm}^{-2}$.
In the Rossi approximation B model,

$$X_{\max} = X_0 \left[\ln\left(\frac{E}{E_c} \right) - 0.5 \right] \simeq 325 \, \mathrm{g\,cm}^{-2}. \tag{4.87}$$

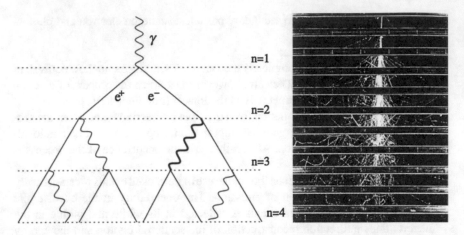

Fig. 4.4 Left: Scheme of the Heitler approximation for the development of an electromagnetic shower. From J. Matthews, Astropart. Phys. 22 (2005) 387. Right: Image of an electromagnetic shower developing through a number of brass plates 1.25 cm thick placed across a cloud chamber (from B. Rossi, "Cosmic rays", McGraw-Hill 1964)

Inverting Eq. (4.85) to obtain the height of X_{max} one gets

$$h_M = -7 \, \text{km} \, \ln \left(\frac{X_{max}}{X} \right). \tag{4.88}$$

Hence,

- Heitler: $h_M \simeq 4.9 \, \text{km}$
- Rossi: $h_M \simeq 8.6 \, \text{km}$
- MC: $h_M \simeq 9.5 \, \text{km}$.

For comparison, the value for a full Monte Carlo simulation is $h_M \simeq 9.5 \, \text{km}$.

41. *Extensive electromagnetic air showers (29).* The main characteristic of an electromagnetic shower (say, initiated by a photon) can be obtained using a simple Heitler model. Let E_0 be the energy of the primary particle and consider that the electrons, positrons and photons in the cascade always interact after traveling a certain atmospheric depth $d = X_0$, and that the energy is always equally shared between the two particles. With this assumptions, we can schematically represent the cascade as in Fig. 4.4, left.

   a. Write the analytical expressions for the number of particles and for the energy of each particle at depth X as a function of d, n and E_0.
   b. The multiplication of the cascade stop when the particles reach a critical energy, E_c (when the decay probability surpasses the interaction probability). Using the expressions obtained in the previous question, write as a

function of E_0, E_c and $\lambda = d/\ln 2$, the expressions, at the shower maximum, for:

  i. the average energy of the particles,
  ii. the number of particles, N_{max},
  iii. the atmospheric depth X_{max}.

---

a. At the n-th generation,

$$X = n \times d \tag{4.89}$$

and the number of produced particles is simply

$$N = 2^n . \tag{4.90}$$

As the energies of the particles are the same at the end of each generation, the energy of each particle is equal is the primary energy divided by the number of particles at this level, i.e.,

$$E_i = \frac{E_0}{2^n} . \tag{4.91}$$

b. i. By construction:

$$E = E_c. \tag{4.92}$$

  ii.

$$N_{max} = \frac{E_0}{E_c} . \tag{4.93}$$

iii. Using Eq. (4.90),

$$N_{max} = 2^{n_{max}} \Leftrightarrow n_{max} = \frac{\ln(N_{max})}{\ln 2} \tag{4.94}$$

where n_{max} is the maximum number of levels.
Since $d = \lambda \ln 2$ the maximum atmospheric depth can be written as

$$X_{max} = n_{max} \times d = \frac{\ln(N_{max})}{\ln 2} d = \lambda \ln \left(\frac{E_0}{E_c} \right) \tag{4.95}$$

where Eq. (4.94) and (4.93) were used to evaluate n_{max} and N_{max}, respectively.

42. *Extensive hadronic air showers (30).* Consider a shower initiated by a proton of energy E_0. We will describe it with a simple Heitler-like model: after each depth d an equal number of pions, n_π, and each of the 3 types is produced: π^0,

π^+, π^-. Neutral pions decay through $\pi^0 \to \gamma\gamma$ and their energy is transferred to the electromagnetic cascade. Only the charged pions will feed the hadronic cascade. We consider that the cascade ends when these particles decay as they reach a given decay energy E_{dec}, after n interactions, originating a muon (plus an undetected neutrino).

a. How many generations are needed to have more that 90 % of the primary energy, E_0 in the electromagnetic component?
b. Assuming the validity of the superposition principle, according to which a nucleus of mass number A and energy E_0 behaves like A nucleons of energy E_0/A, derive expressions for:
    i. the depth where this maximum is reached, X_{max},
    ii. the number of particles at the shower maximum,
    iii. the number of muons produced in the shower, N_μ.

In this case:

$$N_{tot} = n_\pi^n \; ; \; N_{ch} = \left(\frac{2}{3}n_\pi\right)^n \; ; \; E_i = \frac{E_0}{n_\pi^n} \tag{4.96}$$

where N_{ch} is the number of charged particles at the level n and n_π the number of pions produced at each interaction.

a. At each interaction $1/3$ of the energy goes into the electromagnetic channel through the π^0 decay. Therefore the energy that remains for the charged particles is

$$E_{ch} = \left(\frac{2}{3}\right)^n E_0. \tag{4.97}$$

Thus the fraction of electromagnetic energy rises as $E_{em} = E_0 - E_{ch}$. Hence

$$\frac{E_{em}}{E_0} = 1 - \left(\frac{2}{3}\right)^n. \tag{4.98}$$

Taking $E_{em}/E_0 = 0.9$ and inverting Eq. (4.98) to obtain the number of generations (levels), one gets

$$n = \frac{\ln(0.1)}{\ln(2/3)} \simeq 5.7 \text{ generations.} \tag{4.99}$$

b. i. Let us start by evaluating X_{max} for protons. In this case from Eq. (4.96), and recalling that the shower development stops when the energy of the particles reaches E_{dec}, one obtains

$$X_{max} = d \times n_{dec}. \tag{4.100}$$

The maximum number of generations is $E_{\max} = E_0/n_\pi^{n_{dec}}$. Inverting this last expression one gets

$$n_{dec} = \frac{\ln(E_0/E_{dec})}{\ln(n_\pi)}, \tag{4.101}$$

which leads, using Eq. (4.100), to

$$X_{\max} = d\frac{\ln(E_0/E_{dec})}{\ln(n_\pi)}. \tag{4.102}$$

For iron we have 56 nucleons (i.e. the atomic number $A = 26$ protons + 30 neutrons). Using the superposition principle each nucleon carries $E_0/56$ of the primary energy. Substituting in Eq. (4.102),

$$X_{\max} = d\frac{\ln\left(\frac{E_0}{AE_{dec}}\right)}{\ln(n_\pi)} = \frac{d}{\ln(n_\pi)}\left[\ln\left(\frac{E_0}{E_{dec}}\right) - \ln A\right]. \tag{4.103}$$

Notice that the $X_{\max}$ evolution with energy is the same for proton and iron and the curves are separated by a constant term: $\ln(A)$.

ii. Again starting with protons we have for the number of particles at the shower maximum

$$N_{\max} = n_\pi^{n_{dec}} = \frac{E_0}{E_{dec}}. \tag{4.104}$$

For iron primaries,

$$N_{\max} = An_\pi^{n_{dec}}. \tag{4.105}$$

Using the superposition principle and the result of Eq. (4.101) it is easy to see that

$$N_{\max} = An_\pi^{\frac{\ln\left(\frac{E_0}{AE_{dec}}\right)}{\ln(n_\pi)}} = A\frac{E_0}{AE_{dec}} = \frac{E_0}{E_{dec}} \tag{4.106}$$

which means that the number of particles at the shower maximum does not depend on the primary mass composition.

iii. The number of muons in the shower, for this simplified model, is given by

$$N_\mu = N_{ch}|_{X=X_{\max}} = \left(\frac{2}{3}n_\pi\right)^{n_{dec}}. \tag{4.107}$$

Therefore, for proton primaries, using Eq. (4.101),

$$N_\mu = \left(\frac{2}{3}n_\pi\right)^{\frac{\ln(E_0/E_{dec})}{\ln(n_\pi)}} = \left[\left(\frac{2}{3}n_\pi\right)^{\log_{\frac{2}{3}n_\pi}(E_0/E_{dec})}\right]^{\frac{\ln\left(\frac{2}{3}n_\pi\right)}{\ln(n_\pi)}}$$

$$= \left(\frac{E_0}{E_{dec}}\right)^{\frac{\ln\left(\frac{2}{3}n_\pi\right)}{\ln(n_\pi)}} = \left(\frac{E_0}{E_{dec}}\right)^\beta \tag{4.108}$$

where β is a parameter related with the multi-particle production in the hadronic interactions, in particular, the ratio between the hadronic and the electromagnetic component of the interaction.

For iron, using again the superposition principle and the final result of Eq. (4.108), one gets

$$N_\mu = A\left(\frac{E_0/A}{E_{dec}}\right)^\beta = A^{1-\beta}\left(\frac{E_0}{E_{dec}}\right)^\beta. \tag{4.109}$$

43. *Cherenkov telescopes (31).* Suppose you have a Cherenkov telescope with 7 m diameter, and your camera can detect a signal only when you collect 100 photons from a source. Assuming a global efficiency of 0.1 for the acquisition system (including reflectivity of the surface and quantum efficiency of the PMT), what is the minimum energy (neglecting the background) that such a system can detect at a height of 2 km a.s.l.?

Considering a global efficiency of 0.1, the minimum number of Cherenkov photons that must be collected by the telescope in order to detect a signal is $N_\gamma = 1000$. Since the area of the telescope is $38.5\,\text{m}^2$, this corresponds to a minimum density of about 26 photons/m^2 in the Cherenkov "light pool" of the shower. The mean density of Cherenkov photons at a height of 2 km a.s.l. is about 10 photons/m^2, for a primary of 100 GeV, and about 150 photons/m^2 for a primary of 1 TeV (see Chap. 4 of the textbook). Taking the conservative number of 100 photons/m^2 per 1 TeV of primary energy, the minimum energy of the primary yielding the required density of photons is thus of about 260 GeV. This calculation is conservative: in particular, with dedicated triggers, the minimum number of detectable photons can go down by an order of magnitude, albeit with a larger background, and the actual threshold of a Cherenkov telescope with this diameter can go down to $\sim$100 GeV.

44. *Cherenkov telescopes and muon signals (32).* If a muon travels along the axis of a Cherenkov imaging telescope with parabolic mirror, the image of its Cherenkov emission in the focal plane is a circle (the Cherenkov angle can be approximated as a constant, rather small in air: $\theta \simeq 0.02$ rad): in the jargon of gamma-ray astronomers, this topology is called "muon ring". Demonstrate it, considering

Fig. 4.5 Section of a
parabolic reflector, showing
as an example the reflection
from a beam parallel to the
axis of the paraboloid. From
https://amsi.org.au/
ESA_Senior_Years/
SeniorTopic2/2a/
2a_2content_13.html

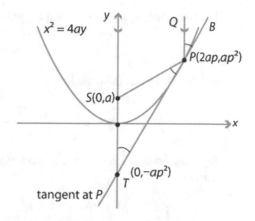

the focal distance of the mirror, R, small with respect to the typical emission
distances of Cherenkov radiation, which are of the order of kilometers.

Let us consider the section of a parabolic mirror as in figure. If the equation of
the parabola is

$$y = \frac{x^2}{4a}$$

($a > 0$), it is known that the light rays parallel to the axis of the parabola are all
reflected in the *focus* of coordinates $S = (0, a)$ (Fig. 4.5).
Our problem here is a bit different, as we deal with a particle traveling along
y towards the vertex and steadily emitting photons along a cone from different
positions along its path. This problem has an axial symmetry and a left-right
symmetry: it is thus enough to demonstrate that all the parallel beams hitting the
parabola on the upper right quadrant are reflected to a single point on the focal
line $y = a$.
To describe the direction of the Cherenkov light beams we consider a generic line
$y = mx + q$, with m fixed, crossing the parabola in a generic point $(2ap, p^2)$
with $p > 0$. Being

$$m_T = \frac{dy}{dx} = \frac{x}{2a} = p$$

the angular coefficient of the line tangent to the parabola in $(2ap, p^2)$, the angular
coefficient m_R of the line describing the reflection of a line of angular coefficient
m over the line of angular coefficient $m_T = p$ is given by the relation

$$\frac{p - m_R}{1 + pm_R} = \frac{m - p}{1 + mp} \implies m_R = \frac{m(p^2 - 1) + 2p}{2mp - p^2 + 1}.$$

Thus the generic lines describing the reflected beams can be written as

$$y = p^2 + m_R(x - 2ap)$$

which all intersect the focal line $y = a$ in the abscissa which solves the equation

$$a = p^2 + \frac{m(p^2 - 1) + 2p}{2mp - p^2 + 1}(x - 2ap).$$

Thus, the point of intersection is (assuming $p \ll 1$, i.e., a small curvature of the mirror)

$$P_F = \left(\frac{a(2mp^3 + 3p^2 + 1) + p^2(-2mp + p^2 - 1)}{m(p^2 - 1) + 2p}, a \right) \simeq \left(-\frac{a}{m}, a \right).$$

With the same procedure (but the calculations are much more tedious, since one cannot use the simplification given by the axial symmetry), one could demonstrate that the Cherenkov image on the focal plane for a muon traveling in a generic direction is a conical section. In particular, the "muon ring" can be an ellipse.

45. *Imaging Air Cherenkov Telescopes: signals from electromagnetic showers (33).*
    In the isothermal approximation, the depth x of the atmosphere at a height h
    (i.e., the amount of atmosphere above h) can be approximated as

$$x \simeq X e^{-h/7\,\mathrm{km}},$$

with $X \simeq 1030$ g/cm^2. If a shower is generated by a gamma ray of $E = 1$ TeV penetrating the atmosphere vertically, considering that the radiation length X_0 of air is approximately 36.6 g/cm^2 (440 m) and its critical energy E_c is about 88 MeV:

a. If 2000 useful Cherenkov photons per radiation length are emitted by charged particles in the visible and near UV, compute the total number N_γ of Cherenkov photons generated by the shower (note: the critical energy is larger than the Cherenkov threshold).

b. Supposing that the Cherenkov photons are all emitted at the center of gravity of the shower—that in the Heitler approximation is just the maximum of the shower minus one radiation length, compute how many photons per square meter arrive to a detector at a height h_d of 2000 m, supposing that the average attenuation length of photons in air is 3 km, and that the light pool can be derived by a opening of $\sim 1.3°$ from the shower maximum (1.3° is the Cherenkov angle and 0.5°, to be added in quadrature, comes from the intrinsic shower spread). Comment on the size of a Cherenkov telescope, considering an average reflectivity of the mirrors (including absorption in transmission) of 70%, and a photodetection efficiency (including all the chains of acquisition) of 20%.

c. Redo the calculations for $E = 50\,\text{GeV}$, and comment.

Like in Problem (4.40), let us compute the depth of the shower maximum, X_{max}, for the two models and afterwards convert it into an altitude using the atmosphere model provided in the problem

$$X_{max} \simeq X e^{-\frac{h}{7\,km}}. \tag{4.110}$$

In the Heitler model,

$$X_{max} = X_0 \left(1 + \frac{\ln(E/E_c)}{\ln 2}\right) \simeq 530\,\text{g cm}^{-2} \tag{4.111}$$

where $E = 1\,\text{TeV}$, $E_c = 88\,\text{MeV}$ and $X_0 = 36.6\,\text{g cm}^{-2}$.
In the Rossi approximation B model,

$$X_{max} = X_0 \left[\ln\left(\frac{E}{E_c}\right) - 0.5\right] \simeq 325\,\text{g cm}^{-2}. \tag{4.112}$$

Inverting Eq. (4.110) to obtain the height of X_{max} one gets

$$h_M = -7\,\text{km}\, \ln\left(\frac{X_{max}}{X}\right). \tag{4.113}$$

Hence,

- Heitler: $h_M \simeq 4.9\,\text{km}$
- Rossi: $h_M \simeq 8.6\,\text{km}$
- MC: $h_M \simeq 9.5\,\text{km}$

where MC is the value for a full Monte Carlo EAS simulation.

a. The total number of Cherenkov photons generated by the shower is

$$N_\gamma^{total} = \left(\frac{E}{E_c}\right) N_\gamma^{Ch} \simeq 2.27 \times 10^7 \text{ photons} \tag{4.114}$$

where N_γ^{Ch} is the number of Cherenkov photons per radiation length and (E/E_c) the total track length in units of radiation length.

b. In this problem we will use the Rossi approximation B model to evaluate the shower main characteristics for a 1 TeV photon induced shower. It shall be assumed that all the photons are coming from the center of gravity of the shower, t_{med}. This quantity can be computed using

Fig. 4.6 Illustration of the problem main variables

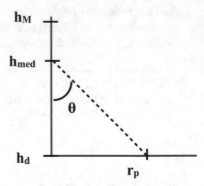

$$t_{med} = t_{max} + 1.7 = \ln\left(\frac{E}{E_c}\right) - 0.5 + 1.7 \simeq 10.5 \qquad (4.115)$$

or, in traversed matter units,

$$X_{med} = t_{med} X_0 \simeq 386 \, g/cm^2 \, . \qquad (4.116)$$

Using the *isothermal* approximation as atmosphere model one obtains for the altitude

$$h_{med} = -7 \ln\left(\frac{X_{med}}{X}\right) \simeq 6.88 \, km \, . \qquad (4.117)$$

Using this altitude as emission point and knowing that the light pool can be derived from an opening angle of $\simeq 1.3°$ (see Fig. 4.6), one gets for r_p

$$r_p = (h_{med} - h_d) \tan \theta \simeq 110.6 \, m. \qquad (4.118)$$

Therefore the number of photons at ground is

$$n_\gamma = \frac{N_\gamma\big|_{ground}}{A} = \frac{N_\gamma \, e^{-(h_{med}-h_d)/3}}{\pi \, r_p^2} = 116 \, photons \, m^{-2} \, . \qquad (4.119)$$

where A is the area of the light pool at ground and $N_\gamma\big|_{ground}$ is the number of Cherenkov photons that reach the ground. N_γ is the total number of photons produced by the shower, calculated in the previous problem, while the exponential term represents the attenuation of these photons while travelling through the atmosphere.

The telescopes of VERITAS and HESS have areas A_{det} of $\simeq 100 \, m^2$. Thus, the number of detected photons is

$$n_\gamma^{det} = n_\gamma \, A_{det} \, \varepsilon_{ref} \, \varepsilon_{acq} \simeq 1600 \, photons \qquad (4.120)$$

where ε_{ref} and ε_{act} are the reflective and acquisition efficiencies, respectively.

c. This problem is solved in the same way by taking into account that the primary energy is now of 50 GeV. We summarize the results in the following table.

E [GeV]	t_{med}	X_{med} [g/cm^2]	h_{med} [km]	r_p [m]	N_γ	n_γ [ph/m^2]	N_γ^{det}
50	7.5	276.0	9.22	163.7	1.1×10^6	1	17
1000	10.5	385.7	6.88	110.6	2.2×10^7	116	1629

For showers induced by photons with $E = 50$ GeV the number of detected Cherenkov photons is extremely low (of the order of the background fluctuations). Therefore these kind of IACT are not suited to measure gamma-ray induced showers below 50 GeV.

46. *Neutrino detectors and the rate of rare events.* The expected rate of UHE neutrinos clearly identified over the background interacting in IceCube from Centaurus A (supposed to be stable) of is 0.46×10^{-3} events/day, and the efficiency for the detection of the interaction with such a strict selection is about 43%. Evaluate the probability to detect at least a clean neutrino in the first year of operation, and in the first 10 years.

The expected number of events in one year is

$$0.46 \times 10^{-3} \times 0.43 \times 365 \simeq 0.094 \, .$$

The probability to see at least one event (i.e., the probability that a Poisson process with expectation value 0.094 results in one or more events) is

$$P_{1 \, year} \simeq 9\% .$$

The expected number of events in 10 years is

$$0.46 \times 10^{-3} \times 0.43 \times 3650 \simeq 0.94 \, .$$

The probability to see at least one event (i.e., the probability that a Poisson process with expectation value 0.94 results in one or more events) is

$$P_{10 \, years} \simeq 61\% .$$

47. *Gravitational waves: different sizes sample different frequencies.* Explain why a larger interferometer is sensitive to gravitational waves of lower frequencies.

In a Michelson interferometer the total phase difference due to passage of a gravitational wave (GW) signal is:

$$\Delta\phi = 4\pi \frac{Lh_+}{\lambda_{laser}} \frac{\sin(\omega_{GW}L/c)}{\omega_{GW}L/c} \cos(\omega_{GW}(t - L/c))$$

where L is the length of the interferometer's arms; ω_{GW} and h_+ refer respectively to the frequency and strain amplitude of the gravitational wave signal.

The goal for a good detection is to have the amplitude of the phase difference as large as possible, so that, given a fixed ω_{GW}, the dependence on L is due to $\sin(\omega_{GW}L/c)$. The optimal length at a given frequency f is:

$$\omega_{GW}\frac{L}{c} = \frac{\pi}{2} \Rightarrow L = \frac{\lambda_{GW}}{4} \Rightarrow L \simeq 750\,\text{km}\left(\frac{100\,\text{Hz}}{f_{GW}}\right).$$

The term $[\sin(\omega_{GW}L/c)]/[(\omega_{GW}L/c)]$ induces a suppression of the sensitivity for $\omega_{GW}L/c \gg 1$ (cutoff frequency).

As expected, the sensitivity of a Michelson interferometer can be increased by increasing the lengths of its arms.

Given the previous equation it is clear that to detect a gravitational wave signal of a frequency of the order of hundreds Hz the optimal length of the arms is hundreds km (not reachable on Earth). The present interferometers employ Fabry-Perot (FP) cavities which bounce the photons back and forth; the amplitude of phase shift due to GW signal in this case is:

$$\mid \Delta\phi_{FP} \mid = \frac{2F}{\pi} \mid \Delta\phi \mid$$

where $\Delta\phi$ is the amplitude phase induced in a standard Michelson interferometer. The factor F is about 400 in the Virgo interferometer.

48. *Energy flux from the first GW signal.* The energy flux (power per unit area) in GWs can be estimated as

$$\Phi \simeq \frac{c^3}{8G}h^2 f^2 \tag{4.121}$$

where $f = \omega/2\pi$ is the frequency of the GW (assumed monochromatic) and h is the RMS amplitude of the strain. GW150914, the first gravitational signal detected, had a peak strain of $h \sim 10^{-21}$ at $f \sim 200$ Hz. Compute the corresponding energy flux.

By inserting the numbers relative to GW150914 in Eq. 4.121 we obtain

$$\Phi \simeq 10^{-3}\text{W/m}^2.$$

This is approximately the electromagnetic energy flux that we receive from the full moon, despite GW150914 being at an estimated distance of ~ 400 Mpc. Guess then why the Moon is so easy to detect while it needs an extremely high technology to "see" gravitational waves.

Chapter 5
Particles and Symmetries

1. *Dirac equation and angular momentum*
   Prove that the Dirac Hamiltonian H_D of a particle in a generic central potential commutes with the total angular momentum operator

$$\mathbf{L} + \frac{1}{2}\boldsymbol{\sigma} .$$

The Dirac Hamiltonian can be written as

$$H_D = \phi(r) - i\mathbf{a}\mathbf{p} + \alpha m$$

where

$$\alpha = \gamma^0 \; ; \; \mathbf{a} = (\gamma^0)^{-1}\gamma$$

(the boldface indicates a 3-vector).
The central potential commutes with any angular operator, and $[\mathbf{L}, \alpha m] = 0$; thus

$$[L, H_D] = -i[\mathbf{L}, \mathbf{a} \cdot \mathbf{p}] .$$

Then, we get

$$[L_j, H_D] = -i[\epsilon_{jkl}r_k p_l, a_i p_i] = -ia_i[\epsilon_{jkl}r_k p_l, p_i]$$

and since $[r_i, p_j] = i\delta_{ij}$

$$- ia_i[\epsilon_{ijk}r_k p_l, p_i] = a_i\epsilon_{jkl}p_l \neq 0 . \tag{5.1}$$

© Springer Nature Switzerland AG 2021
A. De Angelis et al., *Particle and Astroparticle Physics*, Undergraduate Lecture Notes in Physics,
https://doi.org/10.1007/978-3-030-73116-8_5

Spin $\mathbf{S} = \sigma/2$ commutes with the Hamiltonian in the non-relativistic case (the deep reason being that classical mechanics says nothing about spin). What about the Dirac Hamiltonian? One has

$$[S_j, H_D] = -i[S_j, a_i p_i] + m[S_j, \alpha].$$

Since $[S_j, \alpha_i] = \epsilon_{jik} a_k$ and $[S_j, \alpha] = 0$, one has

$$[S_j, H_D] = -ip_i[S_j, a_i] = p_i \epsilon_{jik} a_k = -\epsilon_{jkl} a_i p_l,$$

which is the opposite of the result obtained in (5.1).
Thus

$$\left[\hbar \mathbf{L} + \frac{\sigma}{2}, H_D\right] = 0.$$

2. *Commutation relations (3)*
   Demonstrate that if $\hat{A}$ and $\hat{B}$ are two commuting operators the following relation holds:
$$\exp(\hat{A}) \exp(\hat{B}) = \exp\left(\hat{A} + \hat{B}\right).$$

If $[\hat{A}, \hat{B}] = 0$ then: $\hat{A}, \hat{B}, \exp(\hat{A}), \exp(\hat{B})$ commute.
Let us consider the function

$$f(u) = \exp(u\hat{A}) \exp(u\hat{B}).$$

Its derivatives with respect to u, if $\hat{A}$ and $\hat{B}$ commute, are:

$$\dot{f}(u) = (\hat{A} + \hat{B})f(u)$$
$$\ddot{f}(u) = \frac{1}{2!}(\hat{A} + \hat{B})^2 f(u)$$
$$\dddot{f}(u) = \frac{1}{3!}(\hat{A} + \hat{B})^3 f(u)$$
$$\cdots$$

Let us develop $f(u)$ in Taylor series around $u = 0$ ($f(0) = 1$):

$$f(u) = 1 + (\hat{A} + \hat{B})u + \frac{1}{2!}(\hat{A} + \hat{B})^2 u^2 + \frac{1}{3!}(\hat{A} + \hat{B})^3 u^3 + \cdots$$
$$\equiv \exp[u(\hat{A} + \hat{B})].$$

Then, inserting $u = 1$ one has

$$\exp(\hat{A}) \exp(\hat{B}) = \exp\left(\hat{A} + \hat{B}\right).$$

The same could be verified by "brute force" under the condition $[\hat{A}, \hat{B}] = 0$ using the series expansion of the exponential. On the right side of the equation you get a single infinite series with terms of $(\hat{A} + \hat{B})^j$; on the left side you have two infinite series multiplied together, with terms of $\hat{A}^j$ and $\hat{B}^j$ respectively:

$$(1 + \hat{A} + \frac{1}{2!}\hat{A}^2 + ...)(1 + \hat{B} + \frac{1}{2!}\hat{B}^2 + ...)$$

$$= 1 + (\hat{A} + \hat{B}) + \frac{1}{2!}(\hat{A} + \hat{A}\hat{B} + \hat{B}\hat{A} + \hat{B}^2) + ...$$

$$= 1 + (\hat{A} + \hat{B}) + \frac{1}{2!}(\hat{A} + \hat{B})^2 + ...$$

Our result is a particular case of the so-called Baker-Campbell-Hausdorff formula for the solution for Z of the equation $e^X e^Y = e^Z$:

$$Z = X + Y + \frac{1}{2}[X, Y] + \frac{1}{12}\left([X, [X, Y]] - [Y, [X, Y]]\right) + ...$$

3. *Parity (4)*

Verify explicitly if the spherical harmonics

$$Y_1^{-1}(\theta, \varphi) = \frac{1}{2}\sqrt{\frac{3}{2\pi}} e^{-i\varphi} \sin\theta$$

$$Y_1^0(\theta, \varphi) = \frac{1}{2}\sqrt{\frac{3}{\pi}} \cos\theta$$

$$Y_1^1(\theta, \varphi) = \frac{1}{2}\sqrt{\frac{3}{2\pi}} e^{i\varphi} \sin\theta$$

are eigenstates of the parity operator, and in case they are determine the corresponding eigenvalues.

The parity operator acts on the spherical coordinates by transforming:

$$r \to r$$
$$\theta \to \pi - \theta$$
$$\varphi \to \varphi + \pi.$$

Only the angular part is thus important related to parity symmetry.
All the functions in this exercise change of sign by a parity transformation: they are thus eigenstates of the parity operator with eigenvalue -1.
One has in general for the spherical harmonic $Y_\ell^m(\theta, \varphi)$

$$\hat{P}Y_\ell^m = (-1)^\ell Y_\ell^m \,,$$

where $\hat{P}$ is the parity operator.

4. *Could the neutron be a bound state of electron and proton? (2)*
   The hypothesis that the neutron is a bound state of electron and proton is inconsistent with Heisenberg's uncertainty principle. Why?

   According to the Heisenberg uncertainty principle a relativistic particle confined into a box of length Δx should have a kinetic energy greater than:

   $$E \sim \frac{\hbar c}{\Delta x} \,.$$

   For $\Delta x \sim 1$ fm (the proton size), the kinetic energy of the electron should then be of the order of 100 MeV–200 MeV, which is much higher than the mass difference between the proton and the neutron and also than the measured energies of the electrons in β decay.

   Furthermore, being both the electron and the proton spin-half particles, the electron-proton bound state should have spin 0 or 1 and not one-half as it is the case for the neutron.

5. *Constructing baryons (5)*
   How many different baryon combinations can you make with 1, 2, 3 or 6 different quark flavors? What is the general formula for n flavors?

   A meson is composed by a quark and an anti-quark ($q\bar{q}$) while a baryon is made of three valence quarks (qqq). Noticing that the meson and the baryon must be color neutral we have:

   - 1 flavor (a)

     – Meson: $(a, \bar{a}) \Longrightarrow 1$ meson.
     – Baryon: $(a, a, a) \Longrightarrow 1$ baryon.

   - 2 flavors (a) and (b)

     – Meson: $(a, \bar{a}); (a, \bar{b}); (b, \bar{a}); (b, \bar{b}) \Longrightarrow 4$ mesons.
     – Baryon: $(a, a, a); (a, a, b); (a, b, b); (b, b, b) \Longrightarrow 4$ baryons.

   - 3 flavors (a), (b) and (c)

     – Meson: $(a, \bar{a}); (a, \bar{b}); (a, \bar{c}); (b, \bar{a}); (b, \bar{b}); (b, \bar{c}); (c, \bar{a}); (c, \bar{b}); (c, \bar{c}) \Longrightarrow$ 9 mesons.
     – Baryon: $(a, a, a); (a, a, b); (a, a, c); (a, b, b); (a, b, c); (b, b, b); (b, b, c);$ $(b, c, c); (a, c, c); (c, c, c); \Longrightarrow 10$ baryons.

   For N flavors we will have for the mesons

   - $(q, \bar{q}) \to (N, N) = N^2$.

Table 5.1 Possible combinations to produce baryon with four flavors. The text in bold are the combinations with at least one c-quark. The orange (light grey) cells indicate the baryons with 2 c-quarks and the red (dark grey) with 3 c-quarks

	u	d	s	c
uu	uuu	uud	uus	**uuc**
dd	ddu	ddd	dds	**ddc**
ss	ssu	ssd	sss	**ssc**
cc	**ccu**	**ccd**	ccs	ccc
ud	–	–	uds	**udc**
sc	**scu**	**dsc**	–	–

For a baryon with N flavors we have for the 3 quark the following allowed combinations,

- All quarks have the same flavor $\rightarrow (q, q, q) \rightarrow N$ possible states;
- Two quarks equal then third one must be different $\rightarrow (q_1, q_1, q_3) \rightarrow N(N - 1)$ possible states;
- All quarks have different flavor $\rightarrow (q_1, q_2, q_3) \rightarrow C_3^N$ (number of combinations of 3-to-3) possible states.

Hence, for baryon the general expression is $N + N(N - 1) + C_3^N$.
Using the derived formulas we have for mesons and baryons with 6 flavors, $N = 6$,

- Meson: $N^2 = 6^2 = 36$ possible mesons;
- Baryon: $N + N(N - 1) + C_3^N = 6 + 6 \times 5 + 20 = 56$ possible baryons.

6. *Charmed baryons (6)*
   Using four quarks ($u, d, s,$ and c), construct a table of all the possible baryon species. How many combinations carry a charm of +1? How many carry charm +2, and +3?

We are considering quark combinations for baryons which are made of three quarks. Taking all possible quark combinations (q_1, q_2, q_3), including permutations would lead us to 64 solutions. However, permutations lead to the same baryons and a so a better strategy to build the table would be to consider a di-quark (q_1, q_2) against a single quark (q_3) considering that permutations are not relevant. Still, this table would lead us to 40 combinations.
Taking into consideration the discussions in the previous problem we can build a smaller table: consider only di-quarks with the same flavor against the remainder quark which can have any kind of flavor. By doing so we find the results in Table 5.1. This exercise alone gives us 16 of the 20 allowed combinations. The combinations missing are the ones in which all the quarks of the baryon are different.

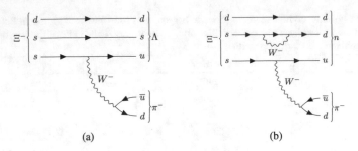

Fig. 5.1 Leading order Feynman diagrams for the process: (a) $\varXi^- \to \varLambda + \pi^-$; (b) $\varXi^- \to n + \pi^-$

We can now simply count the number of baryons with charm:

- $C = +1 \to 6$ baryons;
- $C = +2 \to 3$ baryons;
- $C = +3 \to 1$ baryon.

7. *Exotic hadrons and standard hadrons*
   Indicate the composition in terms of valence quark and antiquark for a baryon, a meson, a tetraquark, and a pentaquark.

- Baryon: qqq;
- Meson: $q\bar{q}$;
- Tetraquark: $q\bar{q}q\bar{q}$;
- Pentaquark: $qqqq\bar{q}$.

8. *Compositeness of quarks? (7)*
   M. Shupe [Phys. Lett. B 611, 87 (1979)] has proposed that all quarks and leptons are composed of two even more elementary constituents: c (with charge $-1/3$) and n (with charge zero) - and their respective antiparticles. You are allowed to combine them in groups of three particles or three antiparticles (ccn, for example, or nnn). Construct all of the eight quarks and leptons in the first generation in this manner. (The other generations are supposed to be excited states.) Notice that each of the quark states admits three possible permutations (ccn, cnc, ncc, for example) - these correspond to the three colors.

- $(n, n, n) \to 0$ charge $\Longrightarrow$ neutrino of the electron;
- $(c, c, c) \to -1$ charge $\Longrightarrow$ electron;
- $(\bar{n}, \bar{n}, \bar{n}) \to 0$ charge $\Longrightarrow$ anti-neutrino of the electron;
- $(\bar{c}, \bar{c}, \bar{c}) \to +1$ charge $\Longrightarrow$ positron;
- $(\bar{c}, \bar{c}, n) \to +2/3$ charge $\Longrightarrow$ up-quark;

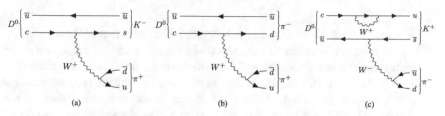

Fig. 5.2 Leading order Feynman diagrams for the process: (a) $D^0 \to K^- + \pi^+$; (b) $D^0 \to \pi^- + \pi^+$; (c) $D^0 \to \pi^- + K^+$

- $(c, c, n) \to -2/3$ charge $\Longrightarrow$ anti-up-quark;
- $(c, n, n) \to -1/3$ charge $\Longrightarrow$ down-quark;
- $(\bar{c}, \bar{n}, \bar{n}) \to +1/3$ charge $\Longrightarrow$ anti-down-quark.

9. *Decays of the Ξ baryon (8)*
   Which decay is more likely:

$$\Xi^- \to \Lambda \pi^- \; ; \; \Xi^- \to n\pi^- \, .$$

Draw the Feynman diagrams at leading order, and explain your answer.

The first decay is more likely, since it involves only one weak decay instead of two (see the Feynman diagrams shown in Fig. 5.1). Another way to see that the second process is more rare is by considering the Ξ^- quark composition (dss, double strangeness), and noticing that the first final state has strangeness 1, while the second has strangeness 0). The phase space for the decays is more or less the same.

10. *Decay of charmed mesons (9)*
    Which decay do you think would be least likely:

$$D^0 \to K^- \pi^+ \; ; \; D^0 \to \pi^- \pi^+ \; ; \; D^0 \to \pi^- K^+ \, .$$

Draw the Feynman diagrams at leading order, and justify your answer.

In Fig. 5.2 the Feynman diagrams corresponding to these decays are shown. The last one is the only one where the two quark flavors change, $\Delta S = -1$, $\Delta C = -1$ (remember by convention the sign of the flavor quantum number of a quark is equal to the sign of its electric charge). Therefore the last decay is the least probable.

11. *Cross sections and isospin (10)*
    Determine the ratio of the following interactions cross sections at the Δ^{++} resonance: $\pi^- p \to K^0 \Sigma^0$; $\pi^- p \to K^+ \Sigma^-$; $\pi^+ p \to K^+ \Sigma^+$.

In strong interactions isospin is conserved, which means that the sum of the isospin quantum numbers of the particles in the initial state is equal to the sum of the isospin quantum numbers of the particles in the final state. Therefore there is one transition amplitude $\mathcal{M}_i$ for each of the channels with the same total isospin quantum numbers. The total transition amplitude that connects the initial and final states is then the sum of all possible isospin transitions amplitudes weighted by the value of the product of the corresponding Clebsch-Gordan coefficients. The cross section is proportional to the square of the total amplitude.

a. For the reaction 1: $\pi^- \, p \; \rightarrow \; K^0 \, \Sigma^0$:

   i. The particle isospin states of each particle are:

$$\pi^- = |1, -1\rangle$$
$$p = \left|\frac{1}{2}, \frac{1}{2}\right\rangle$$
$$K^0 = \left|\frac{1}{2}, -\frac{1}{2}\right\rangle$$
$$\Sigma^0 = |1, 0\rangle \; . \tag{5.2}$$

   ii. The decomposition of the possible isospin states of the initial and final states are using the Clebsch-Gordan tables:

$$\pi^- \, p = |1, -1\rangle \otimes \left|\frac{1}{2}, \frac{1}{2}\right\rangle = \sqrt{\frac{1}{3}}\left|\frac{3}{2}, -\frac{1}{2}\right\rangle - \sqrt{\frac{2}{3}}\left|\frac{1}{2}, -\frac{1}{2}\right\rangle \tag{5.3}$$

$$K^0 \, \Sigma^0 = \left|\frac{1}{2}, -\frac{1}{2}\right\rangle \otimes |1, 0\rangle = \sqrt{\frac{2}{3}}\left|\frac{3}{2}, -\frac{1}{2}\right\rangle + \sqrt{\frac{1}{3}}\left|\frac{1}{2}, -\frac{1}{2}\right\rangle . \tag{5.4}$$

   iii. The total amplitude is

$$\mathcal{M}_1 = \langle K^0 \Sigma^0 \mid \pi^- p\rangle = \frac{\sqrt{2}}{3}\mathcal{M}_{3/2} - \frac{\sqrt{2}}{3}\mathcal{M}_{1/2} . \tag{5.5}$$

The cross section is proportional to

$$\sigma_1 \propto \frac{2}{9}\left|\mathcal{M}_{3/2} - \mathcal{M}_{1/2}\right|^2 . \tag{5.6}$$

b. For the reaction 2: $\pi^- \, p \; \rightarrow \; K^+ \, \Sigma^-$

i. The particle isospin states of each particle are:

$$\pi^- = |1, -1\rangle$$
$$p = \left|\frac{1}{2}, \frac{1}{2}\right\rangle$$
$$K^+ = \left|\frac{1}{2}, \frac{1}{2}\right\rangle$$
$$\Sigma^- = |1, -1\rangle. \qquad (5.7)$$

ii. The decomposition of the possible isospin states of the initial and final states are using the Clebsch-Gordan tables:

$$\pi^- p = |1, -1\rangle \otimes \left|\frac{1}{2}, \frac{1}{2}\right\rangle = \sqrt{\frac{1}{3}}\left|\frac{3}{2}, -\frac{1}{2}\right\rangle - \sqrt{\frac{2}{3}}\left|\frac{1}{2}, -\frac{1}{2}\right\rangle \qquad (5.8)$$

$$K^+ \Sigma^- = \left|\frac{1}{2}, \frac{1}{2}\right\rangle \otimes |1, -1\rangle = \sqrt{\frac{1}{3}}\left|\frac{3}{2}, -\frac{1}{2}\right\rangle - \sqrt{\frac{2}{3}}\left|\frac{1}{2}, -\frac{1}{2}\right\rangle. \qquad (5.9)$$

iii. The total amplitude is

$$\mathcal{M}_2 = \langle K^+ \Sigma^- | \pi^- p \rangle = \frac{1}{3}\mathcal{M}_{3/2} + \frac{2}{3}\mathcal{M}_{1/2}. \qquad (5.10)$$

The cross section is proportional to:

$$\sigma_2 \propto \frac{1}{9}\left|\mathcal{M}_{3/2} + 2\mathcal{M}_{1/2}\right|^2. \qquad (5.11)$$

c. For the reaction 3: $\pi^+ p \rightarrow K^+ \Sigma^+$
   i. The particle isospin states of each particle are

$$\pi^+ = |1, 1\rangle$$
$$p = \left|\frac{1}{2}, \frac{1}{2}\right\rangle$$
$$K^+ = \left|\frac{1}{2}, \frac{1}{2}\right\rangle$$
$$\Sigma^+ = |1, 1\rangle. \qquad (5.12)$$

ii. The decomposition of the possible isospin states of the initial and final states are using the Clebsch-Gordan tables:

$$\pi^+ p = |1, 1\rangle \otimes \left|\frac{1}{2}, \frac{1}{2}\right\rangle = \left|\frac{3}{2}, \frac{3}{2}\right\rangle \qquad (5.13)$$

$$K^+ \Sigma^- = \left|\frac{1}{2}, \frac{1}{2}\right\rangle \otimes |1, 1\rangle = \left|\frac{3}{2}, \frac{3}{2}\right\rangle. \qquad (5.14)$$

iii. The total amplitude is

$$\mathscr{M}_3 = \langle K^+ \Sigma^+ \, | \, \pi^+ p \rangle = \mathscr{M}_{3/2}. \qquad (5.15)$$

The cross section is proportional to

$$\sigma_3 \propto \left|\mathscr{M}_{3/2}\right|^2. \qquad (5.16)$$

At the Δ^{++} resonance (I=3/2 particle), $\mathscr{M}_{3/2} \gg \mathscr{M}_{1/2}$, and thus the relative proportions of the cross sections of the above reactions are

$$\sigma_1 \, : \, \sigma_2 \, : \, \sigma_3 = 2 \, : \, 1 \, : \, 9.$$

12. *Decay branching ratios and isospin (11)*
    Consider the decays of the Σ^{*0} into $\Sigma^+ \pi^-$, $\Sigma^0 \pi^0$ and $\Sigma^- \pi^+$. Determine the ratios between the decay rates in these decay channels.

___

For strong interactions, the evaluation of the decay rates can be done through the conservation of isospin. In fact, the decay rate is related to the amplitude $\mathscr{M}$:

$$\Gamma \propto |\mathscr{M}|^2.$$

The definition of the particle isospin I_3 can be done through the Gell-Mann-Nishijima,

$$I_3 = Q - \frac{1}{2}(B + S), \qquad (5.17)$$

where Q is the particle's charge, B its baryonic number, and S its strangeness. The total isospin I is defined by the taking the family element with higher charge. As such, we can classify the isospin of particles address in this problem as,

$$\pi^+ \to |1, 1\rangle \qquad \pi^0 \to |1, 0\rangle \qquad \pi^- \to |1, -1\rangle \qquad (5.18)$$
$$\Sigma^{*0} \to |1, 0\rangle \qquad \Sigma^+ \to |1, 1\rangle \qquad \Sigma^0 \to |1, 0\rangle \qquad \Sigma^- \to |1, -1\rangle. \qquad (5.19)$$

Using the Clebsch-Gordan coefficients we obtain the following decompositions for the different decay final states:

a) $\Sigma^+ + \pi^-$: $|1, 1\rangle|1, -1\rangle = \dfrac{1}{\sqrt{6}}|2, 0\rangle + \dfrac{1}{\sqrt{2}}|1, 0\rangle + \dfrac{1}{\sqrt{3}}|0, 0\rangle$ (5.20)

b) $\Sigma^0 + \pi^0$: $\qquad\qquad |1, 0\rangle|1, 0\rangle = \dfrac{2}{\sqrt{3}}|2, 0\rangle - \dfrac{1}{\sqrt{3}}|0, 0\rangle$ (5.21)

c) $\Sigma^- + \pi^+$: $|1, -1\rangle|1, 1\rangle = \dfrac{1}{\sqrt{6}}|2, 0\rangle + \dfrac{1}{\sqrt{2}}|1, 0\rangle + \dfrac{1}{\sqrt{3}}|0, 0\rangle$ (5.22)

from which we can, doing the internal product of the initial and final states ($\langle i \mid f\rangle$), obtain the following relations between the amplitudes,

$$\mathcal{M}_a : \mathcal{M}_b : \mathcal{M}_c = 1 : 0 : 1.$$ (5.23)

13. *Quantum numbers and forbidden transitions (12)*
Verify if the following reactions/decays are possible and if not say why:

(a) $pp \to \pi^+\pi^-\pi^0$,
(b) $pp \to ppn$,
(c) $pp \to ppp\bar{p}$,
(d) $p\bar{p} \to \gamma$,
(e) $\pi^- p \to K^0\Lambda$,
(f) $n \to pe^-v$,
(g) $\Lambda \to \pi^- p$,
(h) $e^- \to v_e \gamma$.

---

a. $pp \to \pi^+\pi^-\pi^0$ is impossible: it violates the conservation of the baryon number.

b. $pp \to ppn$ is impossible: it violates the conservation of the baryon number.

c. $pp \to ppp\bar{p}$ is possible.

d. $p\bar{p} \to \gamma$ is impossible: it violates energy-momentum conservation (write the energy-momentum 4-vectors in the center-of-mass system).

e. $\pi^- p \to K^0\Lambda$ is possible via strong interaction, the K^0 meson having quark composition $d\bar{s}$, and the Λ baryon uds.

f. $n \to pe^-v$ is impossible: to conserve the lepton number one must have a $\bar{v}_e$ in the final state.

g. $\Lambda \to \pi^- p$ is possible (weak decay).

h. $e^- \to v_e \gamma$ is impossible: it violates charge conservation.

14. *Width and lifetime of the J/ψ (13)*
The width of the J/ψ meson is $\Gamma \simeq 93$ keV. What is its lifetime? Could you imagine an experiment to measure it directly?

The lifetime of the J/ψ can be computed through the Heseinberg uncertainty principle as

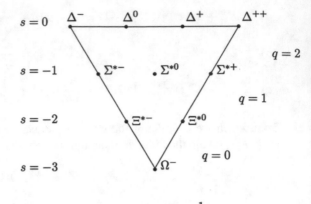

Fig. 5.3 The spin 3/2, parity 1 baryon decuplet. The I_3 axis is the abscissa, while the Y axis is the ordinate. The Ω^- has $Y = 0$, and the Σs have $I_3 = 0$

$$\tau = \frac{\hbar}{\Gamma} \simeq 7.1 \times 10^{-21}. \tag{5.24}$$

A J/ψ would travel in the LAB

$$x = \beta ct = \beta c\gamma\tau \tag{5.25}$$

where $c\tau = 2.1 \times 10^{-12}$ m. Thus, even if the J/ψ would have a boost factor ($\gamma = E/m$) of the order of hundreds the decay length of the particle would still be of the order of the atomic size, and thus, currently, the direct measurement would be basically impossible.

15. Ω^- *mass (14)*
    Verify the relations between the masses of all the particles lying in the fundamental baryon decuplet but the Ω^- and predict the mass of this one. Compare your prediction with the measured values.

The fundamental baryons' decuplet is shown in Fig. 5.3. The abscissa is the z−component of the *isospin*, I_3, while the ordinate is the *hypercharge*.
The baryons with equal hypercharge have the same number of strange quarks and have similar masses. The differences between baryon masses of two consecutive Y levels are similar, and in first approximation are explained attributing to the the strange quark a mass much higher than the masses of the up and down quarks. Then the following equations can be written:

$$M_\Delta - M_\Sigma \simeq M_\Sigma - M_\Xi \tag{5.26}$$
$$M_\Sigma - M_\Xi \simeq M_\Xi - M_\Omega. \tag{5.27}$$

Knowing that the baryon masses are: $M_\Delta \simeq 1232$ MeV, $M_\Sigma \simeq 1384$ MeV and $M_\Xi \simeq 1537$ MeV, one obtains two equations from which the mass of Ω can be extracted:

$$M_\Omega = 2M_\Xi - M_\Sigma \simeq 1682 \, \text{MeV} \tag{5.28}$$

$$M_\Omega = M_\Sigma - M_\Lambda + M_\Xi \simeq 1686 \, \text{MeV} . \tag{5.29}$$

These values are close to one obtained experimentally, $M_\Omega^{\text{exp}} \simeq 1672 \, \text{MeV}$.

16. *Decays and conservation laws (15)*

Is the decay $\pi^0 \to \gamma\gamma\gamma$ possible?

The decay of the π^0 into three photons is forbidden by charge-conjugation invariance. In fact, the photon has charge parity -1 while the π^0 has $+1$. The decay into two photons is thus allowed but not in three.

The photon charge parity -1 can be understood realizing that under charge conjugation charged particles (the sources of the electromagnetic field) change the sign and thus the corresponding electromagnetic field will change also in direction.

17. *Experimental resolution in deep inelastic scattering (16)*

Consider an $e^- p$ deep inelastic scattering experiment where the electron (of mass m_e) scattering angle is $\sim 6°$ and the beam energy is 10 GeV. Make an estimation of the experimental resolution in the measurement of the energy of the scattered electron that is needed to distinguish the elastic channel ($e^- p \to e^- p$) from the first inelastic channel ($e^- p \to e^- p \pi^0$).

Let:

- p_1 and p_3 be, respectively, the 4-momenta of the incoming and outgoing electron;
- p_2 and p_4 the 4-momenta of the incoming (target) and outgoing hadronic system;
- $q = (p_1 - p_3) = (p_4 - p_2)$ the 4-momenta of the exchanged virtual photon,

and note that:

- $p_2^2 = m_p^2$ and $p_4^2 = M^2$, where m_p is the proton mass and M is the invariant mass of the final hadronic state;

$M = m_p$, in the case of elastic scattering;

$M = m_p + m_\pi$ in the case of the first inelastic channel, $e^- p \to e^- p \pi^0$.

In the laboratory reference frame we define: $p_1 = (E, \mathbf{p_1})$; $p_3 = (E', \mathbf{p_3})$; $p_2 = (m_p, 0)$; and θ as the electron scattering angle. Then, by energy-momentum conservation:

$$p_1 + p_2 = p_3 + p_4$$
$$p_4^2 = ((p_1 + p_2) - p_3)^2$$
$$M^2 = ((E + m_p, \mathbf{p_1}) - (E', \mathbf{p_3}))^2$$
$$\frac{1}{2}(M^2 - m_p^2) = E'(E + m_p) - E m_p - m_e^2 + \mathbf{p_1} \cdot \mathbf{p_3}.$$

Neglecting m_e,

$$\frac{1}{2}(M^2 - m_p^2) = E'(E + m_p) - Em_p + EE' \cos\theta$$

and finally

$$E' = \frac{E + \frac{1}{2}(m_p - M^2/m_p)}{1 + E(1 - \cos\theta)/m_p}. \tag{5.30}$$

In the case of elastic scattering $(M = m_p)$ with $\theta \simeq 6°$ and $E = 10$ GeV one obtains

$$E'_{el} = 9.45 \text{ GeV}.$$

In the case of the first inelastic channel $(M = m_p + m_\pi)$ with $\theta \simeq 6°$ and $E = 10$ GeV one obtains

$$E'_{inel} = 9.31 \text{ GeV}.$$

Thus,

$$\Delta(E) = (E'_{el} - E'_{inel}) \simeq 0.14 \text{ GeV} \simeq m_\pi \tag{5.31}$$

as expected. The resolution should be then of the order of $m_\pi/E \simeq 1.5\%$.

18. *Electron-proton elastic and deep inelastic scattering kinematics (17)*
    Consider the e^-p elastic/deep inelastic scattering and deduce the following for-
    mulae:

    a. for elastic and deep inelastic scattering,

    $$Q^2 = 4EE' \sin^2(\theta/2);$$

    b. for elastic scattering,

    $$Q^2 = 2Mv;$$

    c. for elastic and deep inelastic scattering,

    $$Q^2 = xy(s - M^2).$$

In an e^-p elastic or deep inelastic scattering the electron transfers energy-
momentum to the proton through the exchange of a virtual photon. Let p_1 and p_3
be, respectively, the 4-momenta of the incoming and outgoing electron, and p_2
and p_4 the 4-momenta of the incoming (target) and outgoing hadronic system.
Note that the square of the exchanged 4-vector is $q^2 = -Q^2 = (p_1 - p_3)^2 = (p_4 - p_2)^2$ and that $p_2^2 = m_p^2$ and $p_4^2 = M^2$, where m_p is the proton mass and
M is the invariant mass of the final hadronic state ($M = m_p$, in the case of elastic

scattering). Finally, let define in the laboratory reference frame: $p_1 = (E, \mathbf{p_1})$, $p_3 = (E', \mathbf{p_3})$ and $p_2 = (M, 0)$.

a. In the laboratory reference frame

$$Q^2 = -(E - E')^2 - (\mathbf{p_1} - \mathbf{p_3})^2 = -(2m_e^2 - 2EE' + 2\mathbf{p_1} \cdot \mathbf{p_3}) \,.$$

Neglecting m_e,

$$Q^2 = 2EE'(1 - \cos\theta) = 4EE' \sin^2(\theta/2) \,.$$

b.

$$p_4 = p_2 + q \Leftrightarrow p_4^2 = p_2^2 + q^2 + 2p_2 \cdot q$$

(the symbol $\cdot$ between 4-vectors indicates their scalar product). In the case of elastic scattering

$$q^2 = -2p_2 \cdot q \,.$$

Defining the lost energy, ν, as:

$$\nu = \frac{q \cdot p_2}{M}$$

(in the laboratory reference frame $\nu = E - E'$), one obtains

$$Q^2 = 2M\nu \,.$$

c. The square s of the center-of-mass energy is defined as:

$$s = (p_1 + p_2)^2 = 2p_1 \cdot p_2 + M^2 + m_e^2 \,.$$

Neglecting m_e,

$$2p_1 \cdot p_2 = s - M^2.$$

The inelasticity y is defined as:

$$y = \frac{q \cdot p_2}{p_1 \cdot p_2}$$

(in the laboratory reference frame $y = \frac{\nu}{E}$).
Finally, the scaling variable x is defined as:

$$x = \frac{Q^2}{2p_2 \cdot q}$$

(in the laboratory reference frame $x = 1$).

Then,

$$x\,y = \frac{Q^2}{s - M^2}$$

or

$$Q^2 = x\,y(s - M^2).$$

19 *Gottfried sum rule (18)*
Deduce in the framework of the quark-parton model the sum rule

$$\int \frac{1}{x} \left(F_2^{ep}(x) - F_2^{ep}(x) \right) dx = \frac{1}{3} + \frac{2}{3} \int \left(\bar{u}(x) - \bar{d}(x) \right) dx$$

and comment the fact that the value measured in $e^- p$ and $e^- d$ deep inelastic scattering experiments is approximately 1/4.

The form factor F_2 for the electron–proton scattering can be written as a function of the specific quarks PDFs:

$$F_2^{ep}\left(Q^2, x \right) \simeq x \left[\frac{4}{9} \left(u^p(x) + \bar{u}^p(x) \right) + \frac{1}{9} \left(d^p(x) + \bar{d}^p(x) + s^p(x) + \bar{s}^p(x) \right) \right],$$

where the contributions from heavier quarks and antiquarks were neglected as they are strongly suppressed due to their high masses. Similarly, the form factor F_2 for the electron–neutron scattering can be written as:

$$F_2^{en}\left(Q^2, x \right) \simeq x \left[\frac{4}{9} \left(u^n(x) + \bar{u}^n(x) \right) + \frac{1}{9} \left(d^n(x) + \bar{d}^n(x) + s^n(x) + \bar{s}^n(x) \right) \right].$$

The PDFs may be divided into valence and sea. A subscript V or S is added to specify if a given PDF refers to valence or sea. For instances, the u PDF is written as the sum of two PDFs:

$$u(x) = u_V(x) + u_S(x) .$$

Sea quarks and antiquarks with the same flavor are produced in pairs and their PDFs should be similar. In the proton and the neutron there are no valence antiquarks, just sea antiquarks. Then the $\bar{u}$ and the $\bar{d}$ components can then be expressed as:

$$\bar{u}(x) = \bar{u}_S(x) = u_S(x) ; \tag{5.32}$$
$$\bar{d}(x) = \bar{d}_S(x) = d_S(x) . \tag{5.33}$$

The form factor F_2 can then be written as a function of the valence and sea quarks PDFs:

$$F_2^{ep}\left(Q^2, x\right) \simeq x \left[\frac{4}{9}\left(u_V^p(x) + 2\overline{u}^p(x)\right) + \frac{1}{9}\left(d_V^p(x) + 2\overline{d}^p(x) + 2\overline{s}^p(x)\right)\right],$$

$$F_2^{en}\left(Q^2, x\right) \simeq x \left[\frac{4}{9}\left(u_V^n(x) + 2\overline{u}^n(x)\right) + \frac{1}{9}\left(d_V^n(x) + 2\overline{d}^n(x) + 2\overline{s}^n(x)\right)\right].$$

Assuming isospin invariance:

$$u^p(x) = d^n(x) \; ; \; d^p(x) = u^n(x)$$

$$\overline{u}^p(x) = \overline{d}^n(x) \; ; \; \overline{d}^p(x) = \overline{u}^n(x)$$

$$s^p(x) = \overline{s}^p(x) = s^n(x) = \overline{s}^n(x) .$$

Integrating over the full x range the difference between the two F_2 form factor (weighted by $1/x$) one obtains:

$$\int_0^1 \frac{1}{x}\left\{F_2^{ep}\left(Q^2, x\right) - F_2^{en}\left(Q^2, x\right)\right\} dx \simeq \frac{1}{3}\int_0^1 \left\{(u_V(x) - d_V(x))\right\} dx$$
$$+ \frac{2}{3}\int_0^1 \left\{(\overline{u}(x) - \overline{d}(x))\right\} dx. \tag{5.34}$$

The integration over the full x range of the valence quark PDFs of the proton is consistent with the quark model and therefore:

$$\int_0^1 u_V^p(x) \, dx \simeq 2 \; ; \; \int_0^1 d_V^p(x) \, dx \simeq 1. \tag{5.35}$$

Thus,

$$\int_0^1 \frac{1}{x}\left\{F_2^{ep}\left(Q^2, x\right) - F_2^{en}\left(Q^2, x\right)\right\} dx \simeq \frac{1}{3} + \frac{2}{3}\int_0^1 \left\{(\overline{u}(x) - \overline{d}(x))\right\} dx. \tag{5.36}$$

If one assumes also the naïve quark-parton model

$$\overline{u}(x) \simeq \overline{d}(x), \tag{5.37}$$

one gets the Gottfried sum rule and the result would be 1/3, which is however, in strong disagreement with the experimental data (the measured value is 0.235 ± 0.026). There is probably an isospin violation in the sea quark distributions.

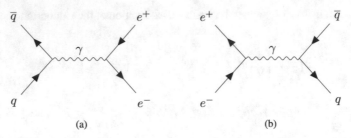

Fig. 5.4 Lowest order Feynman diagrams for the process: (a) $q + \bar{q} \rightarrow e^+ + e^-$; (b) $e^+ + e^- \rightarrow q + \bar{q}$

20 *Number of colors in QCD*
Discuss the number of *colors* in QCD by computing the cross section ratio

$$\frac{\sigma(q\bar{q} \rightarrow e^+e^-)}{\sigma(e^+e^- \rightarrow q\bar{q})} \quad\quad\quad (5.38)$$

where q is a quark flavor. Consider that the masses of the fermions can be neglected and $\sqrt{s} \ll M_W$.

In this problem, the center-of-mass energy is much smaller than the weak interaction bosons mass and thus, the dominant Feynman diagrams for both process are the gamma exchange via an s-channel as shown in Fig. 5.4.
Each of the cross section is proportional to the squared amplitude of the corresponding diagram.
If the number of colors N_C were 1 (that means a world without QCD where the quarks will be not confined) the ratio of the two cross sections will be 1, since the Dirac structure of the above Feynman diagrams is the same, only reversed in time.
The introduction of a number of colors different than 1 make an enormous but, subtle difference. Quarks will have a new quantum number, the color, and:

- The cross section of the process in the denominator ($e^+e^- \rightarrow q\bar{q}$) corresponds in fact to the sum of the cross sections of N_C different processes ($e^+e^- \rightarrow q_i\bar{q}_i$), where the index i runs over the number of colors. In practice this cross section should be multiplied by a N_C factor.
- The cross section of the process in the numerator ($q\bar{q} \rightarrow e^+e^-$) corresponds in fact to the sum of the cross sections of the processes ($q_i\bar{q}_j \rightarrow e^+e^-$), where the index i and j runs over the number of colors. However, as the exchanged gamma has no color, from all the N_C^2 possibles color combinations, only those where $i = j$ have a non-zero amplitude. In practice this cross section should be multiplied by a factor $1/N_C$.

Putting together the two factors and, as in QCD, $N_C = 3$:

$$\frac{\sigma(q\bar{q} \to e^+e^-)}{\sigma(e^+e^- \to q\bar{q})} = \frac{1/3}{3} = \frac{1}{9}. \tag{5.39}$$

21. *Quark-Gluon Plasma tomography*

The hadronic decay of the W-boson, when originated from a top-antitop ($t\bar{t}$) quark pair, can provide a decay chain whose time interval can go from ~ 0.25 fm/c (in its rest frame) up to a few fm/c, depending on the momentum boost provided by the top. Being the W-boson a color neutral particle, such time delay can provide key novel insight into the time structure of the hot and dense colored stated of matter (Quark-Gluon Plasma) that is produced in ultra-relativistic heavy-ion collisions.

a. Consider PbPb collisions at the Large Hadron Collider with a $\sqrt{s_{NN}} = 5.02$ TeV. Knowing that the reconstructed top initiated jet has an average transverse momentum $\langle p_{T,t} \rangle = 115$ GeV/c and the reconstructed hadronic decay of the W boson an average transverse momentum $\langle p_{T,W} \rangle = 95$ GeV/c, calculate the total (average) delay time of this channel.

b. In addition to the top and W boson, the hadronic decay products of the W boson (quark - anti-quark pair) are known to remain in a color-singlet state while propagating through the Quark-Gluon Plasma, increasing the effective delay time. Knowing that this component is fairly independent of the boost and about $\tau_{sing} \simeq 0.34$ fm/c, estimate what would be the minimum transverse momentum that the top would have so that we could reach a total delay of about 1.5 fm/c? Assume that the longitudinal momentum of the top and W boson can be neglected with respect to the transverse momentum component. Compare the obtained value with Fig. 5.5.

---

a. Since we are considering the transverse momentum direction only, the corresponding boost factor along this direction is given by:

$$\gamma_{T,X} = \sqrt{\frac{p_{T,X}^2}{m_X^2} + 1} \ , \quad X = t \text{ or } W , \tag{5.40}$$

where $\gamma = E_T/m$, and $E_T = \sqrt{P_T^2 + m^2}$ is the transverse energy. Notice that, as always, this boost applies to the longitudinal momentum of the particle, which in this case corresponds to the transverse momentum of the top quark relative to the beam direction. The decay times of the top quark and of the W boson are given, respectively, by:

$$\tau_t = 0.15 \text{ fm}/c \,,$$
$$\tau_W = 0.09 \text{ fm}/c \,.$$

In the laboratory frame, the time delay will be boosted with respect to the particle rest frame, yielding:

$$\tau_X^{lab} = \gamma_{T,X} \, \tau_X \,. \tag{5.41}$$

As such, the total average delay time that is possible to reach at the LHC ($\sqrt{s_{NN}} = 5.02$ TeV) is:

$$\langle \tau_{total} \rangle = \left(\frac{\langle p_{T,t} \rangle^2}{m_t^2} + 1 \right)^{1/2} \tau_t + \left(\frac{\langle p_{T,W} \rangle^2}{m_W^2} + 1 \right)^{1/2} \tau_W \simeq 0.32 \text{ fm}/c \,, \tag{5.42}$$

where we used:

$$m_t = 172.76 \text{ GeV}/c^2 \;;\; m_W = 80.38 \text{ GeV}/c^2 \,.$$

b. We start by identifying the dependence of each decay time on the top transverse momentum, $p_{T,t}$. For the top quark, this is simply:

$$\tau_t^{lab} = \gamma_{T,t} \, \tau_t = \left(\frac{p_{T,t}^2}{m_t^2} + 1 \right)^{1/2} \tau_t \,. \tag{5.43}$$

For the W boson, we first need to calculate its transverse momentum when the top decays. In the top rest frame (2-body decay), the kinematics follows:

$$E_W^{cm} = \frac{m_t^2 - m_b^2 + m_W^2}{2m_t} \simeq 105.03 \text{ GeV} \,, \tag{5.44}$$

$$p_W^{cm} = \frac{\sqrt{(m_t^2 - (m_W + m_b)^2)(m_t^2 - (m_W - m_b)^2)}}{2m_t} \simeq 67.60 \text{ GeV}/c \,, \tag{5.45}$$

where the bottom quark mass is $m_b = 4.18$ GeV/c^2. To estimate the W boson momentum in the laboratory frame, it is assumed that the longitudinal momenta of the top and W-boson are negligible (produced at pseudo-rapidity $\eta \simeq 0$). As such, in the following, it is used that $p_W^{cm} \simeq p_{T,W}$ and $\gamma_t \simeq \gamma_{T,t}$. Using Lorentz transformations for the W boson:

$$p_{T,W}^{lab} = \gamma_{T,t} \left(\beta_{T,t} E_W^{cm} + p_W^{cm} \right) \,, \tag{5.46}$$

where

$$\beta_{T,t} = \sqrt{\frac{p_{T,t}^2}{p_{T,t}^2 + m_t^2}} \,. \tag{5.47}$$

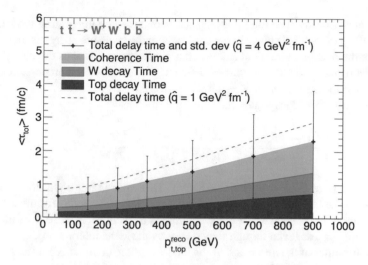

Fig. 5.5 Total delay time and its standard deviation (markers and corresponding error bars). The average contribution of each component (top, W boson and $\tau_{sing} \equiv$ coherence time) is shown as the stacked bands from bottom to top (red, purple and green in color, respectively). From L. Apolinário et al. 2018, Phys. Rev. Lett. 120, 232301

Finally, the W boson decay time in the laboratory frame is given by:

$$\tau_W^{lab} = \gamma_{T,W}\, \tau_W = \left(\frac{(p_{T,W}^{lab})^2}{m_W^2} + 1\right)^{1/2} \tau_W \,. \tag{5.48}$$

The total delay time will now be given by the sum of the three components:

$$\tau_{total} = \tau_t^{lab} + \tau_W^{lab} + \tau_{sing} \,. \tag{5.49}$$

Requiring $\tau_{total} = 1.5$ fm/c yields $p_{T,t} \simeq 564$ GeV/c. Reading from the figure below, one could withdraw 550 GeV/c taking the green line (that includes the 3 delay components) as reference.

22. *Rough estimate of the decay amplitude and of the lifetime of the τ lepton*
    A plausible energy scale for muon and tau decays is the mass of the decaying lepton. Use this to calculate the ratio of the amplitudes:

    $$\frac{\Gamma(\mu \to e\bar{\nu}_e \nu_\mu)}{\Gamma(\tau \to e\bar{\nu}_e \nu_\tau)} \,.$$

    How does this relate to the lifetime of the tau lepton?

    The decay rate is the product of the volume of the phase space times a probability associated to the dynamics of the process, which is the same for both the τ decay and the μ decay. The total phase space is obtained by integrating the spectrum $N(p)dp$ of the final states. In the cases where the available energy is much larger than the masses in the final state such a phase space, and thus the decay rate, is proportional to the fifth power of the energy available in the process, i.e., of the mass of the decaying particle. This is called the Sargent's rule.

The ratio of the decay rates can thus be written as

$$\frac{\Gamma(\mu \rightarrow e\bar{v}_e v_\mu)}{\Gamma(\tau \rightarrow e\bar{v}_e v_\tau)} \simeq \frac{G_F^2 m_\mu^5}{G_F^2 m_\tau^5} = \frac{m_\mu^5}{m_\tau^5} \simeq \frac{106^5}{1777^5} \simeq 7.5 \times 10^{-7}.$$

(masses have been expressed in MeV). This ratio is not equal to the inverse ratio of the lifetimes, as there are more possible decay modes for the τ than for the muon, but it shows that the τ lifetime is a fraction 7.5×10^{-7} of the muon lifetime, or less (even though they decay by the same interaction).

The lifetimes of the muon and of the tau leptons are of about 2.2 μs and of about 0.3 ps respectively (therefore the actual ratio is about 1.4×10^{-7}).

Chapter 6
Interactions and Field Theories

1. *Interaction of spinless particles (1)*. Determine, in the high-energy limit, the electromagnetic differential cross section between two spinless charged nonidentical particles.

A particle with momentum p_1 and another one with a momentum p_2 interact giving origin to a particle with p_3 and another with p_4. Being this an electromagnetic scattering it is mediated by a photon through a t-channel exchange. The $\mathcal{M}$ amplitude can be derived applying the Feynman rules to the diagram in Fig. 6.1,

$$i\mathcal{M} = (ie(p_A + p_C)^\mu \left(\frac{-ig_{\mu\nu}}{q^2} \right) (ie(p_B + p_D)^\nu). \tag{6.1}$$

Neglecting the particle masses as $E \gg m$, the particles initial and final 4-momentum can be written as

$$p_A = (p, p, 0, 0) \tag{6.2}$$
$$p_B = (p, -p, 0, 0) \tag{6.3}$$
$$p_C = (p, p\cos\theta, p\sin\theta, 0) \tag{6.4}$$
$$p_D = (p, -p\cos\theta, -p\sin\theta, 0), \tag{6.5}$$

Fig. 6.1 Feynman diagrams at lowest order for the scattering of two spinless electromagnetic particles

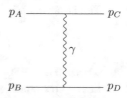

A. De Angelis et al., *Particle and Astroparticle Physics*, Undergraduate Lecture Notes in Physics,
https://doi.org/10.1007/978-3-030-73116-8_6

where p is the particle momentum (and since the mass is neglected is also its energy) and θ is the scattering between the $A - B$ direction and the direction of p_C.
Then

$$(p_A + p_c) = (2p, p(1 + \cos\theta), p\sin\theta, p) \tag{6.6}$$

$$(p_B + p_D) = (2p, -p(1 + \cos\theta), -p\sin\theta, -p) \tag{6.7}$$

$$q = (p_D - p_B) = (0, p(1 - \cos\theta), -p\sin\theta, 0). \tag{6.8}$$

Substituting in Eq. 6.1 we obtain

$$\mathcal{M} = -e^2 \frac{1}{q^2}\left((p_A + p_c)^0(p_B + p_D)^0 - \sum_{i=1}^{3}(p_A + p_c)^i(p_B + p_D)^i\right) \tag{6.9}$$

$$= -e^2 \frac{1}{p^2(1 - \cos\theta)^2 + p^2\sin^2\theta}(4p^2 + p^2(1 + \cos\theta)^2 + p^2\sin^2\theta) \tag{6.10}$$

$$= -e^2 \frac{3 + \cos\theta}{1 - \cos\theta}. \tag{6.11}$$

The differential cross-section for the scattering of two non-identical spinless particles in the center-of-mass frame is given by

$$\frac{d\sigma}{d\Omega} = \frac{|\mathcal{M}|^2}{64\pi^2 s}\frac{|p_C|}{|p_A|} \tag{6.12}$$

where $s = (p_A + p_B)^2$. As we are considering the high energy limit and this is a scattering process $|p_C| = |p_A|$ and we can finally write

$$\frac{d\sigma}{d\Omega} = \frac{\alpha^2}{4s}\frac{(3 + \cos\theta)^2}{(1 - \cos\theta)^2} \tag{6.13}$$

with $\alpha = \frac{e^2}{4\pi}$ being the fine structure constant. Therefore, the cross-section decreases with the square of the center-of-mass energy.

2. *Dirac equation invariance (2).* Show that the Dirac equation written using the covariant derivative is gauge-invariant.

The Dirac equation written using the covariant derivative can be expressed as

$$(i\not{D} - m)\Psi = 0. \tag{6.14}$$

where $\not{D} = \gamma^\mu D_\mu$ and D_μ is the covariant derivative defined as $\partial_\mu + iq A_\mu$. Let us now consider a local phase transformation,

$$\Psi \to e^{iq\lambda(x)}\Psi. \tag{6.15}$$

Substituting in Eq. 6.14 we get

$$i\gamma^\mu(\partial_\mu + iqA_\mu)\Psi - m\Psi = 0 \tag{6.16}$$

$$i\gamma^\mu(\partial_\mu(e^{iq\lambda(x)}\Psi) + iqA_\mu e^{iq\lambda(x)})\Psi - me^{iq\lambda(x)}\Psi = 0 \tag{6.17}$$

$$i\gamma^\mu(e^{iq\lambda(x)}\partial_\mu\Psi + e^{iq\lambda(x)}iqA_\mu + (iq\partial_\mu\lambda)e^{iq\lambda(x)}\Psi + iqA_\mu e^{iq\lambda(x)}\Psi) - me^{iq\lambda(x)}\Psi = 0 \tag{6.18}$$

$$e^{iq\lambda(x)}(i\gamma^\mu(\partial_\mu + iq(A_\mu + \partial_\mu\lambda)) - m)\Psi = 0. \tag{6.19}$$

Noticing that in the electromagnetism gauge the potential vector A_μ transforms as

$$A^\mu \to A^\mu + \partial^\mu\lambda(x) \tag{6.20}$$

so that the physics is the same independently of the reference frame, we finally obtain:

$$(i\gamma^\mu(\partial_\mu + iqA'_\mu) - m)\Psi = 0 \tag{6.21}$$

demonstrating in this way that the Dirac equation using the covariant derivative is invariant under local transformations, i.e. gauge-invariant.

3. *Bilinear covariants (3)*. Show that

a. $\overline{\psi}\psi$ is a scalar;
b. $\overline{\psi}\gamma^5\psi$ is a pseudoscalar;
c. $\overline{\psi}\gamma^\mu\psi$ is a four-vector;
d. $\overline{\psi}\gamma^\mu\gamma^5\psi$ is a pseudo four-vector.

In this solution we will restrict to proper orthochronous (i.e., preserving the direction of time) Lorentz transformations. We first recall some basic results about the algebra of γ matrices and spinors. First, given ψ, a 4-components spinor $\overline{\psi}$ is defined as $\overline{\psi} = \psi^\dagger\gamma^0$. Moreover a Lorentz transformation Λ is represented by a matrix $S(\lambda)$ in such a way that

$$S(\Lambda)^{-1}\gamma^\nu S(\Lambda) = \Lambda^\nu{}_\mu\gamma^\mu \quad \text{and} \quad \psi' = S(\Lambda)\psi,$$

where ψ' is the transformed spinor. Two properties of the gamma matrices that will be useful in what follows are: $(\gamma^\alpha)^\dagger = \gamma^0\gamma^\alpha\gamma^0$, $\gamma^0\gamma^0 = 1$ and $(\gamma^0)^\dagger = \gamma^0$. We would now like to find a convenient way to express the corresponding transformation law for $\overline{\psi}$. To achieve this we first consider the following intermediate result, which can be obtained by considering the adjoint of the first equation above,[1]

---

[1] We use the shorthand $S(\Lambda)^{-\dagger} = (S(\Lambda)^{-1})^\dagger$.

$$S(\Lambda)^\dagger (\gamma^\alpha)^\dagger S(\Lambda)^{-\dagger} = (\Lambda^\alpha{}_\beta \gamma^\beta)^\dagger.$$

Using the first result recalled above this can be rewritten as

$$S(\Lambda)^\dagger \gamma^0 \gamma^\alpha \gamma^0 S(\Lambda)^{-\dagger} = \Lambda^\alpha{}_\beta \gamma^0 \gamma^\beta \gamma^0.$$

We now multiply this equation on the left *and* on the right by γ^0, we use the second results recalled above, and we obtain

$$[\gamma^0 S(\Lambda)^\dagger \gamma^0] \gamma^\alpha [\gamma^0 S(\Lambda)^{-\dagger} \gamma^0] = \Lambda^\alpha{}_\beta \gamma^\beta,$$

from which, using the second and third properties for γ^0 listed above, as well as the law of transformation of the γ matrices, we obtain

$$[\gamma^0 S(\Lambda)^\dagger \gamma^0]^\dagger \gamma^\alpha [\gamma^0 S(\Lambda)^\dagger \gamma^0]^{-1} = S(\Lambda)^{-1} \gamma^\alpha S(\Lambda).$$

By isolating γ^α on the right hand side we finally get

$$[S(\Lambda) \gamma^0 S(\Lambda)^\dagger \gamma^0] \gamma^\alpha [S(\Lambda) \gamma^0 S(\Lambda)^\dagger \gamma^0]^{-1} = \gamma^\alpha.$$

This shows that $S(\Lambda) \gamma^0 S(\Lambda)^\dagger \gamma^0$ commutes with the γs: it must then be a multiple of the identity matrix, from which it is then clear that[2] $S(\Lambda)^\dagger \gamma^0 = \pm \gamma^0 S(\Lambda)^{-1}$. The sign in front of this relation is related to the sign of $\Lambda^0{}_0$, and we can conclude

$$S(\Lambda)^\dagger \gamma^0 = \text{sign}(\Lambda^0{}_0) \gamma^0 S(\Lambda)^{-1}.$$

This result allows us to write the transformation law for $\overline{\psi}$ that we were looking for, as

$$\psi' = S(\Lambda) \overline{\psi} \quad \Rightarrow \quad \overline{\psi}' = \psi^\dagger S(\Lambda)^\dagger \gamma^0 = \text{sign}(\Lambda^0{}_0) \overline{\psi} S(\Lambda)^{-1}.$$

We can now prove the results above. We will consider proper orthochronous Lorentz transformations only, so $\text{sign}(\Lambda^0{}_0) = 1$ below.

a. This comes directly from the two transformations above, as for a proper orthochronus Lorentz transformation we have:

$$\overline{\psi}' \psi' = \overline{\psi} S(\Lambda)^{-1} S(\Lambda) \psi = \overline{\psi} \psi.$$

b. In the second line of the calculation below we will temporarily write S for $S(\Lambda)$. By making use of the definition

$$\gamma^5 = i \gamma^0 \gamma^1 \gamma^2 \gamma^3 = \epsilon_{\mu\nu\rho\sigma} \gamma^\mu \gamma^\nu \gamma^\rho \gamma^\sigma / 4!$$

---

[2]The fact that the proportionality constant is ± 1 follows from the hermiticity of both $S(\Lambda) \gamma^0 S(\Lambda)^\dagger$ and γ^0, added to the fact that $\det(S(\Lambda)^\dagger) = 1$.

we obtain

$$
\begin{aligned}
\overline{\psi}\,'\gamma^5\psi' &= \frac{i}{4!}\epsilon_{\mu\nu\rho\sigma}\overline{\psi}\,'\gamma^\mu\gamma^\nu\gamma^\rho\gamma^\sigma\psi' \\
&= \frac{i}{4!}\epsilon_{\mu\nu\rho\sigma}\overline{\psi}(S^{-1}\gamma^\mu S)(S^{-1}\gamma^\nu S)(S^{-1}\gamma^\rho S)(S^{-1}\gamma^\sigma S)\psi \\
&= \frac{i}{4!}\epsilon_{\mu\nu\rho\sigma}\Lambda^\mu{}_\alpha\Lambda^\nu{}_\beta\Lambda^\rho{}_\eta\Lambda^\sigma{}_\tau\overline{\psi}\gamma^\alpha\gamma^\beta\gamma^\eta\gamma^\tau\psi \\
&= i\frac{\det(\Lambda)}{4!}\epsilon_{\alpha\beta\eta\tau}\overline{\psi}\gamma^\alpha\gamma^\beta\gamma^\eta\gamma^\tau\psi \\
&= \det(\Lambda)(\overline{\psi}\gamma^5\psi).
\end{aligned}
$$

Between the third and the fourth line we make use of the fact that

$$
\epsilon_{\mu\nu\rho\sigma}\Lambda^\mu{}_\alpha\Lambda^\nu{}_\beta\Lambda^\rho{}_\eta\Lambda^\sigma{}_\tau
$$

is a four-indices totally antisymmetric tensor in four dimensions, so it must be a multiple of the totally antisymmetric tensor $\epsilon_{\alpha\beta\eta\tau}$. However, from

$$
\epsilon_{\mu\nu\rho\sigma}\Lambda^\mu{}_\alpha\Lambda^\nu{}_\beta\Lambda^\rho{}_\eta\Lambda^\sigma{}_\tau = k\epsilon_{\alpha\beta\eta\tau},
$$

with k a real constant, by contraction of this equality with $\epsilon^{\alpha\beta\eta\tau}$, and remembering the definition of determinant, we obtain $4!\det(\Lambda) = 4!k$, i.e. $k = \det(\Lambda)$. The first and last lines exactly show the transformation law for a pseudoscalar.

c. The proof goes in a similar way as the first one:

$$
\overline{\psi}'\gamma^\alpha\psi' = \overline{\psi}S(\Lambda)^{-1}\gamma^\alpha S(\Lambda)\psi = \Lambda^\alpha{}_\beta(\overline{\psi}\gamma^\beta\psi),
$$

where in the last step we made use of the transformation law for the γ matrices recalled at the beginning. The equality of the first and last terms is exactly the transformation law for the 4-vector $\overline{\psi}\gamma^\beta\psi$.

d. This result can be shown with very similar methods as those used in the second case. We will also use most of the facts already used there, without further comments.

$$\overline{\psi}\,'\gamma^\omega\gamma^5\psi' = \frac{i}{4!}\epsilon_{\mu\nu\rho\sigma}\overline{\psi}\,'\gamma^\omega\gamma^\mu\gamma^\nu\gamma^\rho\gamma^\sigma\psi'$$

$$= \frac{i}{4!}\epsilon_{\mu\nu\rho\sigma}\overline{\psi}S(\Lambda)^{-1}\gamma^\omega\gamma^\mu\gamma^\nu\gamma^\rho\gamma^\sigma S(\Lambda)\psi$$

$$= \frac{i}{4!}\epsilon_{\mu\nu\rho\sigma}\overline{\psi}(S^{-1}\gamma^\omega S)(S^{-1}\gamma^\mu S)(S^{-1}\gamma^\nu S)(S^{-1}\gamma^\rho S)(S^{-1}\gamma^\sigma S)\psi$$

$$= \frac{i}{4!}\epsilon_{\mu\nu\rho\sigma}\Lambda^\omega{}_\zeta\Lambda^\mu{}_\alpha\Lambda^\nu{}_\beta\Lambda^\rho{}_\eta\Lambda^\sigma{}_\tau\overline{\psi}\gamma^\zeta\gamma^\alpha\gamma^\beta\gamma^\eta\gamma^\tau\psi$$

$$= i\frac{\det(\Lambda)}{4!}\epsilon_{\alpha\beta\eta\tau}\Lambda^\omega{}_\zeta\overline{\psi}\gamma^\alpha\gamma^\beta\gamma^\eta\gamma^\tau\psi$$

$$= \det(\Lambda)\Lambda^\omega{}_\zeta(\overline{\psi}\gamma^\zeta\gamma^5\psi).$$

From the first and last terms we recognize the transformation law of a pseudo-vector.

4. *Chirality and helicity (4)*. Show that the right helicity eigenstate $u_\uparrow$ can be decomposed in the right (u_R) and left (u_L) chiral states as follows:

$$u_\uparrow = \frac{1}{2}\left(1 + \frac{p}{E+m}\right)u_R + \frac{1}{2}\left(1 - \frac{p}{E+m}\right)u_L. \qquad (6.22)$$

The right helicity eigenstate can be written as

$$u_\uparrow = \sqrt{E+m}\begin{pmatrix} \cos\left(\frac{\theta}{2}\right) \\ \sin\left(\frac{\theta}{2}\right)e^{i\phi} \\ \frac{p}{E+m}\cos\left(\frac{\theta}{2}\right) \\ \frac{p}{E+m}\sin\left(\frac{\theta}{2}\right)e^{i\phi} \end{pmatrix}, \qquad (6.23)$$

with p, E and m the particle's momentum, energy and mass, respectively. The angles θ and ϕ are the polar and the azimuthal angles, respectively.

Any eigenstate can be decomposed into its left and right chiral components. By making use of the left and right projector we can write

$$u_\uparrow = \frac{1}{2}(1 + \gamma^5 + 1 - \gamma^5)u_\uparrow$$

$$= \underbrace{\frac{1}{2}(1 + \gamma^5)u_\uparrow}_{P_R} + \underbrace{\frac{1}{2}(1 - \gamma^5)u_\uparrow}_{P_L}. \qquad (6.24)$$

Let us explicitly demonstrate the relation in Eq. 6.22. It is useful to write the P_R and P_L matrices, which in the Dirac-Pauli representation are given by

$$P_R = \frac{1}{2} \begin{pmatrix} 1 & 0 & 1 & 0 \\ 0 & 1 & 0 & 1 \\ 1 & 0 & 1 & 0 \\ 0 & 1 & 0 & 1 \end{pmatrix} \quad \text{and} \quad P_L = \frac{1}{2} \begin{pmatrix} 1 & 0 & -1 & 0 \\ 0 & 1 & 0 & -1 \\ -1 & 0 & 1 & 0 \\ 0 & -1 & 0 & 1 \end{pmatrix}. \quad (6.25)$$

By evaluating $P_R u_\uparrow$ and $P_R u_\uparrow$ we can re-write Eq. 6.24 as

$$u_\uparrow = \underbrace{\frac{1}{2}\left(1 + \frac{p}{E+m}\right)\sqrt{E+m}\begin{pmatrix} \cos\left(\frac{\theta}{2}\right) \\ \sin\left(\frac{\theta}{2}\right)e^{i\phi} \\ \cos\left(\frac{\theta}{2}\right) \\ \sin\left(\frac{\theta}{2}\right)e^{i\phi} \end{pmatrix}}_{u_R} + \underbrace{\frac{1}{2}\left(1 - \frac{p}{E+m}\right)\sqrt{E+m}\begin{pmatrix} \cos\left(\frac{\theta}{2}\right) \\ \sin\left(\frac{\theta}{2}\right)e^{i\phi} \\ -\cos\left(\frac{\theta}{2}\right) \\ -\sin\left(\frac{\theta}{2}\right)e^{i\phi} \end{pmatrix}}_{u_L}$$

$$(6.26)$$

demonstrating in this way the relation presented in Eq. 6.22. The right and left chiral eigenstates, u_R and u_L, respectively, are identified through the application of the right and left projectors, and verifying the following relations:

$$P_R u_R = u_R, \qquad P_R u_L = 0, \qquad P_L u_L = u_L, \qquad \text{and} \qquad P_L u_L = u_L. \quad (6.27)$$

5. *Running electromagnetic coupling (5).* Calculate $\alpha(Q^2)$ for $Q = 1000\,\text{GeV}$.

From

$$\alpha(q^2) \simeq \frac{\alpha(\mu^2)}{1 - \frac{\alpha(\mu^2)}{3\pi}\ln\frac{q^2}{\mu^2}}, \quad (6.28)$$

where $\mu \simeq m_Z$ we get $\alpha(\mu^2) \simeq 0.28$. Plugging these values into the above equation we get

$$\alpha(1000^2\,\text{GeV}^2) \simeq 7.2 \times 10^{-2}.$$

6. *v_μ beams (6).* Consider a beam of v_μ produced through the decay of a primary beam containing pions (90%) and kaons (10%). The primary beam has a momentum of 10 GeV and an intensity of $10^{10}\,\text{s}^{-1}$.

a. Determine the number of pions and kaons that will decay in a tunnel 100 m long.
b. Determine the energy spectrum of the decay products.
c. Calculate the contamination of the v_μ beam, i.e., the fraction of v_e present in that beam.

a. As we want a beam of muon neutrinos, v_μ, then the pions and kaons should be positively charged. Therefore the main decays to be considered are:

- $\pi^+ \to \mu^+ \nu_\mu$ $B.R. \simeq 99.98\%$;
- $K^+ \to \mu^+ \nu_\mu$ $B.R. \simeq 63.55\%$;
- $K^+ \to \pi^+ \pi^0$ $B.R. \simeq 20.66\%$.

Since the primary beam has only 10% of K^+ then the decay $K^+ \to \pi^+ \pi^0$ only contributes with 2% of the overall ν_μ total number and as such it will be neglected. The probability of a particle to decay is given by

$$N(t) = N_0 \left(1 - e^{-\frac{t}{t_0}}\right) \tag{6.29}$$

where t_0 is the particle lifetime in the LAB frame. The particle lifetime in its proper frame, τ, can be related with the one in the LAB through $t_0 = \gamma \tau$, with γ being the particle *boost*. On the other hand, the particle velocity in the LAB is equal to the distance travelled by the particle over the time interval that it took it to cross that distance, i.e. $\beta c = x/t$. Then Eq. 6.29 can be written as

$$N = N_0 \left(1 - \exp \frac{x}{\beta c} \frac{\sqrt{1 - \beta^2}}{\tau}\right), \tag{6.30}$$

making use of the identity $\gamma = (\sqrt{1 - \beta^2})^{-1}$.
Considering the data of the problem:

- $x = 100\,\text{m}$;
- $m_\pi = 0.140\,\text{GeV}/c^2$; $m_K = 0.494\,\text{GeV}/c^2$;
- $N_0^\pi = 0.9 \times 10^{10}\,\text{s}^{-1}$; $N_0^K = 0.1 \times 10^{10}\,\text{s}^{-1}$;
- $\tau_\pi = 2.6 \times 10^{-8}\,\text{s}$; $\tau_K = 1.23 \times 10^{-8}\,\text{s}$,

and evaluating β through

$$\beta_i = \frac{P}{E} = \frac{10}{\sqrt{10^2 + m_i^2}}, \quad i = \pi, K \tag{6.31}$$

then, the number of pions and kaons that will decay can be computed through Eq. 6.30, yielding

$$N_\pi = \qquad\qquad 1.49 \times 10^9 \text{ pions/s}$$
$$N_K = \qquad\qquad 7.41 \times 10^8 \text{ kaons/s}.$$

b. We shall consider a two-body decay in the CM reference frame: $\pi^+ \to \mu^+ \nu_\mu$ and $K^+ \to \mu^+ \nu_\mu$. These reaction will have similar kinematics so let us start by consider the decay of the pion.
   In the CM frame the decay products have the same momentum, irrespective of direction. As such the 4-momentum of all the involved particles can be written as

$$p_\pi = (m_\pi, \mathbf{0}) \quad ; \quad p_\mu = (E_\mu, \mathbf{p}) \quad ; \quad p_\nu = (p, -\mathbf{p}).$$

Through energy-momentum conservation one has

$$p_\pi = p_\mu + p_\nu \Leftrightarrow (p_\pi - p_\nu)^2 = p_\mu^2$$
$$p_\pi^2 + p_\nu^2 - 2p_\pi \cdot p_\nu = m_\mu^2 \Leftrightarrow m_\pi^2 - 2(m_\pi p_\nu) = m_\mu^2$$

$$E_\nu = \frac{m_\pi^2 - m_\mu^2}{2m_\pi} \tag{6.32}$$

and, since $E_\nu = p_\nu = p_\mu$, then

$$E_\mu = \sqrt{E_\nu^2 + m_\mu^2} \, . \tag{6.33}$$

Notice that in the CM frame the sub-products of a two-body decay have a monochromatic energy. In the LAB reference frame, the maximum energy of the muon, E_{max} occurs when it is emitted in the *boost direction*, i.e. the flight direction of the particle that decays, $\theta^* = 0°$. The minimum energy, E_{min}, occurs when the particle is emitted backwards, $\theta^* = 180°$.

To obtain the energy of the decay products in the LAB we just have to apply the Lorentz transformations,

$$E = \gamma_\pi (E^* + \beta_\pi P^* \cos \theta^*) \, , \tag{6.34}$$

where P^* is the total momentum of the particle in the CM. The quantities necessary to change from the CM frame to the LAB can be computed using

$$\gamma_\pi = \frac{E_\pi}{m_\pi} \quad ; \quad \beta_\pi = \frac{p_\pi}{E_\pi} = \frac{\sqrt{E_\pi^2 - m_\pi^2}}{E_\pi} . \tag{6.35}$$

Hence, using the above equations and considerations we can compute the minimum and the maximum energy of the decay sub-products for:

- Pion decay:
  - $E_{max}^\mu(\theta^* = 0°) \simeq 10\,\text{GeV}$;
  - $E_{min}^\mu(\theta^* = 180°) \simeq 5.73\,\text{GeV}$;
  - $E_{max}^\nu(\theta^* = 0°) \simeq 4.22\,\text{GeV}$;
  - $E_{min}^\nu(\theta^* = 180°) \simeq 0\,\text{GeV}$;

- Kaon decay:
  - $E_{max}^\mu(\theta^* = 0°) \simeq 10\,\text{GeV}$;
  - $E_{min}^\mu(\theta^* = 180°) \simeq 0.45\,\text{GeV}$;
  - $E_{max}^\nu(\theta^* = 0°) \simeq 9.53\,\text{GeV}$;
  - $E_{min}^\nu(\theta^* = 180°) \simeq 0\,\text{GeV}$;

The energy spectrum of the decay products in the LAB frame is given by

$$\frac{dN}{dE} = \frac{dN}{d\cos\theta^*} \frac{d\cos\theta^*}{dE} . \tag{6.36}$$

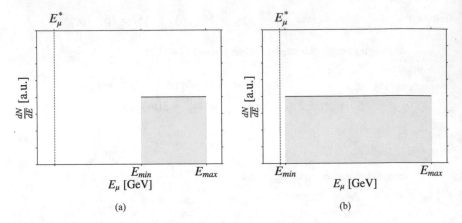

Fig. 6.2 Muon energy spectrum in the LAB (solid line) from the process (a) $\pi^+ \to \mu^+ \nu_\mu$; (b) $K^+ \to \mu^+ \nu_\mu$. The dashed vertical line (red in color) represents the energy of the muon in the decay CM reference frame

In the CM frame the sub-products are produced isotropically and so the term $\frac{dN}{d\cos\theta^*}$ is a constant that we will call k. Deriving Eq. 6.34 as a function of $\cos\theta^*$ we get

$$\frac{dE}{d\cos\theta^*} = \gamma\beta p^* . \qquad (6.37)$$

Therefore

$$\frac{dN}{dE} = k\frac{1}{\gamma\beta p^*} = k' . \qquad (6.38)$$

As k' is a constant, then in the LAB the sub-products energy spectrum will follow an uniform distribution with minimum energy $E_{\min}$ and maximum energy $E_{\max}$. For instance, the muon energy spectrum in LAB for the pion decay and the kaon decay can be seen in Fig. 6.2.

c. The ν_μ beam contamination comes essentially from ν_e. The main channels for getting a ν_e from a π^+ and K^+ beam are:

- $\pi^+ \to e^+ \nu_e$ $B.R. = 1.2 \times 10^{-4}$;
- $K^+ \to e^+ \nu_e$ $B.R. = 1.6 \times 10^{-5}$;
- $K^+ \to \pi^0 e^+ \nu_e$ $B.R. = 5.1 \times 10^{-2}$;
- $K^+ \to \pi^0 \pi^0 e^+ \nu_e$ $B.R. = 2.2 \times 10^{-5}$;
- $K^+ \to \pi^+ \pi^- e^+ \nu_e$ $B.R. = 4.1 \times 10^{-5}$.

Therefore the fraction of electron neutrinos, f_{ν_e}, with respect to the primary beam, taking into consideration that the beam is composed by 90% of pions and only 10% of kaons, is

Table 6.1

	$\nu_\mu\, p$	μ^-	X
Charge	0+1	−1	+2
Baryon number	0+1	0	+1
Lepton number	$L_\mu = +1$	$L_\mu = +1$	$L_\mu = 0$
Energy ($\sqrt{s}$/GeV)	4.43	m_μ	4.43 - m_μ

$$f_{\nu_e} = 0.9(1.2 \times 10^{-4}) + 0.1(0.051 + (1.6 + 2.2 + 4.1) \times 10^{-5})$$
$$\simeq 0.005 = 0.5\% . \tag{6.39}$$

Considering again the positive pion/kaon beam, the main channels to produce muon neutrinos are

- $\pi^+ \to \mu^+ \nu_\mu$ $B.R. = 99.98\%$;
- $K^+ \to \mu^+ \nu_\mu$ $B.R. = 63.55\%$;
- $K^+ \to \mu^+ \nu_\mu$ $B.R. = 3.3\%$.

Therefore, the fraction of muon neutrinos, f_{ν_μ}, with respect to the primary beam is

$$f_{\nu_\mu} = 0.9(0.9998) + 0.1(0.6455 + 0.033) \simeq 0.967 = 96.7\% . \tag{6.40}$$

Hence, the ν_e contamination to the beam of ν_μ is given by the ratio between Eq. 6.39 and 6.40, which is roughly 0.5%.

7. ν_μ *semileptonic interaction (7)*. Consider the process $\nu_\mu + p \to \mu^- + X$.

a. Discuss what X could be (start by computing the available energy in the center-of-mass).

b. Write the amplitude at lowest order for the interaction of the ν_μ with the valence quark d ($\nu_\mu d \longrightarrow \mu^- u$).

c. Compute the effective energy in the center-of-mass for this process supposing that the energy of the ν_μ is 10 GeV and the produced muon takes 5 GeV and is detected at an angle of 10° with the ν_μ beam.

d. Write the cross section of the process $\nu_\mu p \longrightarrow \mu^- X$ as a function of the elementary cross section $\nu_\mu d \to \mu^- u$.

---

a. The X in the reaction $\nu_\mu + p \to \mu^- + X$ can be any particle as long as it does not violate charge, baryon number and energy conservation. Table 6.1 summarizes the conservation of energy and the quantum numbers so that the reaction is possible.

The initial state $\nu_\mu + p$ should be equal to the final state $\mu + X$. The center-of-mass energy, $\sqrt{s}$, was computed noting that the 4-vector of the initial state in the LAB frame is $P_\mu = (E_\nu + m_p, E_\nu)$, and evaluating the $s = P_\mu P^\mu = m_p^2 + 2E_\nu m_p \simeq 4.43\,\text{GeV}$, where E_ν and m_p are the neutrino energy and proton mass, respectively. The maximum energy available for X is obtained subtracting the muon mass to the CM energy (assuming that the muon is produced at rest in the LAB. Therefore X should be at least a baryon, the sum of its charges should be $\sum Q_i = +2$ and the sum of its masses $\sum m_i. = 4.43\,\text{GeV}$ (assume that all particles are produced at rest).

b. Let us consider the elementary process $\nu_\mu(p_1) + d(p_2) \rightarrow \mu^-(p_3) + u(p_4)$. The Feynman diagram for this process is the one shown in Fig. 6.3. The amplitude for this diagram can be written as

$$\mathcal{M} = (-i)\bar{u}(p_3)\left(\frac{-ig}{\sqrt{2}}\right)\gamma^\mu\left(\frac{1-\gamma^5}{2}\right)u(p_1)\left(-i\frac{g_{\mu\nu} - \frac{k_\mu k_\nu}{M_W^2}}{k^2 - M_W^2 + iM_W\Gamma_W}\right)\bar{u}(p_4)\left(\frac{-ig}{\sqrt{2}}\right)\gamma^\nu\left(\frac{1-\gamma^5}{2}\right)u(p_2).$$

(6.41)

with $k = p_3 - p_1$.

c. This reaction can be considered to be a deep inelastic scattering (DIS) process. We know that the energy of the incoming neutrino is $E = 10\,\text{GeV}$, the energy taken by the muon is $E' = 5\,\text{GeV}$, and it deviates from the neutrino initial direction by $\theta = 10°$. In DIS we have the following relations which allows us to computed the square of the momentum transferred, Q^2,

$$Q^2 = 4EE'\sin^2\left(\frac{\theta}{2}\right) = 1.52\,\text{GeV}^2,$$

(6.42)

and the Bjorken-x (the *elasticity* of the scattering process),

$$x = \frac{Q^2}{2m_p(E - E')} = 0.16$$

(6.43)

with m_p being the mass of the target proton. The center-of-mass energy has been computed in the previous problem giving $\sqrt{s} = 4.43\,\text{GeV}$. Thus, finally, the effective center-of-mass energy can be evaluated:

$$\sqrt{\hat{s}} = \sqrt{xs} = 1.78\,\text{GeV}.$$

(6.44)

d. The total cross section of the process $\nu_\mu + p \rightarrow \mu^- + X$ can be obtained evaluating

$$\sigma_{\nu_\mu p \rightarrow \mu^- X} = \int dx\, f_d(x, Q)\, \hat{\sigma}_{\nu_\mu d \rightarrow \mu^- u}(\hat{s}),$$

(6.45)

Fig. 6.3 Feynman diagrams at lowest order for the process $\nu_\mu(p_1) + d(p_2) \rightarrow \mu^-(p_3) + u(p_4)$

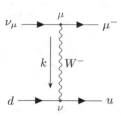

where $\hat{s} = xs$ and $f_d(x, Q)$ is the parton distribution function (PDF) of the d quark in the proton.

8. *Neutrino and anti-neutrino deep inelastic scattering (8).* Determine, in the framework of the quark parton model, the ratio:

$$\frac{\sigma\left(\bar{\nu}_\mu N \longrightarrow \mu^+ X\right)}{\sigma\left(\nu_\mu N \longrightarrow \mu^- X\right)}$$

where N stands for an isoscalar (same number of protons and neutrons) nucleus. Consider that the involved energies are much higher than the particle masses. Take into account only diagrams with valence quarks.

Considering only proton and neutron valence quarks we have only the quarks up and down to participate in theses processes. Through charge conservation we can easily see that the elementary processes are

$$\nu_\mu + N \rightarrow \mu^- + X \Longrightarrow \nu_\mu + d \rightarrow \mu^+ + u \qquad (6.46)$$
$$\bar{\nu}_\mu + N \rightarrow \mu^+ + X \Longrightarrow \bar{\nu}_\mu + u \rightarrow \mu^+ + d . \qquad (6.47)$$

Let us start by evaluating the cross-section of the elementary process $\nu_\mu(p_1) + d(p_2) \rightarrow \mu^-(p_3) + u(p_4)$. The lowest order Feynman diagram for this reaction is shown in Fig. 6.4b and its amplitude can be written as

$$\mathscr{M} = (-i)\left(-i\frac{g}{\sqrt{2}}\right)\bar{u}(p_3)\gamma^\mu\frac{1}{2}(1-\gamma^5)u(p_1)\left(\frac{g_{\mu\nu} - \frac{k_\mu k_\nu}{M_W^2}}{k^2 - M_W^2 + iM_W\Gamma_W}\right)\left(-i\frac{g_W}{\sqrt{2}}\right)\bar{u}(p_4)\gamma^\nu\frac{1}{2}(1-\gamma^5)u(p_2)$$

$$(6.48)$$

with $k = p_3 - p_1$. If $m_q \ll \sqrt{s} \ll M_W$, we can write the above equation as:

$$\mathscr{M} = \frac{g^2}{2M_W^2}\left[\bar{u}(p_3)\gamma^\mu\frac{1}{2}(1-\gamma^5)u(p_1)\right]\left[\bar{u}(p_4)\gamma_\mu\frac{1}{2}(1-\gamma^5)u(p_2)\right]. \quad (6.49)$$

For high-energy neutrino scattering, the quark and neutrino masses can be neglected and the left-handed chiral states ($P_L = \frac{1}{2}(1-\gamma^5)$) are approximately

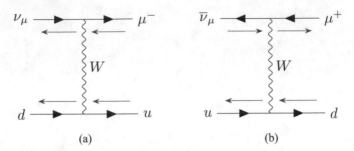

Fig. 6.4 Feynman diagrams at lowest order for process (a) $\nu_\mu + d \to \mu^- + u$; (b) $\bar{\nu}_\mu + u \to \mu^+ + d$. The grey (red in color) arrows represent the non-vanishing helicity states in the calculation of the cross-section

identical to the left-handed helicity states. Thus

$$\mathcal{M} = \frac{g^2}{2M_W^2} J_{u_1 u_3}(\downarrow\downarrow) \cdot J_{u2u4}(\downarrow\downarrow). \tag{6.50}$$

The $\cdot$ symbol between 4-vectors a and b indicates their scalar product $a^\mu b_\mu$. Computing the above inner product between both currents, the amplitude is finally obtained to be

$$\mathcal{M} = \frac{g^2}{M_W^2} \hat{s} , \tag{6.51}$$

where $\hat{s} = (2E)^2$ is the square of the elementary process center-of-mass available energy.

As the fermion masses are being neglected, and we are evaluating this process in the CM reference frame, the differential cross-section is given by

$$\frac{d\sigma}{d\Omega} = \frac{1}{64\pi^2 \hat{s}} \langle |\mathcal{M}|^2 \rangle . \tag{6.52}$$

The spin average matrix can be computed from 6.51 as

$$\langle |\mathcal{M}|^2 \rangle = \frac{1}{2} \left(\frac{g^2}{M_W^2} \hat{s} \right)^2 . \tag{6.53}$$

Notice that as the neutrino is always left-handed, we only have to average over the quark polarizations, thus the $1/2$ factor in the above equation.

The differential cross-section can be then written as

$$\frac{d\sigma}{d\Omega} = \left(\frac{g^2}{8\sqrt{2}\pi M_W^2} \right)^2 \hat{s} = \frac{G_F^2}{4\pi^2} \hat{s} \tag{6.54}$$

and the total cross section can be obtained integrating over $d\Omega$:

$$\sigma_{\nu d \to \mu^- u} = \frac{G_F^2 \hat{s}}{\pi} . \tag{6.55}$$

Let us now repeat the same proceed to evaluate the cross section to the process $\bar{\nu}_\mu(p_1) + u(p_2) \to \mu^+(p_3) + d(p_4)$. The Feynman diagram for this process is the one drawn in Fig. 6.4. Using the same approximations as before we obtain the following amplitude for this process:

$$\mathcal{M} = \frac{g^2}{2M_W^2} \left[v(p_1) \gamma^\mu \frac{1}{2} (1 - \gamma^5) v(p_3) \right] \left[\bar{u}(p_4) \gamma_\mu \frac{1}{2} (1 - \gamma^5) u(p_2) \right] \tag{6.56}$$

$$= \frac{g^2}{2M_W^2} J_{\nu_1 \nu_3}(\uparrow\uparrow) \cdot J_{u2u4}(\downarrow\downarrow) = \frac{1}{2}(1 + \cos\theta) \frac{g^2}{M_W^2} \hat{s} . \tag{6.57}$$

where θ is the polar angle of the μ^+ in the center-of-mass reference frame. We can immediately see that the amplitude in Eq. 6.56 differs from the one in Eq. 6.51 by a factor $\frac{1}{2}(1 + \cos\theta)$. This difference can be understood in terms of the spins of the particles (see the grey (red in color) arrows in Fig. 6.4) and the weak interaction $(V - A)$ nature.

The differential cross-section can be evaluated as before giving:

$$\frac{d\sigma}{d\Omega} = \frac{G_F^2}{16\pi^2}(1 + \cos\theta)\hat{s} . \tag{6.58}$$

The total cross section can be obtained integrating over the solid angle, and noting that

$$\int (1 + \cos\theta)^2 \, d\Omega = \int_0^{2\pi} d\phi \int_{-1}^{+1} (1 + x)^2 dx = \frac{16\pi}{3} , \tag{6.59}$$

with $x = \cos\theta$, we get finally

$$\sigma_{\bar{\nu}_\mu u \to \mu^+ d} = \frac{G_F^2 \hat{s}}{3\pi} . \tag{6.60}$$

Hence, combining Eqs. 6.55 and 6.60, we obtain the ratio

$$\frac{\sigma\left(\bar{\nu}_\mu N \longrightarrow \mu^+ X\right)}{\sigma\left(\nu_\mu N \longrightarrow \mu^- X\right)} = \frac{1}{3} . \tag{6.61}$$

9. *Feynman rules (9).* What is the lowest order diagram for the process $\gamma\gamma \rightarrow$ e^+e^-?

The Breit-Wheeler process is the simplest mechanism by which light can be transformed into matter. Its Feynman diagrams at leading order are shown in Fig. 6.5. Notice that the u-channel appears because the initial momenta of the photons are unknown. Although the pure photon-photon Breit-Wheeler process was one of the first source of pair to be described, its experimental validation in laboratory has to be performed, yet - but the process has been seen at work in the cosmological propagation of gamma rays.

10. *Bhabha scattering (10).* Draw the QED Feynman diagrams at lowest (leading) order for the elastic e^+e^- scattering and discuss why the Bhabha scattering measurements at LEP are done at very small polar angle.

The QED Feynman diagrams at lowest order for the elastic e^+e^- scattering are shown in Fig. 6.6. They can be pictorially regarded as the e^+e^- annihilation into a virtual photon that converts again into an e^+e^- pair (left diagram) and one e^+ (or e^-) that scatters off the other particle by exchanging a virtual photon (right diagram). For simplicity, the initial and final states of the e^+ will be named respectively as 1 and 3, and the initial and final states of the e^- as 2 and 4. In each of these diagrams the photon propagator is $-ig_{\mu\nu}/q^2$, where q^2 is the 4-momentum squared carried by the photon. Since there is energy-momentum conservation at each vertex, one has for q^2 in each of the diagrams,

$$q^2 = (p_1 + p_2)^2 \quad ; \quad q^2 = (p_1 - p_3)^2. \tag{6.62}$$

The above quantities are respectively the s and t Mandelstam variables (s is the square of the center-of-mass energy and t is the square of the variation of the 4-momentum of the initial particles). Neglecting the electron mass ($m_e^2 \ll E$), in the center-of-mass frame we have

$$t \simeq -\frac{s}{2}\,(1 - \cos\theta) \tag{6.63}$$

with θ the angle between the initial and final direction of the electron (or the positron).

The cross-section is obtained by computing the square of the transition amplitude, $|\mathcal{M}|^2$. We can thus expect that for the e^+e^- scattering, a term proportional to $1/(1 - \cos\theta)^2$ will appear in the cross-section. In fact, the QED cross-section at lowest order for the Bhabha scattering is given by:

$$\frac{d\sigma}{d\Omega} = \frac{\alpha^2}{4s}\left(\frac{3 + \cos^2\theta}{1 - \cos\theta}\right)^2 = \frac{\alpha^2}{16\,s}\frac{(3 + \cos^2\theta)^2}{\sin^4(\theta/2)} \tag{6.64}$$

Fig. 6.5 QED Feynman diagrams at lowest order for the Breit-Wheeler process

Fig. 6.6 QED Feynman diagrams at lowest order for the elastic e^+e^- scattering

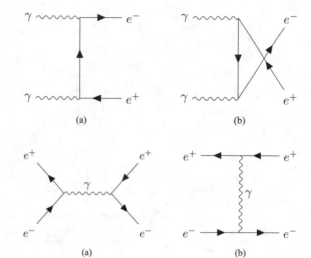

where the relation $\cos(2x) = 1 - 2\sin^2 x$ was used in the last equality. From this expression it is clear that the cross-section increases very steeply with the decrease of the scattering angle. For small values of θ it can be approximated by

$$\frac{d\sigma}{d\Omega} \simeq \frac{16\,\alpha^2}{s}\frac{1}{\theta^4}. \tag{6.65}$$

The divergence of the Bhabha cross-section at leading order with decreasing scattering angle is the reason why at LEP Bhabha scattering measurements are done at very small polar angle. By placing a dedicated detector very close to the colliding beams a large sample of such events are recorded, and the measurements from this process can be performed with very small statistical error.

11. *Bhabha scattering: higher orders (11).* Draw the QED Feynman diagrams at next-to-leading order for the Bhabha scattering.

The next-to-leading (NLO) order diagrams for the $e^+ + e^- \to e^+ + e^-$ are shown in Fig. 6.7. Notice that these diagrams have four coupling instead of the two found at the tree level (see Fig. 6.6).

12. *Compton scattering and Feynman rules (12).* Draw the leading-order Feynman diagram(s) for the Compton scattering $\gamma e^- \to \gamma e^-$ and compute the amplitude for the process.

The lowest order Feynman diagrams for the Compton scattering are the ones shown in Fig. 6.8.

Let us consider the following kinematics: $e^-(p) + \gamma(k) \to e^-(p') + \gamma(k')$. The amplitude for the diagram in Fig. 6.8a is

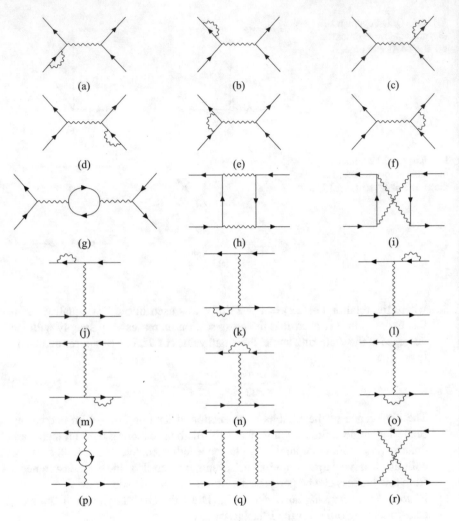

Fig. 6.7 QED Feynman diagrams for the elastic e^+e^- scattering at next-to-leading order

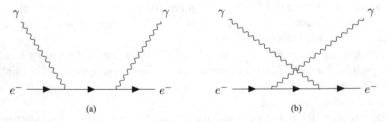

Fig. 6.8 QED Feynman diagrams at lowest order for the electron-photon scattering (Compton effect)

$$i\mathcal{M}_a = \bar{u}(p')(-ie\gamma^\mu)\frac{i(\not{p}+\not{k}+m_e)}{(p+k)^2-m_e^2}(-ie\gamma^\nu)u(p)\varepsilon_\mu^*(k')\varepsilon_\nu(k). \quad (6.66)$$

and in Fig. 6.8b is,

$$i\mathcal{M}_a = \bar{u}(p')(-ie\gamma^\nu)\frac{i(\not{p}'-\not{k}+m_e)}{(p'-k)^2-m_e^2}(-ie\gamma^\mu)u(p)\varepsilon_\mu^*(k')\varepsilon_\nu(k) \quad (6.67)$$

The total amplitude is simply the sum of the two amplitudes, $\mathcal{M}_t = \mathcal{M}_a + \mathcal{M}_b$.

13. *Top pair production (13).* Consider the pair production of top/anti-top quarks at a proton-antiproton collider. Draw the dominant first-order Feynman diagram for this reaction and estimate what should be the minimal beam energy of a collider to make the process happen. Discuss which channels have a clear experimental signature.

In hadron collisions, top quarks are produced dominantly in pairs $(t\bar{t})$ through processes at leading order in QCD. Approximately 85% of the production cross section at the Fermilab Tevatron ($\bar{p}p$ at 1.96 TeV c.m. energy) is from $\bar{q}q$ annihilation, with the remainder from gluon-gluon fusion. The main Feynman diagrams are shown below Fig. 6.9.

At a hadronic collider, the minimum c.m. energy to have a sizable $t\bar{t}$ production is $\sqrt{s} \simeq 2m_t N_p$, where m_t is the top mass and N_p is the number of effective partons in the collision. At Tevatron (the highest energy $\bar{p}p$ accelerator) energies, the number of effective partons was $\simeq 5$; for a top mass of $m_t \sim 173$ GeV the minimum energy was $E_{cm} \simeq \sqrt{s} \simeq 6 \times 2 \times 173\,\mathrm{GeV} \simeq 2\,\mathrm{TeV}$.

14. *c -quark decay (14).* Consider the decay of the c quark. Draw the dominant first-order Feynman diagrams of this decay and express the corresponding decay rates as a function of the muon decay rate and of the Cabibbo angle. Make an

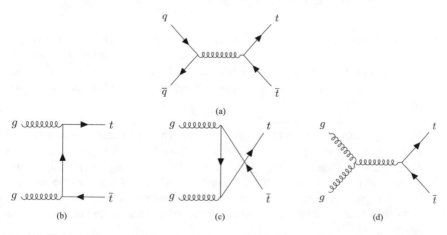

(a)

(b)   (c)   (d)

Fig. 6.9 Diagrams that contribute to the single gluon emission at leading order

Fig. 6.10 Dominant first-order Feynman diagrams for the c decay

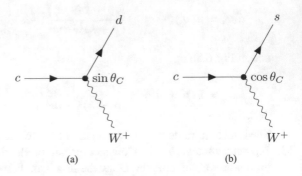

(a)   (b)

estimation of the c quark lifetime knowing that the muon lifetime is about 2.2 μs.

The dominant first-order Feynman diagrams of the c decay are shown in Fig. 6.10 (the left one, subdominant, has an amplitude proportional to $\sin \theta_C$, where θ_C is the Cabibbo angle; the right one has an amplitude proportional to $\cos \theta_C$).
The relative decay rates (neglecting phase space effects, since $m_c \gg m_s, m_d$) are

$$\frac{\Gamma(c \to d)}{\Gamma(c \to s)} \simeq \tan^2(\theta_C) \simeq 0.05.$$

The charm decay width is approximately a factor $(m_c/m_\mu)^5$ larger than the muon decay width.
An estimate of the charm lifetime (neglecting in a first approximation the s mass, since $m_c \gg m_s$) is

$$\tau_c \simeq \tau_\mu \left(\frac{m_\mu}{m_c}\right)^5 \simeq 2.2\mu s \left(\frac{0.105\,\text{GeV}}{1.8\,\text{GeV}}\right)^5 \simeq 1.5\,\text{ps}.$$

15. *Gray disk model in proton–proton interactions (15).* Determine, in the framework of the gray disk model, the mean radius and the opacity of the proton as a function of the c.m. energy (you can use Fig. 6.11 to extract the total proton–proton cross section and the ratio between the elastic and the total proton–proton cross sections).

At low momentum transfer the interaction between two particles can be described in the laboratory reference frame, using quantum mechanics optical models, as one plane wave (associated to the "beam particle") scattering in a diffusion center (associated to the "target" particle). For a short review see Chap. 6 of the textbook. Then, following a semi-classical approach, the scattering amplitude $a(b, s)$ may be written as a function of the impact parameter b and of the center-of-mass energy $\sqrt{(s)}$.

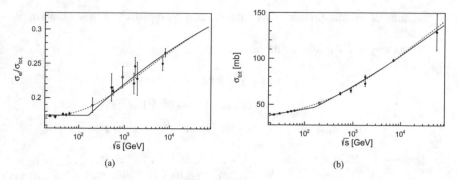

Fig. 6.11 The (a) elastic over total cross-section ratio and (b) total cross-section for proton-proton interactions as a function of the c.m. energy. Points are experimental data and the lines are coming from a fit using a gray disc model. From R. Conceição, J. Dias de Deus, M. Pimenta, Nuclear Physics A 888 (2012) 58

To ensure automatically unitarity, $a(b, s)$ may be parametrized as

$$a(b, s) = \frac{i}{2} (1 - e^{i\chi(b,s)}) \tag{6.68}$$

where

$$\chi(b, s) = \chi_R(b, s) + i\,\chi_I(b, s) \tag{6.69}$$

is called the eikonal function.

It can be shown that the cross sections are related with the eikonal by the following expressions:

$$\sigma_{el}(s) = \int d^2 b \left| 1 - e^{i\chi(b,s)} \right|^2$$

$$\sigma_{tot}(s) = 2 \int d^2 b \left(1 - \cos(\chi_R(b, s))\, e^{-\chi_I(b,s)} \right)$$

$$\sigma_{inel}(s) = \int d^2 b \left(1 - e^{-2\chi_I(b,s)} \right),$$

where the integrations are performed over the target region with a radius R.

In the *gray disk model* the eikonal is imaginary and parametrized as a disk with a radius R and an opacity Ω which is a real number, $0 < \Omega < \infty$:

$$\chi(b, s) = \begin{cases} i\Omega(s)\,, & \text{if } b \leq R\,; \\ 0 & \text{, if } b > R\,. \end{cases} \tag{6.70}$$

The limits $\Omega = 0$ and $\Omega = \infty$ correspond respectively to the total transparency ($\sigma_{tot} = 0$) and to the *black disk* ($\sigma_{inel} = \sigma_{el}$ and $\sigma_{tot} = 2\pi R^2$). In fact, the main features of proton–proton cross sections can be reproduced in such a simple model.

In this way in the gray disk model we have:

$$\sigma_{tot}(s) = 2\pi\left(1 - e^{-\Omega(s)}\right)R^2(s)$$

$$\sigma_{el}(s) = \pi\left(1 - e^{-\Omega(s)}\right)^2 R^2(s)$$

$$\sigma_{inel} = \sigma_{tot} - \sigma_{el} = \pi\left(1 - e^{-2\Omega(s)}\right)R^2(s),$$

and then

$$\frac{\sigma_{el}}{\sigma_{tot}} = \frac{1}{2}\left(1 - e^{-\Omega(s)}\right),\tag{6.71}$$

The opacity and the mean radius of the proton can thus be obtained inverting the above equations:

$$\Omega(s) = -\ln\left(1 - 2\frac{\sigma_{el}(s)}{\sigma_{tot}(s)}\right);$$

$$R(s) = \sqrt{\frac{1}{4\pi}\frac{\sigma_{tot}^2(s)}{\sigma_{el}(s)}}.$$

The ratio of the elastic over the total cross-sections as well as the total cross-section for proton-proton interactions as a function of the c.m. energy are represented in the Fig. 6.12. The ratio is basically constant for $\sqrt{s} \leq 200\,\text{GeV}$ and then grows smoothly being however, at the highest energy shown in figure, still well bellow its asymptotic value of 0.5 ($\Omega = \infty$). Ω is thus constant for $\sqrt{s} \leq 200$ GeV and from the above equations and from the measured values shown in the figure one gets:

$$\Omega_0 \simeq 0.43, \text{ if } \sqrt{s} \leq 200\,\text{GeV}.\tag{6.72}$$

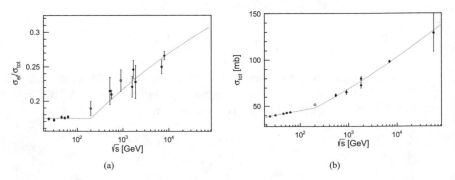

(a)   (b)

Fig. 6.12 (a) Elastic over total cross-section ratio and (b) Total cross section for proton-proton interactions as a function of the c.m. energy. The lines are computed using the parametrizations derived in the exercise

Then, in order to reproduce the experimental data, Ω should grows smoothly and, as a trial, the following parametrization may be used:

$$\Omega(s) = \Omega_0 + \alpha \ln (s/s_0), \text{ if } \sqrt{s} > 200 \text{ GeV.} \tag{6.73}$$

The corresponding parametrization of the ratio of the elastic over the total cross-sections is:

$$\frac{\sigma_{el}}{\sigma_{tot}} = \frac{1}{2} \left(1 - \left(e^{-\Omega_0} \right) \left(\frac{s_0}{s} \right)^\alpha \right). \tag{6.74}$$

Indeed, such parametrization reproduce wells, within experimental errors, the data set (line in the left part of the figure) for $\alpha = 0.044$, $\Omega_0 = 0.43$ and $s_0 = 40\,000 \text{ GeV}^2$.

The evolution of the total proton-proton cross-section as a function of the c.m. energy has also two regimes. For $\sqrt{s} \leq 200 \text{ GeV}$ Ω is constant and the increase of the cross section is just determined by the increase of the mean radius (the so called Geometrical Scale regime). For $\sqrt{s} > 200 \text{ GeV}$ such increase is also determined by the increase of Ω. A logarithmic parametrization for $R(s)$ may also be tried:

$$R(s) = R_0 + \beta \ln (s/s_0). \tag{6.75}$$

Such simple parametrization reproduce well once again the experimental data set (line in the in the right part of the figure) for $R_0 = 4.68$, $\beta = 0.107$ and $s_0 = 200^2 \text{ GeV}^2$.

At very high energies the gray disk may become basically a black disk and thus thereafter the increase of the cross section, will be just determined again by the increase of the mean radius, but limited by the *Froissart Bound* to $\ln^2(s)$.

16. *Helicity of the neutrino.* Is the helicity of a neutrino that travels freely after being produced in a weak decay constant?

The helicity of a neutrino is not a constant of the movement because neutrinos are not massless particles. A left-handed neutrino would become right-handed in an inertial frame that is moving at the speed of light (faster than the speed of the neutrino, whatever its mass is).

17. *Weak interaction in the Yukawa model.* The weak interaction is due to the exchange of $W^\pm$ and Z bosons. In the framework of a Yukawa-like model estimate its range and compare it with the size of a nucleon.

$$R \sim M_W c \sim 3 \times 10^{-16} \text{cm} ,$$

an order of magnitude smaller than the radius of a nucleon.

18. *CP violation and neutron-antineutron oscillations.* Neutron and antineutron are the antiparticle of each other; they are neutral. Explain why $K^0 - \bar{K}^0$ mixing can occur, while the mixing between neutron and antineutron is forbidden.

Kaon decay violates CP conservation and strangeness conservation (we now know that both CP and strangeness are not conserved in weak decays). The neutron-antineutron oscillations would imply that there is baryon number violation. The baryon number conservation is the one that keeps matter stable and its violation has never been observed so far.

19. *CP in neutron β decay.* Apply the parity operator to the β neutron decay $n \rightarrow pe^-\bar{\nu}_e$. Does the resulting process exist in nature? Then, apply the charge conjugation operator. What kind of process do you obtain? What can you conclude about CP in β decay?

The application of the operators P and C to p and e^- gives no particular problem. The problem arises instead for the antineutrino. The parity operator transforms a right-handed $\bar{\nu}_e$ into a left-handed $\bar{\nu}_e$, which does not interact in the Standard Model. After the application of P, C transforms the right-handed antineutrino into a left-handed neutrino, which has the correct helicity. Only the consecutive applications of P and C is therefore possible. The application of CP to the produces the possible process $\bar{n} \rightarrow \bar{p}\nu_e e^+$.

20. K^0 *mixing.* A K^0 beam propagating in vacuum can decay. At a distance d corresponding to 20 times the K_1 lifetime there is a target that absorbs 10% of the incoming beam. If the interaction cross section for $\overline{K}^0$ is three times larger than that of the K^0, calculate the relative amplitudes of K_1 and K_2 in the beam:

a. At $t = 0$;
b. Immediately before the target;
c. Immediately after the target.

Assume low-energy kaons, and neglect relativistic effects.

a. At the beginning we have

$$|K_1\rangle = \frac{1}{\sqrt{2}}(|K^0\rangle + |\overline{K}^0\rangle) \tag{6.76}$$

$$|K_2\rangle = \frac{1}{\sqrt{2}}(|K^0\rangle - |\overline{K}^0\rangle) . \tag{6.77}$$

Therefore the ratio of $K_1/K_2 = 1$.

b. All the K_1 component has decayed before hitting the target, therefore $K_1/K_2 = 0$.

c. Immediately before the target the K_1 component has decayed and the beam is a pure K_2 beam which is a superposition of K^0 and $\overline{K}^0$ states as stated above:

$$|i\rangle = |K_2\rangle = \frac{1}{\sqrt{2}}(|K^0\rangle - |\overline{K}^0\rangle) . \tag{6.78}$$

Immediately after the target the K^0 and $\overline{K}^0$ amplitudes will be reduced by some factor and the final state may be written as:

$$|f\rangle = \frac{1}{\sqrt{2}}(a|K^0\rangle - b|\overline{K}^0\rangle) , \qquad (6.79)$$

or

$$|f\rangle = \frac{a+b}{2\sqrt{2}}(|K^0\rangle - |\overline{K}^0\rangle) + \frac{a-b}{2\sqrt{2}}(|K^0\rangle + |\overline{K}^0\rangle) . \qquad (6.80)$$

In terms of the K_1 and K_2 states:

$$|f\rangle = \frac{1}{2}(a+b)|K_2\rangle + \frac{1}{2}(a-b)|K_1\rangle . \qquad (6.81)$$

The ratio of the amplitudes of $|K_1\rangle$ and $|K_2\rangle$ immediately after the target is then:

$$\frac{K_1}{K_2} = \frac{(a-b)}{(a+b)} . \qquad (6.82)$$

Being the interaction cross section for $\overline{K}^0$ three times larger than of the K^0 then $a = \sqrt{3}\,b$, and:

$$\frac{K_1}{K_2} = \frac{\sqrt{3}-1}{\sqrt{3}+1} \simeq 0.28. \qquad (6.83)$$

21. *The Z boson at LEP.* At the Large Electron-Positron collider (LEP) one of the reactions studied was:

$$e^-(p_1) + e^+(p_2) \rightarrow \mu^-(p_3) + \mu^+(p_4) .$$

a. Draw the Feynman diagrams that contribute in the Standard Model, in lowest order, to this process.
b. Consider now that $\sqrt{s} \simeq m_Z$. Write down the amplitude for the most important diagram. Neglect all fermion masses.
c. Show that if one neglects the fermions masses one can also neglect the terms proportional to the momenta in the numerator of the Feynman propagator for the massive gauge bosons. Write down the simplified amplitude.
d. Show that the electron current obey the relation

$$\overline{v}(p_2)\gamma^\mu(g_V^e - g_A^e\gamma^5)u(p_1) = g_L^e\,\overline{v}(p_2)\gamma^\mu P_L u(p_1) + g_R^e\,\overline{v}(p_2)\gamma^\mu P_R u(p_1)$$

where

$$P_L = \frac{1}{2}(1-\gamma^5), \quad P_R = \frac{1}{2}(1+\gamma^5), \quad g_L^e = g_V^e + g_A^e, \quad g_R^e = g_V^e - g_A^e$$

with similar expressions for the μ current.

e. Use the previous result to write down the non-vanishing helicity amplitudes for this process with the previous assumptions.

f. Calculate the spin averaged amplitude $\langle|\mathcal{M}|^2\rangle$ and write the expression for the differential cross section $d\sigma/d\Omega$ in the CM frame.

g. In the LEP experiment an important quantity that was measured was the Forward-Backward (FB) asymmetry defined by

$$A_{FB} = \frac{\int_0^{\pi/2} d\Omega \, d\sigma/d\Omega - \int_{\pi/2}^{\pi} d\Omega \, d\sigma/d\Omega}{\int_0^{\pi/2} d\Omega \, d\sigma/d\Omega + \int_{\pi/2}^{\pi} d\Omega \, d\sigma/d\Omega}.$$

Explain the name and why it might be useful. The following expressions are useful:

$$(g_L^e)^2(g_L^\mu)^2 + (g_R^e)^2(g_R^\mu)^2 + (g_L^e)^2(g_R^\mu)^2 + (g_R^e)^2(g_L^\mu)^2 = 4\left((g_V^e)^2 + (g_A^e)^2\right)\left((g_V^\mu)^2 + (g_A^\mu)^2\right)$$

$$(g_L^e)^2(g_L^\mu)^2 + (g_R^e)^2(g_R^\mu)^2 - (g_L^e)^2(g_R^\mu)^2 - (g_R^e)^2(g_L^\mu)^2 = 16 g_V^e g_A^e g_V^\mu g_A^\mu.$$

a. The possible lowest order Feynman diagrams are the ones displayed in Fig. 6.13.

b. At $\sqrt{s} \simeq M_Z$ the diagram in Fig. 6.13b dominates because of the Z denominator,

$$\frac{1}{s - M_Z^2 + i M_Z \Gamma_Z} \simeq \frac{1}{i M_Z \Gamma_Z} \qquad (6.84)$$

with $\Gamma_Z \ll M_Z$.

Fig. 6.13 Tree-level Feynman diagram for the process $e^+e^- \to \mu^+\mu^-$

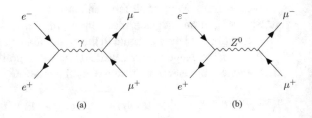

Then

$$\mathcal{M} \simeq \mathcal{M}_b = (-i)\left(-i\frac{g}{\cos\theta_W}\right)^2 \bar{v}(p_2)\gamma^\mu(g_V^e - g_A^e\gamma^5)u(p_1)$$

$$\left(-i\frac{\left(g_{\mu\nu} - \frac{q_\mu q_\nu}{M_Z^2}\right)}{s - M_Z^2 + iM_Z\Gamma_Z}\right)\bar{u}(p_3)\gamma^\nu(g_V^\mu - g_A\gamma^5)v(p_4). \tag{6.85}$$

c. Considering, for instance, the term in the electron line, we see that it would be proportional to

$$X = \bar{v}(p_2)\gamma^\mu(g_V^e - g_A^e\gamma^5)u(p_1)q_\mu \tag{6.86}$$

with $q = p_1 + p_2$. Therefore:

$$\begin{aligned}
X &= \bar{v}(p_2)\not{p}_1(g_V^e - g_A^e\gamma^5)u(p_1) + \bar{v}(p_2)\not{p}_2(g_V^e - g_A^e\gamma^5)u(p_1) \\
&= \bar{v}(p_2)(g_V^e + g_A^e\gamma^5)\not{p}_1 u(p_1) + \bar{v}(p_2)\not{p}_2(g_V^e - g_A^e\gamma^5)u(p_1) \\
&= 0
\end{aligned}$$

because, according to the Dirac equation, for massless fermions we have $\not{p}_1 u(p_1) = 0$ and $\bar{v}(p_2)\not{p}_2 = 0$. So Eq. 6.85 can be simplified into

$$\mathcal{M} = \left(\frac{g}{\cos\theta_W}\right)^2 \frac{1}{iM_Z\Gamma_Z}\bar{v}(p_2)\gamma^\mu(g_V^e + g_A^e\gamma^5)u(p_1)\bar{u}(p_3)\gamma_\mu(g_V^\mu + g_A^\mu\gamma^5)v(p_4). \tag{6.87}$$

d. Taking into account the fact that $1 = P_L + P_R$, the electron current of in $\mathcal{M}$ can be written as

$$\bar{v}(p_2)\gamma^\mu(g_V^e + g_A^e\gamma^5)u(p_1) = \bar{v}(p_2)\gamma^\mu(g_V^e + g_A^e\gamma^5)(P_L + P_R)u(p_1).$$

Since $\gamma^5 P_L = -P_L$ and $\gamma^5 P_R = P_R$, we can re-write the above expression as

$$\begin{aligned}
\bar{v}(p_2)\gamma^\mu(g_V^e + g_A^e\gamma^5)u(p_1) &= \bar{v}(p_2)\gamma^\mu\left[(g_V^e + g_A^e)P_L + (g_V^e - g_A^e)P_R\right]u(p_1) \\
&= g_L^e\bar{v}(p_2)\gamma^\mu P_L u(p_1) + g_R^e\bar{v}(p_2)\gamma^\mu P_R u(p_1) \tag{6.88}
\end{aligned}$$

with $g_L^e = g_V^e + g_A^e$ and $g_R^e = g_V^e - g_A^e$.

e. Using the results in the previous problem we can write

$$\mathcal{M} = C_Z\left[g_L^e\bar{v}(p_2)\gamma^\mu P_L u(p_1) + g_R^e\bar{v}(p_2)\gamma^\mu P_R u(p_1)\right]$$
$$\left[g_L^\mu\bar{u}(p_3)\gamma_\mu P_L v(p_4) + g_R^\mu\bar{u}(p_3)\gamma_\mu P_R v(p_4)\right] \tag{6.89}$$

with $C_Z = \left(\frac{g}{\cos\theta_W}\right)^2 \frac{1}{iM_Z\Gamma_Z}$. We therefore have:

$$
\begin{aligned}
\mathcal{M} =\ & C_Z\, g_L^e\, g_L^\mu\, \bar{v}(p_2)\, \gamma^\mu\, P_L\, u(p_1)\, \bar{u}(p_3)\, \gamma_\mu\, P_L\, v(p_4) \\
&+ C_Z\, g_L^e\, g_R^\mu\, \bar{v}(p_2)\, \gamma^\mu\, P_L\, u(p_1)\, \bar{u}(p_3)\, \gamma_\mu\, P_R\, v(p_4) \\
&+ C_Z\, g_R^e\, g_L^\mu\, \bar{v}(p_2)\, \gamma^\mu\, P_R\, u(p_1)\, \bar{u}(p_3)\, \gamma_\mu\, P_L\, v(p_4) \\
&+ C_Z\, g_R^e\, g_R^\mu\, \bar{v}(p_2)\, \gamma^\mu\, P_R\, u(p_1)\, \bar{u}(p_3)\, \gamma_\mu\, P_R\, v(p_4),
\end{aligned}
\tag{6.90}
$$

which can be written as

$$
\mathcal{M} = \mathcal{M}(\downarrow\uparrow;\downarrow\uparrow) + \mathcal{M}(\downarrow\uparrow;\uparrow\downarrow) + \mathcal{M}(\uparrow\downarrow;\downarrow\uparrow) + \mathcal{M}(\uparrow\downarrow;\uparrow\downarrow) \tag{6.91}
$$

with the notation $\mathcal{M}(h_{e^-} h_{e^+}; h_{\mu^-} h_{\mu^+})$.
We obtain therefore:

$$
\begin{aligned}
\mathcal{M}(\downarrow\uparrow;\downarrow\uparrow) &= C_Z\, g_L^e\, g_L^\mu\, J_{u1v2}(\downarrow\uparrow) \cdot J_{u3v4}(\downarrow\uparrow) \\
&= C_Z\, g_L^e\, g_L^\mu\, s\, (0,-1,i,0) \cdot (0,-\cos\theta,-i,\sin\theta) \\
&= C_Z\, g_L^e\, g_L^\mu\, s\, (1+\cos\theta)
\end{aligned}
\tag{6.92}
$$

$$
\begin{aligned}
\mathcal{M}(\downarrow\uparrow;\uparrow\downarrow) &= C_Z\, g_L^e\, g_R^\mu\, J_{u1v2}(\downarrow\uparrow) \cdot J_{u3v4}(\uparrow\downarrow) \\
&= C_Z\, g_L^e\, g_R^\mu\, s\, (0,-1,i,0) \cdot (0,-\cos\theta,i,\sin\theta) \\
&= C_Z\, g_L^e\, g_L^\mu\, s\, (1-\cos\theta)
\end{aligned}
\tag{6.93}
$$

$$
\begin{aligned}
\mathcal{M}(\uparrow\downarrow;\downarrow\uparrow) &= C_Z\, g_R^e\, g_L^\mu\, J_{u1v2}(\uparrow\downarrow) \cdot J_{u3v4}(\downarrow\uparrow) \\
&= C_Z\, g_R^e\, g_L^\mu\, s\, (0,-1,-i,0) \cdot (0,-\cos\theta,-i,\sin\theta) \\
&= C_Z\, g_R^e\, g_L^\mu\, s\, (1-\cos\theta)
\end{aligned}
\tag{6.94}
$$

$$
\begin{aligned}
\mathcal{M}(\uparrow\downarrow;\downarrow\uparrow) &= C_Z\, g_R^e\, g_R^\mu\, J_{u1v2}(\uparrow\downarrow) \cdot J_{u3v4}(\downarrow\uparrow) \\
&= C_Z\, g_R^e\, g_L^\mu\, s\, (0,-1,-i,0) \cdot (0,-\cos\theta,i,\sin\theta) \\
&= C_Z\, g_R^e\, g_R^\mu\, s\, (1+\cos\theta).
\end{aligned}
\tag{6.95}
$$

f. The differential cross section is given by

$$
\frac{d\sigma}{d\Omega} = \frac{1}{64\pi^2 s} \langle |\mathcal{M}|^2 \rangle \tag{6.96}
$$

with

$$\langle |\mathcal{M}|^2 \rangle = \frac{1}{4}\left[|\mathcal{M}\ (\downarrow\uparrow;\ \downarrow\uparrow)|^2 + |\mathcal{M}\ (\downarrow\uparrow;\ \uparrow\downarrow)|^2 + |\mathcal{M}\ (\uparrow\downarrow;\ \downarrow\uparrow)|^2 + |\mathcal{M}\ (\uparrow\downarrow;\ \uparrow\downarrow)|^2\right] \tag{6.97}$$

$$= \frac{1}{4}c_Z^2\,s^2\left[(1+\cos\theta)^2((g_L^e)^2\,(g_L^\mu)^2) + (g_R^e)^2\,(g_R^\mu)^2) + (1-\cos\theta)^2((g_L^e)^2\,(g_R^\mu)^2) + (g_R^e)^2\,(g_L^\mu)^2)\right] \tag{6.98}$$

$$= \frac{1}{4}c_Z^2\,s^2\left[(1+\cos^2\theta)((g_L^e)^2\,(g_L^\mu)^2 + (g_R^e)^2\,(g_R^\mu)^2 + (g_L^e)^2\,(g_R^\mu)^2 + (g_R^e)^2\,(g_L^\mu)^2) \right.$$
$$\left. +2\cos\theta((g_L^e)^2\,(g_L^\mu)^2 + (g_R^e)^2\,(g_R^\mu)^2 - (g_L^e)^2\,(g_R^\mu)^2 + (g_R^e)^2\,(g_L^\mu)^2)\right]. \tag{6.99}$$

g. The integral $\int_0^{\pi/2} d\Omega$ selects events, particles, emitted to the forward region of the interaction while $\int_{\pi/2}^{\pi} d\Omega$ selects events going into the backward region, hence the name *Forward-Backward asymmetry, A_{FB}.*
Since

$$\int_0^{\pi/2} d\Omega(1+\cos^2\theta) = \int_{\pi/2}^{\pi} d\Omega(1+\cos^2\theta) \tag{6.100}$$

and

$$\int_0^{\pi} d\Omega \cos\theta = -\int_{-\pi/2}^{\pi} d\Omega \cos\theta, \tag{6.101}$$

then A_{FB} is proportional to the terms in $\cos\theta$. These are proportional to

$$A_{FB} \propto (g_L^e)^2\,(g_L^\mu)^2 + (g_R^e)^2\,(g_R^\mu)^2 - (g_L^e)^2\,(g_R^\mu)^2 + (g_R^e)^2\,(g_L^\mu)^2 \tag{6.102}$$
$$\propto g_V^e\,g_A^e\,g_V^\mu\,g_A^\mu. \tag{6.103}$$

It is, thus, sensitive to the sign of the Standard Model couplings, while the total cross-section is not. Hence, the quantity A_{FB} can be used to test the SM couplings.

22. *Cross sections at LEP and the number of light neutrino families.* At LEP final states with electromagnetic and strong interactions were detected. Taking the $\sqrt{s} = m_Z$:

a. Compute at the first order the ratio

$$\frac{\sigma(e^-e^+ \to \nu\bar{\nu})}{\sigma(e^-e^+ \to \mu^+\mu^-)}$$

b. Compute at the first order the ratio

$$\frac{\sigma(e^-e^+ \to hadrons)}{\sigma(e^-e^+ \to \mu^+\mu^-)}$$

c. Discuss how it was possible to determine in LEP that the number of light neutrino family is 2.9840 ± 0.0082.

---

a. Let us start by noting that at the Z resonance, $\sqrt{s} = m_Z$, the dominant diagrams are Z-exchange via s-channel,

$$\frac{\sigma(e^+e^- \to \nu\bar{\nu})}{\sigma(e^+e^- \to \mu^+\mu^-)} \propto$$

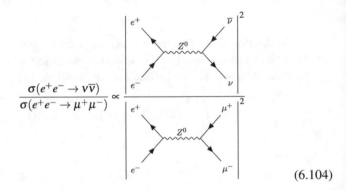

(6.104)

To compute this ratio it is useful to note that the only quantity that is different is the Z vertex of the final state products,

$$-i\frac{g}{\cos\theta_W}\gamma^\mu \left(g_V^f - g_A^f\gamma^5\right) = -i\frac{g}{\cos\theta_W}\gamma^\mu \left(\frac{1}{2}f_L(1-\gamma^5) + f_R(1+\gamma^5)\right), \quad (6.105)$$

with $f_L = \frac{1}{2}(g_V + g_A)$ and $f_R = \frac{1}{2}(g_V - g_A)$. Therefore the current of $\mathcal{M}$ can be split into

$$\overline{\Psi}_L\gamma^\mu(g_V + g_A)\Psi_L + \overline{\Psi}_R\gamma^\mu(g_V - g_A)\Psi_R. \quad (6.106)$$

Again, noting that the Dirac structure for the two processes is the same, and considering the ration between them, we can write:

$$\sigma(Z \to f\bar{f}) \propto |\mathcal{M}|^2 \propto (g_V^f + g_A^f)^2 + (g_V^f - g_A^f)^2 = (g_V^f)^2 + (g_A^f)^2. \quad (6.107)$$

Using

$$g_V^f = \frac{1}{2}T_f^3 - Q_f\sin^2\theta_W \quad ; \quad g_A^f = \frac{1}{2}T_f^3 \quad ; \quad \sin^2\theta_W \simeq 0.231, \quad (6.108)$$

where T_f^3 is the weak isospin projection and θ_W is the Weinberg angle. We can then build the following table for the Z coupling with different fermions (see Table 6.2).

Table 6.2

f	T_f^3	Q	g_A	g_V	$g_V^2 + g_A^2$
$\nu_e,\ \nu_\mu,\ \nu_\tau$	1/2	0	0.25	0.25	0.125
$e,\ \mu,\ \tau$	−1/2	−1	−0.25	−0.02	0.06
$u,\ c,\ t$	1/2	2/3	0.25	0.096	0.07
$d,\ s,\ b$	−1/2	−1/3	−0.25	−0.17	0.09

We can now evaluate

$$\frac{\sigma(e^+e^- \to \nu\bar{\nu})}{\sigma(e^+e^- \to \mu^+\mu^-)} = \frac{3 \times \left[(g_V^\nu)^2 + (g_A^\nu)^2\right]}{(g_V^\mu)^2 + (g_A^\mu)^2} = \frac{3 \times 0.125}{0.06} = 6.25,$$

(6.109)

with N_ν being the number of neutrino flavors.

b. Using the same approach as in the previous problem, and noting that

$$\frac{\sigma(e^+e^- \to q\bar{q})}{\sigma(e^+e^- \to \mu^+\mu^-)} \propto$$

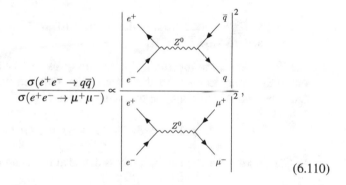

(6.110)

we can compute the ratio with respect to hadrons ($q\bar{q}$) as

$$\frac{\sigma(e^+e^- \to \text{hadrons})}{\sigma(e^+e^- \to \mu^+\mu^-)} = \frac{N_c N_q^- \alpha + N_c N_q^+ \beta}{\gamma} = \frac{3 \times 3 \times 0.09 + 3 \times 2 \times 0.07}{0.06} = 20.5$$

(6.111)

where α, β and γ are the couplings for the negatively charged quarks, the positive ones, and muons, respectively. The variable $N_c = 3$ is related with the number of *colors* that a quark carries. Finally N_q^- and N_q^+ are the number of quark families. Note that the Z mass is smaller than the top quark mass, and therefore, it cannot decay into this quark, i.e., $N_q^+ = 2$.

c. Due to the neutrino very low cross-section, neutrinos cannot be detected at the LEP detectors and thus the process $Z \to \nu\bar{\nu}$ cannot be directly observed. However, the total width of the Z boson is

$$\Gamma_Z = 3\Gamma_{l\bar{l}} + \Gamma_{\text{hadrons}} + N_\nu \Gamma_{\nu\bar{\nu}}$$

(6.112)

where it was used the lepton universality present in the weak interactions. Hence, N_ν, the number of neutrino families, can be measured using,

$$N_\nu = \frac{\Gamma_Z - \Gamma_{l\bar{l}} - \Gamma_{\text{hadrons}}}{\Gamma_{\nu\bar{\nu}}^{\text{SM}}} = \frac{\Gamma_{\text{invisible}}}{\Gamma_{\nu\bar{\nu}}^{\text{SM}}} \qquad (6.113)$$

where

- $\Gamma_Z \rightarrow$ is the total width of Z and is proportional to its lifetime and can be measured through any decay mode ($\Gamma(Z \rightarrow f\bar{f})$);
- $\Gamma_{l\bar{l}} \rightarrow$ is the partial with of $Z \rightarrow l\bar{l}$, with $l = e$, μ, τ (any lepton);
- $\Gamma_{\text{hadrons}} \rightarrow$ is the partial width of the Z decaying into $q\bar{q}$ (final states of hadronic nature).

The partial widths can be measured through the measurement of the process cross-section,

$$\sigma(e^+e^- \rightarrow Z \rightarrow f\bar{f}) \propto \Gamma(Z \rightarrow f\bar{f}) \equiv \Gamma_{f\bar{f}}. \qquad (6.114)$$

Finally, $\Gamma_{\nu\bar{\nu}}^{\text{SM}}$ is the partial width for the process $Z \rightarrow \nu_i\bar{\nu}_i$ as predicted by the Standard Model.

23. *The W^- boson at the Tevatron.* The Tevatron was an accelerator where protons and antiprotons beams with the same energy collided head-on. Consider the detection of a W^- through the following process:

$$p + \bar{p} \rightarrow W^- \rightarrow e^-(p_3) + \bar{\nu}_e(p_4).$$

a. Considering the elementary process that is at its origin at the quark level:

$$q(p_1) + \bar{q}'(p_2) \rightarrow W^- \rightarrow e^-(p_3). + \bar{\nu}_e(p_4),$$

identify the valence quarks q and $\bar{q}'$.
b. Draw the Feynman diagram(s) that contribute in the Standard Model, at lowest order, to this process.
c. Write down the amplitude. Neglect all fermion masses.
d. Show that if one neglects the fermion masses one can also neglect the terms proportional to the momenta in the numerator of the Feynman propagator for the massive gauge bosons. Write down the simplified amplitude.
e. Write down the non-vanishing helicity amplitudes for this process with the previous assumptions.
f. Calculate the spin-averaged amplitude $\langle |\mathcal{M}|^2 \rangle$ and write the expression for the differential cross section $d\sigma/d\Omega$ in the CM frame in terms of the energy in the CM frame for this elementary process $\hat{s}$.
g. Evaluate the total cross section in the CM frame $\sigma(\hat{s})$.

Fig. 6.14 Lowest order
Feynman diagram for the
process $\bar{u} + d \to e^- + \bar{\nu}_e$

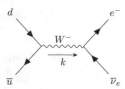

h. Write the amplitude for the process $p\bar{p} \to W \to e^-\bar{\nu}_e$ as a function of the elementary cross-section, calculated above, and taking into account only the PDFs of the valence quarks. Note: no calculations are necessary.

---

a. Taking into consideration the charge, Q, conservation, one must have $Q[q] - Q[q'] = -1$. The possible values for the quark charges are $-1/3$ and $+2/3$. Moreover, since we want only valence quarks of the proton and the anti-proton we have only $u, d, \bar{u}$ and $\bar{d}$ quarks. Therefore the solution is

$$Q[q] = -\frac{1}{3} \;\Rightarrow\; q = d \;\;; \;\; Q[q'] = \frac{2}{3} \;\Rightarrow\; q' = u \,.$$

b. The lowest order Feynman diagram for the process $d(p_1) + \bar{u}(p_2) \to e^-(p_3) + \bar{\nu}_e(p_4)$ is the one shown in Fig. 6.14, where $k = p_1 + p_2 = p_3 + p_4$.

c. The amplitude for the diagram displayed in Fig. 6.14 is

$$\mathcal{M} = (-i)\left(-\frac{ig}{\sqrt{2}}\right)^2 \bar{v}(p_2)\gamma^\mu P_L u(p_1) \frac{-i\left(g_{\mu\nu} - \frac{k_\mu k_\nu}{M_W^2}\right)}{k^2 - M_W^2 + iM_W\Gamma_W}\bar{u}(p_3)\gamma^\nu P_L v(p_4) \quad (6.115)$$

with $P_L = \frac{1-\gamma^5}{2}$.

d. Let us look, for instance, to the electron-neutrino line. We have one term,

$$\begin{aligned}
X &= \bar{u}(p_3)\gamma^\nu P_L v(p_4)k_\nu \\
&= \bar{u}(p_3)\gamma^\nu P_L v(p_4)(p_3 + p_4)_\nu \\
&= \bar{u}(p_3)\slashed{p}_3 P_L v(p_4) + \bar{u}(p_3)\slashed{p}_4 P_L v(p_4) \\
&= \bar{u}(p_3)\slashed{p}_3 P_L v(p_4) + \bar{u}(p_3)P_R\slashed{p}_4 v(p_4) \,, \quad (6.116)
\end{aligned}$$

where we have used $\slashed{p}P_L = P_R\slashed{p}$. For massless fermions the Dirac equation yields $\slashed{p}_4 v(p_4) = 0$ and $\bar{u}(p_3)\slashed{p}_3 = 0$, making Eq. 6.116 as $X = 0$. As such, the term proportional to $\frac{k_\mu k_\nu}{M_W^2}$ vanishes. We can, thus, write the amplitude as

$$\mathcal{M} = \frac{g^2}{2} \frac{1}{k^2 - M_W^2 + i M_W \Gamma_W} \bar{v}(p_2) \gamma^\mu P_L u(p_1) \bar{u}(p_3) \gamma_\mu P_L v(p_4) .$$

(6.117)

e. As the W only couples to left-handed currents we have only one non-zero helicity amplitude. Using the notation $\mathcal{M}(h_d h_{\bar{u}}; h_{e^-} h_{\bar{\nu}_e})$, we have,

$$\mathcal{M}(\downarrow\uparrow;\downarrow\uparrow) \equiv$$

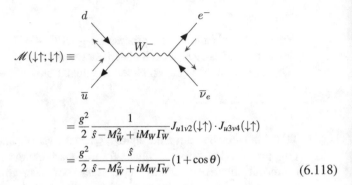

$$= \frac{g^2}{2} \frac{1}{\hat{s} - M_W^2 + i M_W \Gamma_W} J_{u1v2}(\downarrow\uparrow) \cdot J_{u3v4}(\downarrow\uparrow)$$

$$= \frac{g^2}{2} \frac{\hat{s}}{\hat{s} - M_W^2 + i M_W \Gamma_W} (1 + \cos\theta)$$

(6.118)

where $\hat{s} = (p_1 + p_2)^2$ is the square of the center-of-mass energy carried by the elementary particles participating in the interaction.

f. The spin averaged amplitude can be computed as

$$\langle |\mathcal{M}|^2 \rangle = \frac{1}{4} |\mathcal{M}(\downarrow\uparrow; \downarrow\uparrow)|^2$$

(6.119)

$$= \frac{g^4}{8} \left| \frac{\hat{s}}{\hat{s} - M_W^2 + i M_W \Gamma_W} \right|^2 (1 + \cos\theta)^2$$

(6.120)

(the factor 1/4 appears because each fermion entering in the process can have two spin combinations).

The differential cross-section can then be computing through

$$\frac{d\hat{\sigma}}{d\Omega} = \frac{1}{64\pi\hat{s}} \frac{|\mathbf{p}_3|}{|\mathbf{p}_1|} \langle |\mathcal{M}|^2 \rangle .$$

(6.121)

In the CM frame, neglecting fermion massess, we have $|\mathbf{p}_1| = |\mathbf{p}_3|$. Inserting Eq. 6.120 in Eq. 6.121 we get,

$$\frac{d\hat{\sigma}}{d\Omega} = \frac{g^4}{512\pi\hat{s}} \left| \frac{\hat{s}}{\hat{s} - M_W^2 + i M_W \Gamma_W} \right|^2 (1 + \cos\theta)^2 b .$$

(6.122)

g. Integrating Eq. 6.122 over the solid angle,

$$\int \frac{d\hat{\sigma}}{d\Omega} d\Omega = \frac{g^4}{512\pi\hat{s}} \left| \frac{\hat{s}}{\hat{s} - M_W^2 + i M_W \Gamma_W} \right|^2 \int (1 + \cos\theta)^2 d\Omega .$$

(6.123)

Using

$$\int d\Omega (1 + \cos\theta)^2 = 2\pi \int_{-1}^{1} dx (1+x)^2$$

$$= 2\pi \int_{-1}^{1} dx (1 + 2x + x^2)$$

$$= 2\pi \left(2 + \frac{2}{3} \right) = \frac{16\pi}{3} \tag{6.124}$$

we finally obtain the following expression for the elementary cross-section,

$$\hat{\sigma}(\hat{s}) = \frac{g^4}{96\pi\hat{s}} \left| \frac{\hat{s}}{\hat{s} - M_W^2 + i M_W \Gamma_W} \right|^2. \tag{6.125}$$

h. Let us consider the reaction $p + \overline{p} \to W^- \to e^- + \overline{\nu}_e$ with momentum p_1 for the proton and p_2 for the anti-proton. For the elementary process $d + \overline{u} \to W^- \to e^- + \overline{\nu}_e$, we take d with momentum $\hat{p}_1$ and $\overline{u}$ with $\hat{p}_2$. We have then,

$$\hat{p}_1 = x_1 p_1 \tag{6.126}$$
$$\hat{p}_2 = x_2 p_2 \tag{6.127}$$

where x_1 and x_2 are the fractions of the and anti-proton carried by the quarks. Again, neglecting the fermion masses,

$$\hat{s} = (\hat{p}_1 + \hat{p}_2)^2 = 2\hat{p}_1 \cdot \hat{p}_2 = x_1 x_2 2 p_1 \cdot p_2 = x_1 x_2 s. \tag{6.128}$$

Hence, we can write the $p + \overline{p} \to W^- \to e^- + \overline{\nu}_e$ cross-section as

$$\sigma(s) = \int_0^1 dx_1 \int_0^1 dx_2 \, f_d(x_1) \, f_{\overline{u}}(x_2) \, \hat{s}(x_1 x_2 \, s) \tag{6.129}$$

with $f_d(x_1)$ and $f_{\overline{u}}(x_2)$ being the quark and anti-quark parton distribution functions.

24. *Top quark at the Tevatron.* The top quark, that was missing to complete the quark sector of the standard model, was discovered at the Tevatron. Consider the single top production in the collision of $p\overline{p}$ at $\sqrt{s} = 1.96$ TeV.

a. Draw the first order Feynman diagram in the s-channel for the single top production at Tevatron.
b. Describe the possible final states that could be observed experimentally and discuss which channels have a clear experimental signature.

Fig. 6.15 Feynman
diagrams at lowest order for
the single top production at
Tevatron

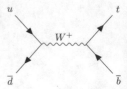

a. The main Feynman diagram at first order for the single top production is the
   one shown in Fig. 6.15. The u quark is more likely to come from the proton
   valence while the $\bar{d}$ from the anti-proton valence quarks.
b. The top quark decays, almost exclusively, to W^+b. The W boson has two
   main decay channels: leptonic ($W^+ \to l_i^+ \bar{\nu}_i$, with $i = e, \mu, \tau$) and hadronic
   ($W^+ \to q_i \bar{q}_j$, with $i = u, c$ and $j = d, s$). There is a substantial source of
   hadronic background in pp collisions from the underlying event and pileup.
   These will contaminate the W decay's hadronic products, making it a less
   precise channel than the leptonic decay. As such, in a pp collision, the W
   decay that yields isolated leptons is more straightforward to be identified
   experimentally.

25. *Angular ordering of gluon emission in QCD.* In the context of $e^+e^- \to$ hadrons,
    single soft gluon emission can be factorized out from the $q\bar{q}$ matrix element and
    interpreted as an additional correction. Multiple gluon emissions can be added by
    going higher orders in perturbation theory, or, alternatively, by adopting a proba-
    bilistic picture based on the recursive application of single gluon emissions. One
    simply has to guarantee that there is no interference between successive emis-
    sions that can break this probabilistic picture. It is possible to show that, in the
    soft gluon emission, successive emissions follow angular ordering, a property
    widely used in Monte Carlo event generators. It follows that the double differ-
    ential emission probability of single gluon from a $q\bar{q}$ antenna singlet (produced
    from a photon), in the so-called soft limit, is given by:

    $$dS = dS_1 + dS_2, \tag{6.130}$$

    where

    $$dS_i = \frac{\alpha_s C_F}{\pi} \frac{dE_k}{E_k} \frac{d(\cos\theta_{ik})}{1 - \cos\theta_{ik}} \Theta(\theta_{12} > \theta_{ik}). \tag{6.131}$$

    E_k is the energy of the radiated gluon, and the remaining kinematic variables
    are explicitly shown in Fig. 6.16.
    To demonstrate this result:

    a. Show that, in the soft limit, the averaged squared matrix element of the full
       process can be written as:

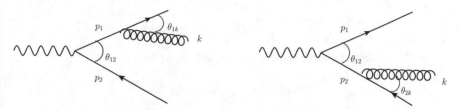

Fig. 6.16 Diagrams that contribute to the single gluon emission at leading order

$$\langle |M_{q\bar{q}g}|^2 \rangle \simeq \langle |M_{q\bar{q}}|^2 \rangle \, C_F \, g_s^2 \, \frac{2 p_1 \cdot p_2}{(p_1 \cdot k)(p_2 \cdot k)} \tag{6.132}$$

where $M_{q\bar{q}}$ is the matrix element of the process $\gamma \to q\bar{q}$.

b. Consider now the triple differential gluon emission probability, where it was added the corresponding phase space, $d^3k/(2E_k(2\pi)^3)$

$$dS = \frac{\alpha_s C_F}{\pi} E_k d E_k \frac{p_1 \cdot p_2}{(p_1 \cdot k)(p_2 \cdot k)} d\Omega, \quad \text{where } d\Omega = d(\cos\theta_k)\frac{d\phi_k}{2\pi}. \tag{6.133}$$

Show that, without assuming the center-of-mass frame, this would yield:

$$dS = \frac{\alpha_s C_F}{\pi} \frac{d E_k}{E_k} \frac{1 - \cos\theta_{12}}{(1 - \cos\theta_{1k})(1 - \cos\theta_{2k})} d\Omega . \tag{6.134}$$

c. Separate them into the angular pattern of the gluon radiation from each leg, such that:

$$W = \frac{1 - \cos\theta_{12}}{(1 - \cos\theta_{1k})(1 - \cos\theta_{2k})} = \frac{W_1 + W_2}{2}, \tag{6.135}$$

where

$$W_1 = W + \frac{1}{1 - \cos\theta_{1k}} - \frac{1}{1 - \cos\theta_{2k}}. \tag{6.136}$$

Defining your kinematics in $3D$ space, integrate each contribution independently over the azimuthal angle.

Suggestion: for W_1, you can fix the quark along the $\mathbf{e}_z$ and the anti-quark along the $x - z$ plane (the gluon will be given as a function of this reference frame). From here, you can relate θ_{2k} with ϕ_{1k} to finally integrate over ϕ_{1k}.

d. Using the results from the previous question, identify the S_1 and S_2 components from Eq. (6.130)

a. The matrix element for the $\gamma \to q\bar{q}$ process is given by

$$M_{q\bar{q}} = \bar{u}_a(p_1)i.e.\gamma_\mu \delta_{ab} v_b(p_2) \tag{6.137}$$

where a and b are the colors of the quark and anti-quark in the fundamental representation. By adding a gluon, the matrix element will have 2 contributions from the 2 possible diagrams shown in Fig. 6.16. These will yield:

$$
\begin{aligned}
M_{q\bar{q}g} &= \bar{u}_a(p_1)(-ig_s)\not{\epsilon}(k)t^A_{aa'}\frac{i(\not{p}_1 + \not{k})}{(p_1+k)^2}i.e.\gamma_\mu \delta_{a'b}v_b(p_2) \\
&\quad + \bar{u}_a(p_1)i.e.\gamma_\mu \delta_{ab'}\frac{-i(\not{p}_2+\not{k})}{(p_2+k)^2}(-ig_s)\not{\epsilon}(k)t^A_{b'b}v_b(p_2) = \\
&= i.e.g_s\bar{u}(p_1)\left[\not{\epsilon}(k)t^A\frac{(\not{p}_1+\not{k})}{(p_1+k)^2}\gamma_\mu - \gamma_\mu\frac{i(\not{p}_2+\not{k})}{(p_2+k)^2}\not{\epsilon}(k)t^A\right]v(p_2)
\end{aligned}
\tag{6.138}
$$

where g_s is the strong coupling constant, A the color of the gluon in the adjoint representation, t^A_{ij} the SU(3) group matrices and ε the gluon polarization. In the last equality we omit the color indices to improve readability. Focusing on the Dirac structure of the first term:

$$\bar{u}(p_1)\not{\epsilon}(\not{p}_1 + \not{k}) = 2\varepsilon \cdot (p_1 + k)\bar{u}(p_1) - \bar{u}(p_1)(\not{p}_1 + \not{k})\not{\epsilon} \simeq 2\varepsilon \cdot p_1\bar{u}(p_1). \tag{6.139}$$

In the first equation we used the property of the Dirac γ matrices:

$$\gamma_\mu\gamma_\nu = 2g_{\mu\nu} - \gamma_\nu\gamma_\mu \tag{6.140}$$

and in the second equality we used the fact that the gluon carries a small momentum fraction of the parent parton. In particular:

$$
\begin{aligned}
\varepsilon \cdot k &= 0 \\
\bar{u}(p_1)\not{p}_1 &\sim 0 \quad \text{(high-energy limit)} \\
k_\mu &\ll p_\mu \quad \text{(soft limit)}.
\end{aligned}
\tag{6.141}
$$

Doing the same for the second term we get:

$$M_{q\bar{q}g} = i.e.g_s\bar{u}(p_1)\gamma_\mu t^A v(p_2)\left(\frac{p_1 \cdot \varepsilon}{p_1 \cdot k} - \frac{p_2 \cdot \epsilon}{p_2 \cdot k}\right). \tag{6.142}$$

Summing over polarization, spin and color states we finally arrive to:

$$\langle |M_{q\bar{q}g}|^2 \rangle = \langle |M_{q\bar{q}}|^2 \rangle C_F g_s^2 \frac{2p_1 \cdot p_2}{(p_1 \cdot k)(p_2 \cdot k)}, \tag{6.143}$$

where we have used the color algebra identities:

$$\sum_{a,b} |M_{q\bar{q}}|^2 = N_c$$

$$\sum_{a,b,A} t_{ab}^A t_{ba}^A = N_c C_F \tag{6.144}$$

being $N_c = 3$ and $C_F = (N_c^2 - 1)/(2N_c)$ the Casimir factor, and the polarization sum result, that, in the soft limit, goes as:

$$\sum_{pol} \varepsilon_\mu(k)\varepsilon_\nu(k) = -g_{\mu\nu} + \mathcal{O}(k_\mu, k_\nu). \tag{6.145}$$

We thus found an expression where we can factor out the process that is independent of the gluon, $M_{q\bar{q}}$.

b. Defining the kinematics in the laboratory frame, it follows:

$$p_1 \cdot p_2 = E_1 E_2 - \mathbf{p_1} \cdot \mathbf{p_2},$$
$$p_1 \cdot k = E_1 E_k - \mathbf{p_1} \cdot \mathbf{k}, \tag{6.146}$$
$$p_2 \cdot k = E_2 E_k - \mathbf{p_2} \cdot \mathbf{k},$$

where the 3D momenta of each particle are represented in boldface. The emission probability will follow:

$$dS = \frac{\alpha_s C_F}{\pi} E_k dE_k \frac{p_1 \cdot p_2}{(p_1 \cdot k)(p_2 \cdot k)} d\Omega$$
$$= \frac{\alpha_s C_F}{\pi} \frac{dE_k}{E_k} \frac{1 - (\mathbf{p_1} \cdot \mathbf{p_2})/(E_1 E_2)}{(1 - (\mathbf{p_1} \cdot \mathbf{k})/(E_1 E_k))(1 - (\mathbf{p_2} \cdot \mathbf{k})/(E_1 E_k))} d\Omega. \tag{6.147}$$

In the high energy limit (massless particles), the ratios of the 3D momenta by the corresponding energies will simply yield:

$$\frac{\mathbf{p_i} \cdot \mathbf{p_j}}{E_i E_j} \simeq \cos\theta_{ij}. \tag{6.148}$$

Finally:

$$dS = \frac{\alpha_s C_F}{\pi} \frac{dE_k}{E_k} \frac{1 - \cos\theta_{12}}{(1 - \cos\theta_{1k})(1 - \cos\theta_{2k})} d\Omega \tag{6.149}$$

Fig. 6.17 Scheme of the
Euclidean coordinates used
in the solution of this
problem

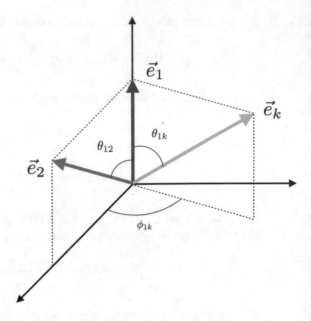

c. Separating the angular part as suggested yield:

$$W_1 = \frac{1 - \cos\theta_{12}}{(1 - \cos\theta_{1k})(1 - \cos\theta_{2k})} + \frac{1}{1 - \cos\theta_{1k}} - \frac{1}{1 - \cos\theta_{2k}}. \quad (6.150)$$

$$W_2 = \frac{1 - \cos\theta_{12}}{(1 - \cos\theta_{1k})(1 - \cos\theta_{2k})} + \frac{1}{1 - \cos\theta_{2k}} - \frac{1}{1 - \cos\theta_{1k}}. \quad (6.151)$$

We start by integrating W_1, over its azimuthal phase space included in $d\Omega = d(\cos\theta_{1k})d\phi_{1k}/(2\pi)$. To do so, one has to write the θ_{2k} as a function of the azimuthal angle ϕ_{1k}. For that, we fix the direction of the quark over the z axis, and the anti-quark in the $x - z$ plane (see Fig. 6.17). The Euclidean coordinates of the unitary vectors of each particle direction can thus be written as:

$$\mathbf{e}_1 = (0, 0, 1)$$
$$\mathbf{e}_2 = (\sin\theta_{12}, 0, \cos\theta_{12}) \quad (6.152)$$
$$\mathbf{e}_k = (\sin\theta_{1k}\cos\phi_{1k}, \sin\theta_{1k}\sin\phi_{1k}, \cos\theta_{1k})$$

Using these results, one has the identity:

$$\mathbf{e}_2 \cdot \mathbf{e}_k = \cos\theta_{2k} = \sin\theta_{12}\sin\theta_{1k}\cos\phi_{1k} + \cos\theta_{12}\cos\theta_{1k}. \quad (6.153)$$

The W_1 component can then be expressed as:

$$W_1 = \frac{1}{1 - \cos\theta_{1k}} + \left(\frac{1 - \cos\theta_{12}}{1 - \cos\theta_{1k}} - 1\right)\frac{1}{a - b\cos\phi_{1k}} \quad (6.154)$$

where we introduced:

$$a = 1 - \cos\theta_{12}\cos\theta_{1k} \quad (6.155)$$
$$b = \sin\theta_{12}\sin\theta_{1k}. \quad (6.156)$$

To integrate over ϕ_{1k} we note:

$$\cos\phi_{1k} = \frac{e^{i\phi_{1k}} - e^{-i\phi_{1k}}}{2} = \frac{z^2 + 1}{2z} \quad (6.157)$$

where we made the substitution $z = e^{i\phi_{1k}}$. The integral over the azimuthal angle ϕ_{1k} of the second term of W_1 can finally be computed. In particular:

$$\mathscr{I} \equiv \int_0^{2\pi} \frac{d\phi_{1k}}{2\pi}\frac{1}{a - b\cos\phi_{1k}} = \frac{1}{\pi i}\oint \frac{dz}{2az - bz^2 - b} = \frac{1}{|\cos\theta_{12} - \cos\theta_{1k}|} \quad (6.158)$$

where we made use of the same substitution as before, $z = e^{i\phi_{1k}}$. The last equality can be found by using the Residues theorem. Putting everything together we find the result for W_1 to be:

$$\int_0^{2\pi} \frac{d\phi_{1k}}{2\pi} W_1 = \frac{1}{1 - \cos\theta_{1k}} + \left(\frac{1 - \cos\theta_{12}}{1 - \cos\theta_{1k}} - 1\right)\mathscr{I}$$
$$= \frac{1}{1 - \cos\theta_{1k}}\left(1 - \frac{\cos\theta_{1k} - \cos\theta_{12}}{|\cos\theta_{1k} - \cos\theta_{12}|}\right) \quad (6.159)$$
$$= \frac{1}{1 - \cos\theta_{1k}}\begin{cases} 2 & \text{if } \theta_{1k} < \theta_{12} \\ 0 & \text{otherwise.} \end{cases}$$

For the W_2 term we just note that the results are the same exchanging $1 \leftrightarrow 2$. As such:

$$\int_0^{2\pi} \frac{d\phi_{1k}}{2\pi} W_2 = \frac{1}{1 - \cos\theta_{2k}}\begin{cases} 2 & \text{if } \theta_{2k} < \theta_{12} \\ 0 & \text{otherwise.} \end{cases} \quad (6.160)$$

We have thus identified the Θ function needed to suppress radiation at angles larger than the previous emission, thus yielding the angular ordering property:

$$\int_0^{2\pi} \frac{d\phi_{1k}}{2\pi} W_i = \frac{2}{1 - \cos\theta_{ik}}\Theta(\theta_{ik} - \theta_{12}) \equiv \frac{2}{1 - \cos\theta_{ik}}\Theta(\theta_{ik} < \theta_{12}) , \quad i = 1, 2. \quad (6.161)$$

d. To finalize the identification with the emission spectrum of each quark and anti-quark separately, we just use the results of W_1 and W_2:

$$dS = \frac{\alpha_s C_F}{\pi} \frac{dE_k}{E_k} \frac{W_1 + W_2}{2} d\Omega \equiv dS_1 + dS_2 \qquad (6.162)$$

where:

$$dS_i = \frac{\alpha_s C_F}{\pi} \frac{dE_k}{E_k} \frac{d(\cos\theta_{ik})}{1 - \cos\theta_{ik}} \Theta(\theta_{12} > \theta_{ik}) \quad , \quad i = 1, 2. \qquad (6.163)$$

Chapter 7
The Higgs Mechanism and the Standard Model of Particle Physics

1. *Symmetry breaking introducing a real scalar field (1).* Consider a simple model where a real scalar field Φ is introduced being the Lagrangian of such field:

$$L = \frac{1}{2}\left(\partial_\mu \Phi\right)^2 - \frac{1}{2}\mu^2 \Phi^2 - \frac{1}{4}\lambda\Phi^4.$$

Discuss the particle spectrum originated by a small quantum perturbation around the minimum of the potential (vacuum) for the cases $\mu^2 > 0$ and $\mu^2 < 0$. Can such model accommodate a Goldstone boson?

From the Lagrangian of the considered model:

$$L = \frac{1}{2}(\partial_\mu \phi^2(x)) - \frac{1}{2}\mu^2\phi^2(x) - \frac{1}{4}\lambda\phi^4(x), \tag{7.1}$$

one can identify the potential as given by:

$$V = \frac{1}{2}\mu^2\phi^2(x) + \frac{1}{4}\lambda\phi^4(x). \tag{7.2}$$

To determine the minimum of the potential, one has:

$$\frac{\partial V}{\partial \phi(x)} = \mu^2\phi(x) + \lambda\phi^3(x) = 0 \Rightarrow \phi_{min}(x) = 0 \vee \mu^2 + \lambda\phi_{min}^2(x) = 0. \tag{7.3}$$

The solution corresponding to an existing field is then given by:

$$\phi_{min}^2 = -\frac{\mu^2}{\lambda} \tag{7.4}$$

© Springer Nature Switzerland AG 2021
A. De Angelis et al., *Particle and Astroparticle Physics*, Undergraduate Lecture Notes in Physics,
https://doi.org/10.1007/978-3-030-73116-8_7

that depends only on the two constants μ and λ. A real minimum of the potential can only exist if $\mu^2 < 0$, being this solution the only possibility to apply small quantum perturbations.

Selecting this condition, $\mu^2 < 0$, it follows:

$$\phi(x) \simeq \phi_{min} + h(x) \tag{7.5}$$

where $h(x)$ is considered to be a small perturbation around the vacuum solution, i.e., $h(x) \ll \phi_{min}$. Substituting this into the Lagrangian and working each term independently, one has:

$$\frac{1}{2}(\partial_\mu \phi^2(x)) = \frac{1}{2}(\partial_\mu \phi_{min} + \partial_\mu h(x))^2 = \frac{1}{2}(\partial_\mu h(x))^2 \tag{7.6}$$

$$\frac{1}{2}\mu^2 \phi^2(x) = \frac{1}{2}\mu^2(\phi_{min}^2 + h(x))^2 = \frac{1}{2}\mu^2 \phi_{min}^2 + \frac{1}{2}\mu^2 h^2(x) + \mu^2 \phi_{min} h(x) \tag{7.7}$$

$$\frac{1}{2}\mu^2 \phi^4(x) = \frac{1}{2}(\partial_\mu \phi_{min} + \partial_\mu h(x))^4$$
$$= \frac{1}{4}\lambda \phi_{min}^4 + \frac{6}{4}\lambda \phi_{min}^2 h^2(x) + \lambda \phi_{min}^3 h(x) + \mathcal{O}(h^3(x)) \tag{7.8}$$

where we neglected higher order terms of the perturbation field (in particular, $h^3(x)$ and $h^4(x)$ terms). Adding the two terms of the potential:

$$V = \underbrace{\frac{1}{2}\mu^2 \phi^2 + \frac{1}{4}\lambda \phi_{min}^4}_{V_{min}=0} + \frac{1}{2}\mu^2 h^2(x) - \frac{3}{2}\mu^2 h^2(x) + \underbrace{\mu^2 \phi_{min} h(x) - \mu^2 \phi_{min} h(x)}_{=0}, \tag{7.9}$$

where we have used the minimum condition, Eq. (7.4). Summing all terms, the perturbed Lagrangian is finally given by:

$$L = \frac{1}{2}(\partial_\mu h(x))^2 - \frac{1}{2}(-2\mu^2)h^2(x). \tag{7.10}$$

Since $\mu^2 < 0$, we can identify the mass of the particle as being $m = \sqrt{-2\mu^2}$. Although we can produce massive scalar fields, this model cannot accommodate a Goldstone boson as the only symmetry present in the Lagrangian is discrete ($\phi(x) \to -\phi(x)$). For a Goldstone boson to appear, there must be a continuous symmetry, for instance, if $\phi(x)$ were an imaginary field.

2. *Handling left and right projection operators (2).* Demonstrate that

$$\left(\frac{1}{2}(1-\gamma^5)\right)\left(\frac{1}{2}(1-\gamma^5)\right) = \left(\frac{1}{2}(1-\gamma^5)\right);$$

$$\left(\frac{1}{2}(1-\gamma^5)\right)\left(\frac{1}{2}(1+\gamma^5)\right) = 0.$$

To prove this result we use the property that $\gamma^5\gamma^5 = 1$, where 1 is in this case the 4×4 identity matrix (we shall use later in the similar way the 0 symbol). By direct calculation:

$$\left(\frac{1}{2}(1-\gamma^5)\right)\left(\frac{1}{2}(1-\gamma^5)\right) = \left(\frac{1}{2}(1-\gamma^5)\right)\left(\frac{1}{2}(1-\gamma^5)\right)$$

$$= \frac{1}{2}\cdot\frac{1}{2}(1-\gamma^5-\gamma^5+\gamma^5\gamma^5)$$

$$= \frac{1}{2}\cdot\frac{1}{2}(1-\gamma^5-\gamma^5+1)$$

$$= \frac{1}{2}\cdot\frac{1}{2}\cdot 2(1-\gamma^5)$$

$$= \frac{1}{2}(1-\gamma^5).$$

The second result follows, in a similar way, from

$$(1-\gamma^5)(1+\gamma^5) = (1+\gamma^5-\gamma^5-\gamma^5\gamma^5) = 0,$$

as the last term is $\gamma^5\gamma^5 = 1$.

3. *Fermion mass terms (3).* Show that the fermion mass term $\mathscr{L}_f = -m_f \bar{f} f$ is gauge invariant in QED but not in $\mathrm{SU}(2)_L \otimes \mathrm{U}(1)_Y$.

The mass term $-m_f \bar{f} f$ can be decomposed in terms which combine the left and the right chiral eigenstates, f_L, f_R and their adjoint $\bar{f}_L$ and $\bar{f}_R$ defined as:

$$f_L = \frac{1}{2}\left(1-\gamma^5\right) f \;;\; f_R = \frac{1}{2}\left(1+\gamma^5\right) f \tag{7.11}$$

$$\bar{f}_L = \bar{f}\frac{1}{2}\left(1+\gamma^5\right) \;;\; \bar{f}_R = \bar{f}\frac{1}{2}\left(1-\gamma^5\right). \tag{7.12}$$

Indeed, using the following identities:

$$1 = \frac{1}{2}\left(1 - \gamma^5\right) + \frac{1}{2}\left(1 + \gamma^5\right); \tag{7.13}$$

$$\left(1 - \gamma^5\right) = \frac{1}{2}\left(1 - \gamma^5\right)\left(1 - \gamma^5\right); \tag{7.14}$$

$$\left(1 + \gamma^5\right) = \frac{1}{2}\left(1 + \gamma^5\right)\left(1 + \gamma^5\right), \tag{7.15}$$

one obtains:

$$
\begin{aligned}
-m_f \bar{f} f &= -m_f \bar{f}\left(\frac{1}{2}(1 - \gamma^5) + \frac{1}{2}(1 + \gamma^5)\right)f \\
&= -m_f\left(\bar{f}\frac{1}{2}(1 - \gamma^5)\frac{1}{2}(1 - \gamma^5)f + \bar{f}\frac{1}{2}(1 + \gamma^5)\frac{1}{2}(1 + \gamma^5)f\right)
\end{aligned} \tag{7.16}
$$

and

$$ - m_f \bar{f} f = -m_f(\bar{f}_R f_L + \bar{f}_L f_R). \tag{7.17}$$

In QED this term is invariant under the $U(1)_{QED}$ phase transformations but not under the $SU(2)_L$ isospin symmetry transformations, since f_L is a member of an $SU(2)_L$ doublet while f_R is a singlet (there are no right Dirac fermions in the Standard Model).

4. *Mass of the photon (4).* **Show that in the Standard Model of particle physics (SM) the diagonalization of the $(W3_\mu, B_\mu)$ mass matrix ensures the existence of a massless photon.**

Let us call W^1, W^2. W^0 the three gauge fields of $SU(2)$. We call $W^a_{\mu\nu}$ ($a = 1, ..., 3$) the field tensors of $SU(2)$ and $B_{\mu\nu}$ the field tensor of $U(1)$. Notice that $B_{\mu\nu}$ is not equal to $F_{\mu\nu}$,(the electromagnetic tensor) as W^0 is not the Z field: since we use a tensor product of the two spaces, in general the neutral field B can mix to the neutral field W^0, and the photon and Z states are a linear combination of the two.

The electromagnetic current,

$$j^{em}_\mu = \bar{e}\gamma_\mu e = \bar{e}_L\gamma_\mu e_L + \bar{e}_R\gamma_\mu e_R$$

(where L and R are respectively the left and right component of the electron), is invariant under $U(1)_Q$, the gauge group of QED associated to the electromagnetic charge, since it contains the full e field (the electromagnetic field treats equally both chiralities). It is clearly not invariant under $SU(2)_L$, because of the R term.

After spontaneous symmetry breaking (SSB), the Standard Model Lagrangian can be approximated around its minimum as

$$\mathscr{L}_{free} = \text{constant} + \text{kinetic terms} +$$

$$+ \frac{1}{2} \left(-2\mu^2 \right) h^2 \tag{7.18}$$

$$+ \frac{1}{2} \left(\frac{1}{4} g^2 v^2 \right) W_\mu^1 \, W^{1\mu} + \frac{1}{2} \left(\frac{1}{4} g^2 v^2 \right) W_\mu^2 \, W^{2\mu} \tag{7.19}$$

$$+ \frac{1}{8} v^2 \left(W^{3\mu} \ B^\mu \right) \begin{pmatrix} g^2 & -gg'Y \\ -gg'Y & g'^2 \end{pmatrix} \begin{pmatrix} W_\mu^3 \\ B_\mu \end{pmatrix} .$$

$$+ \mathscr{O}(3) , \tag{7.20}$$

where h is the field around the minimum.

- As usual in the SSB, the h field acquires mass (7.19); we shall call the corresponding particle H. This is the famous SM Higgs boson, and its mass is

$$m_H = \sqrt{-2\mu^2} = \sqrt{2\lambda} v . \tag{7.21}$$

- We now analyze the term (7.19). Two massive charged bosons W^1 and W^2 have the same mass $gv/2$. We have seen, however, that physical states of integer charge ± 1 can be constructed by a linear combination of them:

$$W^\pm = \sqrt{\frac{1}{2}} (W^1 \pm i W^2) .$$

The mass is as well

$$M_{W^\pm} = \frac{1}{2} gv , \tag{7.22}$$

and these states correspond naturally to the charged current vectors.
- Finally, let us analyze the term (7.20).
Here the fields W^3 and B couple through a non-diagonal matrix; they thus are not mass eigenstates. The physical fields can be obtained by an appropriate rotation diagonalizing the mass matrix

$$M = \begin{pmatrix} g^2 & -gg'Y \\ -gg'Y & g'^2 \end{pmatrix} .$$

For $Y = \pm 1$ the determinant of the matrix is 0, and when we shall diagonalize one of the two eigenstates will be massless. If we introduce the fields A_μ and Z_μ defined as

$$A_\mu = \sin\theta_W W^0_\mu + \cos\theta_W B_\mu \tag{7.23}$$

$$Z_\mu = \cos\theta_W W^0_\mu - \sin\theta_W B_\mu , \tag{7.24}$$

where the angle θ_W, called the Weinberg angle, parametrizes the electroweak mixing:

$$\tan\theta_W = \frac{g'}{g} , \tag{7.25}$$

the term (7.20) becomes

$$\frac{1}{8} v^2 (A^\mu \ Z^\mu) \begin{pmatrix} 0 & 0 \\ 0 & g^2 + g'^2 \end{pmatrix} \begin{pmatrix} A_\mu \\ Z_\mu \end{pmatrix} . \tag{7.26}$$

A_μ is then massless (we can identify it with the photon).

5. *Fermion couplings (5).* Verify that choosing the weak hypercharge according to the Gell-Mann Nishijima formula ($Y = 2(Q - I_W^{(3)})$) ensures the right couplings of all fermions with the electroweak neutral bosons (Z, γ). Remember that $e = g \sin\theta_W = g' \cos\theta_W$.

The electroweak neutral bosons (Z, γ) interactions can be written considering the left and right helicity states of the fermions. Indeed, for a generic fermion f,

$$\overline{\psi}_f \gamma^\mu (g_V^f - g_A^f \gamma^5) \psi_f$$
$$= \overline{\psi}_f \gamma^\mu \left[\frac{1}{2}(g_V^f + g_A^f)(1 - \gamma^5) + \frac{1}{2}(g_V^f - g_A^f)(1 + \gamma^5) \right] \psi_f$$
$$= \overline{\psi}_{fL} \gamma^\mu g_L \psi_{fL} + \overline{\psi}_{fR} \gamma^\mu g_R \psi_{fR} \tag{7.27}$$

where the left and right couplings g_L and g_R are given by

$$g_L = \frac{1}{2}(g_V + g_A) \tag{7.28}$$

$$g_R = \frac{1}{2}(g_V - g_A) . \tag{7.29}$$

In QED $g_L = g_R$ and thus parity P is conserved. In fact, there is a remarkable agreement between the theoretical predictions and the experiment as it is demonstrated in the study of the anomalous magnetic momentum of the electron ($g - 2$), where high order corrections are computed up to the eighth-order (891 diagrams) and the most significant tenth-order terms.

In the electroweak unification the physical currents of QED and the weak neutral currents can be written as

$$j_{em}^{\mu} = j_Y^{\mu} \cos \theta_W + j_3^{\mu} \sin \theta_W$$
$$j_Z^{\mu} = -j_Y^{\mu} \sin \theta_W + j_3^{\mu} \cos \theta_W . \tag{7.30}$$

Knowing the fermion charge Q_f and weak isospin $I_W^{(3)}$ it is possible to obtain the weak hypercharge, Y using the Gell-Mann Nishijima formula

$$Y = 2 \left(Q - I_W^{(3)} \right) . \tag{7.31}$$

As such we can build Table 7.2. Notice that from $I_W^{(3)}$ we can evaluate the left state hypercharge Y_L while the right state Y_R is obtained considering $I_W^{(3)} = 0$.

Let us start by checking the electromagnetic couplings, C_{em}. In this unified model the electromagnetic current can be written as

$$j_{em}^{\mu} = C_{em} \bar{u}_L \gamma^{\mu} u_R + C_{em} \bar{u}_R \gamma^{\mu} u_R = j_Y^{\mu} \cos \theta_W + j_3^{\mu} \sin \theta_W . \tag{7.32}$$

Thus, we can find the electromagnetic couplings for the fermions through the following equations,

$$\bar{u}_L \gamma^{\mu} u_L : C_{em} = \frac{1}{2} g' Y_L \cos \theta_W + I_W^{(3)} g \sin \theta_W \tag{7.33}$$

$$\bar{u}_R \gamma^{\mu} u_R : C_{em} = \frac{1}{2} g' Y_R \cos \theta_W . \tag{7.34}$$

Noting that $g' \cos \theta_W = g \sin \theta_W$ and that $g \sin \theta_W = e$ we can write the above equations as

$$\bar{u}_L \gamma^{\mu} u_L : C_{em} = \left(\frac{1}{2} Y_L + I_W^{(3)} \right) e \tag{7.35}$$

$$\bar{u}_R \gamma^{\mu} u_R : C_{em} = \left(\frac{1}{2} Y_R \right) e . \tag{7.36}$$

Table 7.1 Relative branching fractions of the Z into $f\bar{f}$ pair: predictions at leading order from the SM (for $\sin^2 \theta_W = 0.23$) are compared to experimental results

Particle	g_V	g_A	Predicted (%)	Experimental (%)
Neutrinos (all)	1/4	1/4	20.5	(20.00 ± 0.06)
Electron	$-1/4 + \sin^2 \theta_W$	$-1/4$	3.4	(3.363 ± 0.004)
Muon	$-1/4 + \sin^2 \theta_W$	$-1/4$	3.4	(3.366 ± 0.007)
Tau	$-1/4 + \sin^2 \theta_W$	$-1/4$	3.4	(3.367 ± 0.008)
Down-type quarks d, s, b	$-1/4 + 1/3 \sin^2 \theta_W$	$-1/4$	15.2	(15.6 ± 0.4)
Up-type quarks u, c	$1/4 - 2/3 \sin^2 \theta_W$	$1/4$	11.8	(11.6 ± 0.6)

Table 7.2 Fermions charge, Q_f, weak isospin $I_W^{(3)}$ and left and right hypercharges, Y_L and Y_R, respectively

Fermion	Q_f	$I_W^{(3)}$	Y_L	Y_R
$\nu_e,\ \nu_\mu,\ \nu_\tau$	0	$+\frac{1}{2}$	-1	0
$e^-,\ \mu^-,\ \tau^-$	-1	$-\frac{1}{2}$	-1	-2
$u,\ c,\ t$	$+\frac{2}{3}$	$+\frac{1}{2}$	$+\frac{1}{3}$	$+\frac{4}{3}$
$d,\ s,\ b$	$-\frac{1}{3}$	$-\frac{1}{2}$	$+\frac{1}{3}$	$-\frac{2}{3}$

Using the values for Y and $I_W^{(3)}$ given in Table 7.2 and inserting them in the previous equations it is easy to verify that $C_{em} = eQ_f$, i.e., the coupling of the photon to fermions. Let us know verify the coupling for the neutral weak current. From Eq. 7.30 we have

$$j_Z^\mu = -\frac{1}{2}g' \sin\theta_W \left[Y_L \bar{u}_L \gamma^\mu u_L + Y_R \bar{u}_R \gamma^\mu u_R \right] + I_W^{(3)} g \cos \left[Y_L \bar{u}_L \gamma^\mu u_L \right].$$

(7.37)

We will show that the use of the Gell-Mann Nishijima formula leads to the couplings measured experimentally by substituting this equation in the neutral weak current above equation. We have

$$j_Z^\mu = -g' \sin\theta_W \left[\left(Q + I_W^{(3)} \right) \bar{u}_L \gamma^\mu u_L + Q \bar{u}_R \gamma^\mu u_R \right] + I_W^{(3)} g \cos\theta_W \left[Y_L \bar{u}_L \gamma^\mu u_L \right]$$

$$j_Z^\mu = \left[-\left(Q + I_W^{(3)} \right) \frac{\sin^2\theta_W}{\cos\theta_W} + I_W^{(3)} \cos\theta_W \right] \bar{u}_L \gamma^\mu u_L - g \left[\frac{\sin^2\theta_W}{\cos\theta_W} Q \right] \bar{u}_R \gamma^\mu u_R \quad (7.38)$$

where the relation $g \sin\theta_W = g' \cos\theta_W$ was used. By taking the coupling to the Z boson as

$$g_Z = \frac{g}{\cos\theta_W} = \frac{e}{\sin\theta_W \cos\theta_W}$$

(7.39)

then the neutral current can be expressed as

$$j_Z^\mu = g_Z(g_L \bar{u}_L \gamma^\mu u_L + g_R \bar{u}_R \gamma^\mu u_R)$$

(7.40)

where the quantities g_L and g_R can be identified as

$$g_L = I_W^{(3)} - Q \sin^2\theta_W$$

(7.41)

$$g_R = -Q \sin^2\theta_W.$$

(7.42)

This shows that the Z weak boson couples to both left and right chiral states but with different strengths. As shown before, the left and right couplings can be written in terms of the vector and axial-vector couplings as

$$g_V = \frac{1}{2}(g_L + g_R) = \frac{1}{2}I_W^{(3)} - Q \sin^2 \theta_W , \tag{7.43}$$

$$g_A = \frac{1}{2}(g_L - g_R) = \frac{1}{2}I_W^{(3)} . \tag{7.44}$$

Using the fermion charge and weak isopsin, given in Table 7.2, and taking $\sin^2 \theta_W = 0.23$, it is now a trivial exercise to see that we can obtain the weak couplings between the Z and the fermions, displayed in Table 7.1.

6. $\sin^2 \theta_W$ (6). Determine the value of $\sin^2 \theta_W$ from the experimental measurements of:

a. G_F and M_W;
b. M_W and M_Z.

---

a. The sine of the Weinberg angle is the ratio between the electromagnetic and weak coupling constants,

$$\sin \theta_W = \frac{g_e}{g_W} . \tag{7.45}$$

The weak interaction coupling constant can be related to the Fermi coupling constant, G_F, and the W boson mass, through

$$G_F = \frac{\sqrt{2}}{8} \left(\frac{g_W}{M_W c^2} \right)^2 (\hbar c)^3 . \tag{7.46}$$

Combining Eqs. 7.45 and 7.46 we can write:

$$\sin^2 \theta_W = \frac{\sqrt{s}(\hbar c)^3 g_e^2}{8 G_F (M_W c^2)^2} . \tag{7.47}$$

The experimental values for the Fermi coupling constant and the mass of the W boson are

$$\frac{G_F}{(\hbar c)^3} = 1.16637 \times 10^{-6} \, \text{GeV}^{-2}$$

$$M_W = 80.398 \, \text{GeV}/c^2.$$

From Eq. 7.47 we get
$$\sin^2 \theta_W = 0.215017.$$

b. The masses of the weak interaction force carriers can be related through

$$M_W = M_Z \cos \theta_W . \tag{7.48}$$

Expanding the above equation we get

$$\left(\frac{M_W}{M_Z}\right)^2 = \cos^2\theta_W = 1 - \sin^2\theta_W \Leftrightarrow \sin^2\theta_W = 1 - \left(\frac{M_W}{M_Z}\right)^2. \quad (7.49)$$

As the experimental value of the Z-boson is $M_Z = 91.187\,\text{GeV}/c^2$, then from Eq. 7.49 we obtain

$$\sin^2\theta_W = 0.222646.$$

7. *W decays (7).* Compute at leading order the ratio of the probabilities that a $W^\pm$ boson decays into leptons to the probability that it decays into hadrons.

The W coupling to fermions is $-i\frac{g}{\sqrt{s}}\gamma^\mu\frac{(1-\gamma^5)}{2}$, and thus it is the same to all fermions. Considering for instance the W^- decay, we have:

$$W^- \to e^-\bar{\nu}_e \qquad\qquad\qquad (7.50)$$
$$W^- \to \mu^-\bar{\nu}_\mu \qquad\qquad\qquad (7.51)$$
$$W^- \to \tau^-\bar{\nu}_\tau \qquad\qquad\qquad (7.52)$$
$$W^- \to d\bar{u} \qquad\qquad\qquad (7.53)$$
$$W^- \to s\bar{c}. \qquad\qquad\qquad (7.54)$$

Notice, that the $W^- \to b\bar{t}$ decay is not allowed due to kinematic constraints as $m_W < m_t$ and if you neglect the mass of the quarks, the sum of the probabilities of the decays into $q\bar{q}$ is independent of the mixing between the families, as the CKM matrix is unitary. Therefore,

$$\frac{\Gamma(W^- \to l\bar{l})}{\Gamma(W^- \to q\bar{q})} = \frac{3}{2\times 3} \qquad\qquad (7.55)$$

where the factor 3 in the denominator is the number of *colors* in QCD. The same argument can be used for W^+, which means that

$$\frac{\Gamma(W^\pm \to l\bar{l})}{\Gamma(W^\pm \to q\bar{q})} = \frac{1}{2}. \qquad\qquad (7.56)$$

8. *Glashow, Iliopoulos, and Maiani (GIM) mechanism (8).* Justify the GIM mechanism in the framework of the Standard Model.

Nicola Cabibbo suggested in 1963 that the quark u, d and s strong eigenstates (the ones known by then) were not the same than the corresponding weak eigenstates, but rather there was a mixing between the d and s quarks which have the same electromagnetic charge of $-1/3$:

$$d' = d \cos\theta_C + s \sin\theta_C, \tag{7.57}$$

where the angle θ_C was designated as the Cabibbo angle.

This mechanism explained the suppression of some decay channels of strange particles as, for example, $K^- \to \mu^- \bar{\nu}_\mu$ or $\Lambda \to p\,e^-\,\bar{\nu}_e$. From these measurements $\theta_C \simeq 13°$.

However, flavor changing neutral currents (FCNC) which are experimentally strongly suppressed (as the decay $K^0 \to \mu^-\mu^+$) were allowed in the Cabibbo model.

Glashow, Iliopoulos, and Maiani proposed in 1970 the introduction of the quark charm, the c, with the same quantum numbers of the u quark (electromagnetic charge of $2/3$). This model is known nowadays as the GIM mechanism. The quarks were then organized in two families (two $SU(2)_L$ doublets) and the weak currents symmetrized.

The mixing between the d and s quarks is in the GIM mechanism expressed by two orthogonal combinations:

$$d' = d\,\cos\theta_C + s\sin\theta_C$$
$$s' = -d\sin\theta_C + s\cos\theta_C.$$

or, in matrix form:

$$\begin{pmatrix} d' \\ s' \end{pmatrix} = V_C \begin{pmatrix} d \\ s \end{pmatrix} = \begin{pmatrix} \cos\theta_c & \sin\theta_c \\ -\sin\theta_c & \cos\theta_c \end{pmatrix} \begin{pmatrix} d \\ s \end{pmatrix}.$$

where V_C is a 2×2 rotation matrix.

This model ensures the cancellation of leading order and loop diagrams as it is exemplified in Fig. 7.1 for one loop $K^0 \to \mu^-\mu^+$ diagrams. This cancellation would be perfect if the masses of u and c quarks were equal. As the c mass is higher than u mass the sum of such diagrams leads to terms proportional to $m_c^2/m_{Z,W}^2$.

The J/ψ meson was discovered in 1974 confirming the existence of the c quark. The GIM mechanism was converted in a building block of the Standard Model.

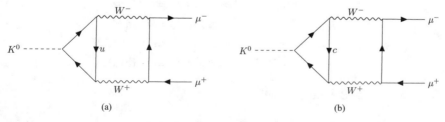

Fig. 7.1 Feynman diagrams at lowest order for process $K_L^0 \to \mu^+\mu^-$

The GIM mechanism was generalized to three families with the introduction of two new quarks, the bottom, b, and the top, t, by Makoto Kobayashi and Toshihide Maskawa. The mixing matrix, known as the CKM matrix, is:

$$V_{CKM} = \begin{pmatrix} V_{ud} & V_{us} & V_{ub} \\ V_{cd} & V_{cs} & V_{cb} \\ V_{td} & V_{ts} & V_{tb} \end{pmatrix}.$$

The introduction of the third generation allow CP symmetry violation in the Standard Model without, however, explaining the measured magnitude of such violation.

9. *Fourth family exclusion limit at LEP (9).* One of the first results of LEP was the exclusion at a level of 5σ of a fourth family of light neutrinos. Estimate the number of hadronic events that had to be detected to establish such a limit taking into account only statistical errors. Consider that the radiative corrections at $\sqrt{s} = m_Z$ reduced the cross-section, evaluated at tree-level, by $\sim 27\%$.

As seen in exercise 6, the Z decay cross-section is proportional to its decay width. In fact, for $\sqrt{s} \simeq m_Z$, the dominant Feynman diagram is the Z exchange through the s-channel, and the cross-section of the process $e^+ + e^- \to f\bar{f}$, where f is a fermion, is given by

$$\sigma(e^+e^- \to f\bar{f}) = \frac{12\pi}{m_Z^2} \frac{\Gamma_{e^+e^-} \Gamma_{f\bar{f}}}{\Gamma_Z^2}. \tag{7.58}$$

It was also seen that the decay widths are proportional to the sum of the squares of the vectorial and axial Z couplings, i.e.,

$$\Gamma_f \propto (g_V^f)^2 + (g_A^f)^2. \tag{7.59}$$

According to the Standard Model, the leptonic, hadronic and *invisible* partial widths can be calculated as

$$\Gamma_l \equiv \Gamma_e = \Gamma_\mu = \Gamma_\tau = 84\,\text{MeV} \tag{7.60}$$
$$\Gamma_u = \Gamma_c = 296\,\text{MeV} \tag{7.61}$$
$$\Gamma_d = \Gamma_s = \Gamma_b = 374\,\text{MeV} \tag{7.62}$$
$$\Gamma_{\text{had}} = \Gamma_u + \Gamma_d + \Gamma_s + \Gamma_c + \Gamma_b = 1714\,\text{MeV} \tag{7.63}$$
$$\Gamma_\nu = 167.3\,\text{MeV} \tag{7.64}$$
$$\Gamma_{\text{inv}}|_{N_\nu=3} = N_\nu \times \Gamma_\nu = 501.9\,\text{MeV} \tag{7.65}$$

and the total Z width is simply

$$\Gamma_Z = \Gamma_h + 3\Gamma_l + N_\nu \Gamma_\nu. \tag{7.66}$$

Notice that, according to Eq. 7.58, the partial width depend on the total width Γ_Z, and therefore on the number of neutrino families N_ν.

From now one let us consider only the process $e^+ + e^- \rightarrow q\bar{q}$, which is related to Γ_{had}. The cross section of this process can be computed using the previous equations, considering 3 and 4 families of neutrinos, yielding $\sigma_3 \simeq 30\,\text{nb}$ and $\sigma_4 \simeq 27\,\text{nb}$, respectively.

In order to perform a measurement with a significance of 5σ, the error in the cross section has to be five times smaller than the difference between the theoretical expectations for a contiguous number of neutrino families, i.e.,

$$|\sigma_3 - \sigma_4| \geq 5\varepsilon(\sigma_{\text{exp}}) \tag{7.67}$$

where σ_{exp} is the error on the cross section measurement.

The cross-section measurement is proportional to the number of events $\sigma_{\text{exp}} = kN$, while the error on the number of events, N, is $\sqrt{N}$. As such, the error on the measured cross-section is

$$\varepsilon(\sigma_{\text{exp}}) = \sqrt{\left(\frac{\partial \sigma_{\text{exp}}}{\partial N}\right)^2 (\varepsilon(N))^2} = k\varepsilon(N) = \frac{\sigma_{\text{exp}}}{N}\sqrt{N} = \frac{\sigma_{\text{exp}}}{\sqrt{N}}. \tag{7.68}$$

Substituting Eq. 7.68 into 7.67, and assuming that we are testing the 3 neutrino family hypothesis, $\sigma_{\text{exp}} \rightarrow \sigma_3$ we get

$$|\sigma_3 - \sigma_4| \geq 5\frac{\sigma_3}{\sqrt{N}} \quad \Leftrightarrow \quad N \geq 25\left(\frac{\sigma_3}{\sigma_3 - \sigma_4}\right)^2. \tag{7.69}$$

Plugging in the values obtained previously for σ_3 and σ_4 we obtain finally

$$N \gtrsim 2\,500. \tag{7.70}$$

10. *Higgs decays into ZZ (10).* The Higgs boson was first observed in the $H \rightarrow \gamma\gamma$ and $H \rightarrow ZZ \rightarrow 4\,leptons$ decay channels. Compute the branching fraction of $ZZ \rightarrow \mu^+\mu^-\mu^+\mu^-$ normalized to $ZZ \rightarrow anything$.

The amplitude of the decay of the Z into each flavor-antiflavor pair kinematically accessible is, at leading order,

$$\Gamma_Z \rightarrow f\bar{f} \propto \left(g_{Af}^2 + g_{Vf}^2\right),$$

where g_A and g_V are respectively the axial and vector coupling of the flavor (in the case of quarks a factor of 3 should be inserted in order to take into account the number of colors). As a consequence, summing over all flavors, the branching fraction of the Z into muon pairs is

$$\frac{\Gamma_Z \to \mu^+\mu^-}{\Gamma_Z \to anything} \simeq 0.034 \,.$$

The probability to have both Z decaying into $\mu^+\mu^-$ is thus

$$Br(ZZ \to \mu^+\mu^-\mu^+\mu^-) \simeq 0.034^2 \simeq 1.2 \times 10^{-3} \,.$$

This final state has a low probability, but it is relatively clean from the experimental point of view: it is thus a "golden" channel, despite its small frequency with respect to, say, hadronic channels (which are affected by a larger background).

11. *Higgs decays into $\gamma\gamma$ (11).* Draw the lowest order Feynman diagram for the decay of the Higgs boson into $\gamma\gamma$ and discuss why this channel was a golden channel in the discovery of the Higgs boson.

The dominant diagram for the Higgs particle decay into $\gamma\gamma$ is shown in Fig. 7.2. The lowest possible order for the process is at one loop.

You could have any (charged) particles in the loop, but the Higgs particle couples to mass, and thus prefers the heavy particles: this implies that the dominant contributions come from t and W (which are virtually produced), and b.

The $\gamma\gamma$ decay, although not the favorite from the point of view of the branching fraction (about 0.2%, compared to 58% and 21% respectively for the two dominant decay channels, $\bar{b}b$ and WW, which are possible at first order), is a golden channel (and indeed it provided the first evidence for the Higgs) because its topology is experimentally easier to detect than a hadronic topology. The background for all processes producing two photons is quite low compared to the hadronic backgrounds, and thus the small branching fraction is traded for the relative lack of background. This fact makes $\gamma\gamma$ channel one of the best for Higgs detection.

12. *Top production at the Tevatron and at the LHC.* In March 1995, the CDF group and its rival experiment D0 at Fermilab jointly reported the discovery of the top quark at a mass of (176 ± 18) GeV. One year before, the CDF group had submitted an article suggesting tentative evidence for the existence of top with a mass of about 175 GeV.

Fig. 7.2 Dominant Feynman diagram for the Higgs decay into $\gamma\gamma$

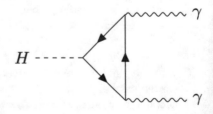

Indirect determinations by LEP made before the actual discovery indicated a mass of (165 ± 20) GeV, well in agreement with the value found later from the measurements following the discovery.

The top quark was later copiously produced at LHC.

Which physical processes are dominant in the production of the top quark at the Tevatron and at the LHC? How was it possible to determine the top mass at LEP without actually producing the particle?

In hadron collisions, top quarks are produced dominantly in pairs ($t\bar{t}$) through processes at leading order in QCD. Approximately 85% of the production cross section at the Tevatron ($\bar{p}p$ at 1.96 TeV c.m. energy) is from $\bar{q}q$ annihilation, with the remainder from gluon-gluon fusion, while at LHC (pp) energies about 90% of the production is from the latter process at $\sqrt{s} \simeq 14$ TeV. Electroweak single top-quark production mechanisms, namely from $\bar{q}q' \to t\bar{b}$ and $qb \to tq'$, mediated by virtual $s-$channel and $t-$channel W bosons, and Wt associated production from b-gluon fusion, lead to cross section about one order of magnitude smaller at LHC energies.

In the years before the top quark discovery, it was realized that some precision measurements of the electroweak vector boson masses and couplings at LEP are very sensitive to the value of the top quark mass; even if not produced, a virtual production of top quarks could influence the actual value of variables well measured and well calculated within the Standard Model of particle physics (in particular, the Z mass and the Z decay amplitude in bb pairs). The top quark properties were measurable even if the particle could not be directly detected. The world average value of the top quark mass is presently of (172.8 ± 0.3) GeV, and the accuracy is dominated by LHC measurements.

13. *Top mass measurement at the LHC.* Which decay channels of the t quark are simplest to measure at the LHC? Which detector aspects are crucial for a precise mass measurement?

The main decay channel of the top quark is into bW^+.

The b-quark can be tagged combining the information collected from the microvertex detector (as b-hadrons can live enough to travel a few microns before decaying) and the morphology of the hadronic jet reconstructed using the overall detectors of the experiment, namely the hadronic calorimeters. If fact, as the b-quark has a large mass the jets produced in its decay are wider.

The W can decay into leptons or quarks. The simplest signature for a W is the leptonic decay into electrons or muons:

$$t \to b\,W^+ \to b\,\mu^+\nu_\mu \tag{7.71}$$

$$t \to b\,W^+ \to b\,e^+\nu_e . \tag{7.72}$$

The collision between two protons generates many unwanted hadronic particles (*underlying event*), but leptons, more scarce in hadronic interactions, can be easily identified and connected with the interaction of interest. The muons are detected by muon chambers, placed at the periphery of the experiments, and shielded by iron or lead walls, while electrons (positrons) can be detected using electromagnetic calorimeters. Both particles, being charged, also deposit energy in the silicon tracker. While for electrons, its energy can be estimated using the calorimeters, for muons, its energy is obtained essentially looking at the curvature induced by the experiment magnetic field.

At the LHC, the best measurement of the mass of the top is coming from the study of the $t\bar{t}$ production where the top quarks decay in the above channels; final states with two high-p_T b-jets and two high-p_T leptons and missing energy (neutrinos) are the golden experimental signature. The mass is obtained from a global fit of the particles kinematical variables assuming that in the final state there are two W bosons and two top quarks.

Chapter 8
The Standard Model of Cosmology and the Dark Universe

1. *Cosmological principle and Hubble law (1)*
   Show that the Hubble law does not contradict the Cosmological principle (all points in space and time are equivalent).

   _____

   The cosmological principle expresses the fact that, on large enough scales, the matter distribution in the Universe is homogeneous and isotropic. The fact that the Hubble law does not contradict this statement can be shown by proving that, given a homogeneous and isotropic distribution of matter, the Hubble law results as a consequence.
   If we consider a spacetime with homogeneous and isotropic spatial sections, in comoving coordinates the metric has the form

   $$ds^2 = dt^2 - a^2(t)\left\{d\chi^2 + \Sigma^2(\chi)(d\theta^2 + \sin^2\theta d\phi^2)\right\},$$

   where $\Sigma^2(\chi) = \sin^2\chi$ (resp. χ^2, $\sinh^2\chi$), depending on the closed (respectively flat, open) character of the spatial sections. If we call D the distance between an observer in the origin of the above reference system and another point, we measure $D = D(t)$ along a surface of constant time t; by also choosing to consider the radial distance (i.e., constant θ and ϕ), we end up with the three conditions $dt = d\theta = d\phi = 0$, so that $D(t) = a(t)\chi$. For the observed velocity $V(t)$ of a body at a constant value of the coordinate χ we then obtain

   $$V(t) = \frac{dD(t)}{dt} = \dot{a}(t)\chi = \frac{\dot{a}(t)}{a(t)}D(t).$$

   In the above relation $\dot{a}(t)/a(t) = H(t)$ is the Hubble constant, so that the above relation gives the Hubble law $V(t) = H(t)D(t)$. This shows that the Hubble law is consistent with a homogeneous and isotropic cosmological model that respects the cosmological principle.

© Springer Nature Switzerland AG 2021
A. De Angelis et al., *Particle and Astroparticle Physics*, Undergraduate Lecture Notes in Physics,
https://doi.org/10.1007/978-3-030-73116-8_8

2. *Olbers' Paradox (2)*

Why is the night dark? Does the existence of interstellar dust (explanation studied by Olbers himself) solve the paradox?

In modern cosmology we can find many reasons why the Olbers' Paradox cannot arise:

- Redshift: due to cosmological expansion light is redshifted.
- Finite speed of light: we can see only the stars whose light had the time to reach us.
- Finite age of stars: stars can only shine for a finite amount of time. For an infinite Universe, this means that stars should have switched on at different times depending on their distance from Earth to be shining all at the same time and fill the Universe with light. This extreme coincidence clearly implies a strong violation of the Copernican principle.
- Finite age of the Universe: The Universe has a "beginning", the Big Bang. In any direction we can see up to a finite distance in space (i.e. time).

An interstellar medium absorbing light (solution studied by Olbers himself) does not solve the problem, as it would heat up, reach the equilibrium and irradiate the same amount of light.

3. *Steady state Universe (3)*

In a steady state Universe with Hubble law, matter has to be permanently created. Compute in that scenario the creation rate of matter.

A steady state Universe must appear identical in any direction and any time. A consequence is that the matter density ρ and the Hubble parameter must be constant in time:

$$H(t) = H_0 \; ; \; \rho(t) = \rho(0) \,. \tag{8.1}$$

A constant Hubble parameter means an Universe in an exponential expansion:

$$\dot{a} = a H_0 \Longrightarrow a(t) \propto e^{H_0 t} \,.$$

The volume of a spherical region will then increases exponentially, and to keep in it a constant density the mass contained in it will also has to increase:

$$M = \rho(0) V \; ; \; \dot{M} = \rho(0) \dot{V} = 3 H_0 \rho(0) V.$$

Then, the rate of mass creation in a steady Universe is:

$$\frac{\dot{M}}{V} = 3 H_0 \rho(0) \simeq 6 \times 10^{-28} \, \mathrm{kg \, km^{-3} \, yr^{-1}}, \tag{8.2}$$

which means the creation of about one proton per year per cubic kilometer.

4. *Blackbody distribution and the Cosmic Microwave Background (4)*

In 1965 Penzias and Wilson discovered that nowadays the Universe is filled with a cosmic microwave background which follows an almost perfect Planck blackbody formula. Show that the blackbody form of the energy density of the background photons was preserved during the expansion and the cooling that had occurred in the Universe after photon decoupling.

For an adiabatically expanding Universe filled with radiation one can find:

$$\frac{d}{dt}\left(\epsilon\, a^4\right) = \frac{d}{dt}\left(T^4 a^4\right) = 0 \quad \Longrightarrow \quad T(a) = T_0 a^{-1}. \tag{8.3}$$

Moreover, using the redshift formula $z = \lambda'/\lambda - 1$ and $1 + z = 1/a$, we have that the frequency of the radiation shifts towards red as:

$$\nu(a) = \nu_0 a^{-1}. \tag{8.4}$$

Substituting these expression into the Planck blackbody formula

$$\epsilon_\gamma(\nu)d\nu = \frac{8\pi h}{c^3}\frac{\nu^3\, d\nu}{e^{-\frac{h\nu}{k_B T}} - 1},$$

we have that:

$$\epsilon_\gamma(\nu; a) = \frac{8\pi h}{c^3}\frac{\nu^3(a)}{e^{-\frac{h\nu(a)}{k_B T(a)}} - 1} = \frac{8\pi h}{c^3}\frac{\nu_0^3}{e^{-\frac{h\nu_0}{k_B T_0}} - 1}\, a^{-3} = \epsilon_\gamma(\nu)a^{-3}. \tag{8.5}$$

This confirms that the shape of the blackbody is conserved by the expansion, while its energy redshifts away.

5. *The CMB and our body (5)*

If CMB photons are absorbed by the human body (which is a reasonable assumption), what is the power received by a human in space because of CMB?

If you are lazy as the authors of this book, you can assume your surface area to be $A \sim 1$ m^2 – but if you are not lazy you can do better.
The number of photons per second you absorb is

$$\frac{dN_\gamma}{dt} = n_\gamma c A \sim 4 \times 10^8 \text{m}^{-3} \times 3 \times 10^8 \text{m/s} \times 1\text{m}^2 \simeq 1.23 \times 10^{17}\text{s}^{-1}.$$

(b) What is the approximate rate, in watts, at which you would absorb radiative energy from
The mean energy per CMB photon is $\langle E_\gamma \rangle \sim 10^{-22}$ J, and thus the rate of absorption of energy is

$$\frac{dE}{dt} \sim \langle E_\gamma \rangle \frac{dN_\gamma}{dt} \sim 12.5\,\mu W\,.$$

CMB is thus not a very powerful microwave oven.

6. *CMB, infrared and visible photons (6)*
Estimate the number of near-visible photons (λ from $0.3\,\mu m$ to $1\,\mu m$) in a cubic centimeter of interstellar space. Estimate the number of far-infrared photons in the region of λ from $1000\,\mu m$ to $1\,\mu m$.

VISIBLE/NIR: An accurate calculation would require the knowledge of the radiation sources distribution and a proper integration of their spectrum. Anyways, we can obtain a fair estimate with some crude approximations. Considering a point located within the Milky Way's disk, the main sources of diffuse visible/near IR photons are stars. In the far IR ($20 - 500\,\mu m$), on the contrary, the dominant source is radiation reprocessed by dust. At longer wavelengths, the CMB becomes the most relevant contribution.

For the visible part of the spectrum, we can reasonably assume that the Milky Way is dominated by stars with $R_\star \simeq 0.75 R_\odot$ and $T_\star \simeq 4000\,K$ and that the average number of stars is $n_\star \simeq 0.24\,pc^{-3}$ (roughly corresponding to 1 star in every sphere with radius of $1\,pc$). Using the Stefan-Boltzmann's Law, the average stellar luminosity is:

$$L_\star = 4\pi R_\star^2 \sigma T_\star^4 = 5.03 \times 10^{32}\,erg\,s^{-1}, \tag{8.6}$$

where $\sigma = 5.67 \times 10^{-5}\,erg\,cm^{-2}\,s^{-1}\,K^{-4}$ is the Stefan-Boltzmann's constant. In a random point of the interstellar space, the stellar radiation field energy flux would be:

$$F_\star = \int_V \frac{n_\star L_\star}{4\pi r^2}dV = 4\pi \int_0^{D_{max}} \frac{n_\star L_\star}{4\pi r^2}r^2 dr. \tag{8.7}$$

Assuming $D_{max} = 100\,pc$ (approximately the scale height of the disk and a distance where dust absorption is not yet too severe, thus justifying the use of spherical symmetry), we get:

$$F_\star = \frac{3}{4\pi}\frac{L_\star \times 100\,pc}{(1\,pc)^3} = 1.26 \times 10^{-3}\,erg\,cm^{-2}\,s^{-1}. \tag{8.8}$$

In the assumption of isotropic radiation, the associated energy density is:

$$\varepsilon_\star = \frac{4\pi}{c}F_\star = 5.30 \times 10^{-13}\,erg\,cm^{-3} = 0.331\,eV\,cm^{-3}, \tag{8.9}$$

that, with an average photon energy of $\langle h\nu \rangle = 1.35\,eV$ (the most common energy for photons emitted by a thermal source with temperature equal to $T_\star$), yields an estimate of:

$$n_\gamma^{VIS} = 0.24\,cm^{-3}. \tag{8.10}$$

To place this number in context, the density of visible photons due to Solar radiation at the distance of the Earth is $n_{\gamma,\odot}^{VIS} = 1.42 \times 10^7 \, cm^{-3}$. From our point of view, the Sun is approximately 10^{12} times brighter than a barely visible star for the naked eye. Given the 8 orders of magnitude difference between Solar energy density and interstellar energy density, the estimate above is as acceptable as the statement that a good naked eye should expect to see something between 1000 and 10000 stars, in dark nights across the whole sky.

Far IR: The dominant source of far IR radiation is interstellar dust, which absorbs UV and visible star light, re-emitting it in the IR. The average dust temperature is $T_{dust} \simeq 10 \, K$, with a maximum photon emission at approximately $\lambda \simeq 367 \, \mu m$. Dust is concentrated within a thin layer of the disk, but, contrary to stellar light, its emission can travel much longer distances within the disk itself. We can estimate (an upper limit to) the number of FIR photons assuming that the dust radiation energy density cannot be higher than the stellar radiation energy density. The energy of the most numerous dust-emitted photons is $\langle h\nu \rangle \simeq 3.38 \, meV$, which yields:

$$n_\gamma^{FIR} \leq 97.94 \, cm^{-3},$$

showing that, in both cases, the CMB is dominating the stage in terms of photon number densities.

7. *Requirements for a cosmic neutrino background detector (7)*
   Let the typical energy of a neutrino in the Cosmic Neutrino Background be ~ 0.2 meV. What is the approximate interaction cross-section for cosmic neutrinos? How far would typically a cosmic neutrino travel in ice before interacting?

   From a dimensional analysis we can approximate the cross section to be $\sigma \sim G_F^2 \times E_\nu^2$, where G_F is the Fermi constant and E_ν is the energy of the cosmic neutrinos. So numerically we get

   $$\sigma \sim 5.44 \times 10^{-54} eV^{-2} = 2.1 \times 10^{-63} \, cm^2.$$

   where we have used $(1eV)^{-1} = 197 \times 10^{-7} \, cm$.
   The average path the cosmic neutrinos travel in ice is $\lambda = 1/(n_{ice}\sigma)$, where n_{ice} is the number density of ice ($\rho_{ice} = 0.917 \, g/cm^3$). The number density can be evaluated from the mass density as follows:

   $$n_{ice} = \frac{\rho_{ice} N_A}{M_{H_2O}},$$

   where $N_A = 6.022 \times 10^{23} mol^{-1}$ is the Avogadro's number and $M_{H_2O} \simeq 18 \, g/mol$ is the molar density of H_2O. Using these numbers we get

   $$n_{ice} \simeq 3.06 \times 10^{22} \, cm^{-3}.$$

Thus the average path is

$$\lambda \simeq 1.5 \times 10^{40} \text{ cm.}$$

8. *Dark Matter and mini-BHs (8)*

   If Black Holes (BHs) of mass $10^{-8} M_\odot$ made up all the dark matter in the halo of our Galaxy, how far away would the nearest such BH on average? How frequently would you expect such a BH to pass within 1 AU of the Sun?

   Method 1. The virial mass of a galaxy can be defined as the mass M_{200} enclosed in the radius R_{200} within which the mean density is 200 times the critical density; for the Milky Way, $R_{200} = 305$ kpc and $M_{200} = 1.4 \times 10^{12} M_\odot$.
   The number of mini-BHs, with $M_{BH} = 10^{-8} M_\odot$, to make up the total dark matter (DM) mass is:

   $$N_{BH} = \frac{M_{200}}{M_{BH}} = \frac{1.4 \times 10^{12} M_\odot}{10^{-8} M_\odot} \simeq 1.4 \times 10^{20}. \tag{8.11}$$

   We can calculate the halo volume modeling it as a cylinder with a R_{200} radius and a $h = 2000$ ly $= 613$ pc height:

   $$V = \pi R_{200}^2 h = \pi (305 \text{ kpc})^2 \, 613 \text{ pc} \simeq 1.79 \times 10^{14} \text{ pc}^3. \tag{8.12}$$

   The average "unitary" volume for each BH is:

   $$V_U = \frac{V}{N_{BH}} = \frac{3.6 \times 10^{14} \text{pc}^3}{1.4 \times 10^{20}} \simeq 1.3 \times 10^{-6} \text{ pc}^3, \tag{8.13}$$

   which means that, on average, the distance from the closest object would be half of the diagonal of a cube with volume V_U:

   $$d_{\min} = \frac{\sqrt{3}}{2} \sqrt[3]{V_U} = \frac{\sqrt{3}}{2} \sqrt[3]{2.6 \times 10^{-6} \text{ pc}^3}$$
   $$\simeq 0.0095 \text{ pc} \simeq 1960 \text{ AU} \simeq 2.9 \times 10^{11} \text{ km.} \tag{8.14}$$

   We consider the velocity dispersion in the Milky Way Halo, assuming that these DM objects would move with same average velocity as the other objects in the galaxy, $\sigma \sim 200$ km/s. The time it would take for an object to cover the distance up to 1 AU from the Sun would be of the order of:

   $$t \sim \frac{d_{\min}}{\sigma} \simeq 46.5 \text{ yr.} \tag{8.15}$$

   Method 2. Considering the average DM density in the region around the Solar System to be:

   $$\rho_{local}^{DM} \sim 0.4 \, \frac{\text{GeV}}{\text{cm}^3} \simeq 3.59 \times 10^{-58} \frac{M_\odot}{\text{cm}^3}. \tag{8.16}$$

In order to make up a $10^{-8} M_\odot$ mass, the volume required is:

$$V_U' = \frac{10^{-8} M_\odot}{\rho_{local}^{DM}} \simeq 9.5 \times 10^{-7} \ \text{pc}^3. \tag{8.17}$$

Then, similarly to *Method 1*:

$$d_{min}' = \frac{\sqrt{3}}{2} \sqrt[3]{V_U'} = \frac{\sqrt{3}}{2} \sqrt[3]{9.5 \times 10^{-7} \ \text{pc}^3}$$
$$\simeq 0.0085 \ \text{pc} \simeq 1750 \ \text{AU} \simeq 2.6 \times 10^{11} \ \text{km} \tag{8.18}$$

$$t' \sim \frac{d_{min}}{\sigma} \simeq 41.2 \ \text{yr} \tag{8.19}$$

which is not so far from the time $t \simeq 46.5$ yr calculated previously.

9. *Nucleosynthesis and neutron lifetime (9)*

The value of the neutron lifetime, which is unusually long for weak decay processes (why?), is determinant in the evolution of the Universe. Discuss what would have been the primordial fraction of He if the neutron lifetime would have been one tenth of its real value.

The neutron lifetime $\tau_n = 885.7$ s is so long because of the weakness of the weak force responsible for the decay and the small difference between neutron and proton masses. Indeed, by dimensional analysis one can observe that the decay rate is very small:

$$\Gamma \sim G_F^2 (m_n - m_p)^5 \ , \tag{8.20}$$

where $G_F \simeq 10^{-5} \ \text{GeV}^{-2}$ is the Fermi constant and $(m_n - m_p) \simeq 10^{-3}$ is the difference between neutron and proton masses. Since

$$\frac{n_n}{n_p} \sim 0.2 e^{-t/\tau_n},$$

if the neutron lifetime had been one tenth of its real value,

$$\tau_n' = \frac{1}{10} \tau_n \ , \tag{8.21}$$

the relative abundance of neutrons and protons would have been much smaller due to a faster decay:

$$\frac{n_n'}{n_p'} \simeq 0.2 e^{-t/\tau_n'} = \simeq 0.2 e^{-10t/\tau_n} = 0.2 \left(\frac{n_n/n_p}{0.2} \right)^{10} \simeq 5.6 \times 10^{-3} \ , \tag{8.22}$$

where we used the real value of neutron-proton ratio $n_n/n_p \simeq 0.14$. Using this and the fact that

$$Y_p = 1 - \frac{\rho\,(\text{H})}{\rho\,(\text{Baryons})} = 1 - \frac{n_p - n_n}{n_p + n_n} = \frac{2\frac{n_n}{n_p}}{1 + \frac{n_n}{n_p}},$$

one can see that helium abundance would have been accordingly smaller:

$$Y'_p = \frac{2n'_n/n'_p}{1 + n'_n/n'_p} \simeq 0.01 . \tag{8.23}$$

10. *GPS time corrections (10)*
    Identical clocks situated in a GPS satellite and at the Earth surface have different periods due general relativity effects. Compute the time difference in one day between one clock situated in a satellite in a circular orbit around Earth with a period of 12h and one clock situated on the Equator at the Earth surface. Consider that Earth has a spherical symmetry and use the Schwarzschild metric.

Period and distance of a satellite in a circular orbit are related by:

$$T = 2\pi\sqrt{\frac{r^3}{GM}} , \tag{8.24}$$

where G is the gravitational constant and M is the mass of the Earth. If $T = 12\,\text{h}$ then $r \simeq 2.66 \times 10^4$ km. According to the formulas for the relative time stretching. the difference between two identical clocks at radial distance r one from the other is:

$$\frac{\Delta\tau}{\tau_*} = \left(1 - \frac{r_S}{r}\right)^{-1} , \tag{8.25}$$

where r_S is the Schwarzschild radius and is given by the equation

$$M = \frac{r_S c^2}{2G} . \tag{8.26}$$

Using the numerical values of r and r_S one finds that the clock on the satellite is 4×10^{-10} times faster than the other one. In a whole day, the accumulated delay will be:

$$\Delta\tau \simeq 38 \ \mu s . \tag{8.27}$$

11. *Asymptotically matter-dominated Universe (11)*
    Consider a Universe composed only by matter and radiation. Show that whatever would have been the initial proportion between the matter and the radiation energy densities this Universe will be asymptotically matter dominated.

Let us imagine a two components Universe, filled by matter and radiation. Let us assume that these two components are not interacting. This last hypothesis allows us to write two independent energy conservation laws for matter and radiation. If ρ_r is the radiation density, and ρ_m is the matter density, the corresponding conservation laws imply

$$\rho_r \propto a^{-4} \quad \text{and} \quad \rho_m \propto a^{-3}, \tag{8.28}$$

where a is the scale factor of the Friedmann metric. In the two component Universe the Friedmann equation reads

$$\frac{\dot{a}^2}{a^2} = \frac{8\pi G}{3}(\rho_r + \rho_m).$$

Let us now consider a Universe that is initially radiation dominated. In this case we can neglect ρ_m with respect to ρ_r, and we can write with good approximation

$$\frac{\dot{a}^2}{a^2} = \frac{8\pi G}{3}\rho_r \propto a^{-4}$$

for the Friedmann equation above. Then, $a \propto t^{1/2}$, and, using (8.28) this shows that, in this epoch, the radiation density behaves as $\rho_r \propto t^{-2}$, while the matter density evolves as $\rho_m \propto t^{-3/2}$. Thus the radiation content dilutes faster as the Universe expands in the radiation dominated epoch. Eventually the radiation will dilute so much that the radiation domination condition will not be satisfied anymore. After enough time, the matter content will become dominant. In this other epoch, the Universe evolution will be governed by the Friedmann equation

$$\frac{\dot{a}^2}{a^2} = \frac{8\pi G}{3}\rho_m \propto a^{-3},$$

in which we are now neglecting ρ_r with respect to ρ_m. Accordingly, the scale factor now evolves as $a \propto t^{2/3}$, and the radiation and matter densities scale as $\rho_r \propto t^{-2-2/3}$ and $\rho_m \propto t^{-2}$, respectively.

As we could anticipate from the conservation laws, radiation will continue to become less and less important for the evolution of the Universe, so that the Universe is eventually matter dominated. Of course, the same conclusion would have been reached if we would have started with a matter dominated Universe, as we would have been from the beginning in the second regime. In presence of matter and radiation, the Universe will then become, eventually, matter dominated.

12. *Cosmological distances (12)*

Consider a light source at a redshift of $z = 2$ in an Einstein-de Sitter Universe.

(a) How far has the light from this object traveled to reach us? (b) How distant is this object today?

In an Einstein-de Sitter critical Universe, we can set $\Omega_{0\gamma} = \Omega_{0K} = \Omega_{0\Lambda} = 0$, while $\Omega_{0m} = 1$. The evolution equation for the Hubble parameter, therefore, becomes:

$$\frac{H}{H_0} = a^{-3/2}, \tag{8.29}$$

which implies $\dot{a} \propto a^{-1/2}$, whose solutions have the expected behaviour of the scale parameter $a(t) \propto t^{2/3}$. The answers sought by the problem are:

(a) the light travel distance:

$$D_T = D_H \int_0^2 a \frac{H_0}{H} dz = \frac{c}{H_0} \int_0^2 \frac{dz}{(1+z)^{5/2}}. \tag{8.30}$$

The above integral can be easily solved with the variable substitution $x = 1 + z$ ($dx = dz$), which leads to:

$$D_T = \frac{c}{H_0} \int_1^3 \frac{dx}{x^{5/2}} = \frac{c}{H_0} \left[\frac{2}{3} - \frac{2}{9\sqrt{3}} \right]. \tag{8.31}$$

Using $H_0 = 70 \, \text{km s}^{-1} \text{Mpc}^{-1}$, we get $D_T \simeq 2306 \, \text{Mpc}$;

(b) the proper distance:

$$D_P = D_H \int_0^2 \frac{H_0}{H} dz = \frac{c}{H_0} \int_0^2 \frac{dz}{(1+z)^{3/2}}. \tag{8.32}$$

Again, we can apply the variable substitution $x = 1 + z$:

$$D_P = \frac{c}{H_0} \int_1^3 \frac{dx}{x^{3/2}} = \frac{c}{H_0} \left[2 - \frac{2}{\sqrt{3}} \right], \tag{8.33}$$

which, under the same cosmological parameters, gives $D_P \simeq 3620 \, \text{Mpc}$.

13. *Decoupling (13)*

What are the characteristic temperatures (or energies) at which (a) neutrinos decouple; (b) electron-positron pairs annihilate; (c) protons and neutrons drop out of equilibrium; (d) light atomic nuclei form, (e) neutral hydrogen atoms form, (f) photons decouple from baryonic matter?

(a) Neutrinos are scattered by their interactions with electrons and positrons, such as the reaction $e^- e^+ \longleftrightarrow \nu_e + \bar{\nu}_e$; $e^- + e^+ \longleftrightarrow \nu_e + \bar{\nu}_e$.

The approximate rate of these interactions is set by the number density of electrons and positrons, the averaged product of the cross section for interaction and the velocity of the particles. The number density n of the relativistic electrons and positrons depends on the cube of the temperature T. The product of the cross section and velocity for weak interactions for temperatures (energies) below W/Z boson masses (~ 100 GeV) is given approximately by $\langle \sigma v \rangle \sim G_F^2 T^2 \langle \sigma v \rangle \sim G_F^2 T^2$, where G_F is Fermi's constant. Putting it all together, the rate of weak interactions is

$$\Gamma = n \langle \sigma v \rangle \sim G_F^2 T^5 .$$

This can be compared to the expansion rate which is given by the Hubble parameter H:

$$H = \sqrt{\frac{8\pi}{3} G \rho},$$

where G is the gravitational constant and ρ is the energy density of the Universe. At this point in cosmic history, the energy density is dominated by radiation, so that $\rho \propto T^4$. As the rate of weak interaction depends more strongly on temperature, it will fall more quickly as the Universe cools. Thus when the two rates are approximately equal (dropping terms of order unity, including an effective degeneracy term which counts the number of states of particles which are interacting) gives the approximate temperature at which neutrinos decouple:

$$G_F^2 T^5 \sim \sqrt{G T^4}.$$

Solving for temperature gives

$$T \sim \left(\frac{\sqrt{G}}{G_F^2} \right)^{1/3} \sim 1 \text{ MeV} \sim 10^{10} \text{ K}.$$

Neutrino decoupling is expected to have left behind a cosmic neutrino background, analogous to the cosmic microwave background radiation of electromagnetic radiation which was emitted at a much later epoch. The detection of the neutrino background is far beyond the capabilities of the present generation of neutrino detectors. The time since the big bang, in standard cosmology, is about 1 s.

(b) The relevant temperature at which electrons and positrons remain in thermal equilibrium (energy of photons is still high enough to produce electron-positron pairs is):

$$T \sim 2m_e \sim 1 \text{ MeV} ;$$

we called m_e the electron mass.

(c) Protons and neutrons drop out of equilibrium when the temperature of the Universe falls below their mass difference, i.e., again when

$$T \lesssim 1 \text{ MeV}.$$

(d) Stable light nuclei form when the temperature of the Universe falls below the binding energies, i.e., when

$$T \sim (10 - 0.1) \text{ MeV}.$$

Neutrons and protons can then combine to form D and He nuclei.

(e) Neutral hydrogen forms when the temperature of the Universe falls below the binging energy of the electron to the nucleus:

$$T \sim 13.6 \text{ eV} \simeq 140\,000 \text{ K}.$$

(f) Below a temperature

$$T \sim 0.4 \text{ eV} \simeq 4000 \text{ K}$$

(corresponding in standard cosmology to an age of about 370 ky, or to a redshift $z \sim 1100$), photons are no longer in thermal equilibrium with matter. They cannot easily induce shifts in the electron energy levels of light nuclei and the Universe becomes transparent. Recombination (the word is misleading, since the Big Bang theory does not posit that protons and electrons had been combined before, but the name exists for historical reasons since before the Big Bang hypothesis) lasts for about 100 ky, during which Universe becomes more and more transparent to photons; the photons of the CMB originate at this time. The radius of the Universe is about 14 Mpc at this time, and the baryonic matter density is approximately a billion times higher than today.

14. *Evolution of momentum (14)*

    How does the momentum of a free particle evolve with redshift (or scale factor)?

    Let us consider the motion of a free particle in a space-time with metric:

    $$ds^2 = -c^2 dt^2 + a^2(t) \left[\frac{dr^2}{1 - kr^2} + r^2 d\Omega^2 \right], \tag{8.34}$$

    where, as usual, we define $d\Omega^2 = d\theta^2 + \sin^2\theta d\phi^2$ and we use the curvature parameter k. Since in the particle's frame it is $ds^2 = -c^2 d\tau^2$, with τ expressing the particle's proper time, we have:

    $$\gamma d\tau = dt \tag{8.35}$$

and:

$$-c^2 d\tau^2 = -c^2 dt^2 + a^2(t)\left[\frac{dr^2}{1 - kr^2} + r^2 d\Omega^2\right]. \tag{8.36}$$

Observing the particle along the direction of its motion (i.e., by taking $d\Omega = 0$), we have:

$$c^2\gamma^{-2}dt^2 = c^2 dt^2 - \frac{a^2(t)dr^2}{1 - kr^2}, \tag{8.37}$$

hence, given that $\gamma = (1 - v/c)^{-1/2}$, we have:

$$(c^2 - v^2)dt^2 = c^2 dt^2 - \frac{a^2(t)dr^2}{1 - kr^2}, \tag{8.38}$$

which eventually yields:

$$v = \frac{a(t)}{\sqrt{1 - kr^2}}\frac{dr}{dt}. \tag{8.39}$$

Using the space component of the momentum, we obtain:

$$p = mv = m\frac{a(t)}{\sqrt{1 - kr^2}}\frac{dr}{dt}, \tag{8.40}$$

which implies a proportionality between momentum and scale factor that, in the case of a flat spacetime with $k = 0$, simply reduces to $p = ma(t)dr/dt$.

15. ΛCDM and distances (15)

Estimate the expected apparent magnitude of a type Ia supernova (absolute magnitude $M \simeq -19$ at a redshift $z = 1$ in the ΛCDM Universe.

Remembering that we define the luminosity distance as the distance that enters the flux-luminosity conversion formula:

$$L_{em} = 4\pi D_L^2 F_{obs}, \tag{8.41}$$

were L_{em} is the luminosity emitted in the source frame, while F_{obs} is the flux that we observe, we can estimate the luminosity distance of the source in the ΛCDM scenario:

$$D_L = (1 + z)D_P = 2\frac{c}{H_0}\int_0^1 \frac{dz}{\sqrt{\Omega_\Lambda + \Omega_m(1 + z)^3}}, \tag{8.42}$$

where D_P is the proper distance and we assumed a flat Universe with negligible radiation contribution ($\Omega_K = \Omega_\gamma = 0$). The integral term has no simple analytical solution, but, in the assumption of $\Omega_m = 0.307$ and $\Omega_\Lambda = 0.693$, taking $H_0 = 70\,km\,s^{-1}\,Mpc^{-1}$, it yields $D_L = 6580\,Mpc$.

Using a total absolute magnitude (i.e., integrated over all the emitted frequencies) of $M = -19$, we can introduce the luminosity distance in the distance modulus:

$$m - M = 5\log\left(\frac{D_L}{1\text{pc}}\right) - 5. \tag{8.43}$$

The equation above solves in $m \simeq 21.8$, a quite nice target for HST or large ground based telescopes.

16. *Flatness of the Early Universe (16)*
   The present experimental data indicate a value for a total energy density of the Universe compatible with one within a few per mil. How close to one should have been at the scale of the electroweak symmetry breaking.

We can use the definition

$$\Omega = \frac{8\pi G}{3H^2}\rho \tag{8.44}$$

to recast the first Friedmann equation as:

$$\rho a^2\left(\frac{1}{\Omega} - 1\right) = -\frac{3Kc^2}{8\pi G} = \text{const.} \tag{8.45}$$

Therefore the energy density Ω evolves in time as:

$$\Omega - 1 \sim \frac{1}{\rho a^2} \sim \begin{cases} a & \text{(Matter domination)} \\ a^2 & \text{(Radiation domination)} \end{cases}, \tag{8.46}$$

where we have taken into account that $\Omega - 1 \ll 1$. We know that at the epoch of nucleosynthesis, $\Omega_n - 1 \lesssim 10^{-12}$. Therefore we can follow the evolution of Ω in the radiation dominated era from nucleosynthesis back to electroweak scale using:

$$\Omega_{EW} - 1 = (\Omega_n - 1)a^2 = (\Omega_n - 1)(1 + z)^{-2}, \tag{8.47}$$

where z is the redshift interval between electroweak and nucleosynthesis epochs, which is around 10^{18}. Then finally:

$$\Omega_{EW} - 1 \lesssim 10^{-30}. \tag{8.48}$$

17. *Virial theorem.*
   A cluster of galaxies, called Abell 2715 (at a redshift $\simeq 0.114$), contains about 200 galaxies, each the mass of the Milky Way. The average distance of the galaxies from the center of the cluster is 1 Mpc. If Abell 2715 is a virialized system, what is the approximate average velocity of the galaxies with respect to the center? The Milky Way galaxy has a mass of about 2.0×10^{42} kg.

If Abell 2715 is a virialized system, then its potential energy E_p will be twice its entire kinetic energy E_k. If it has 200 galaxies, each of mass 2×10^{42} kg, then the mass of the entire cluster is $200 \times 2 \times 10^{42} = 4 \times 10^{44}$ kg. So on average, each galaxy will have potential and kinetic energies

$$\langle E_p \rangle = \frac{GMm}{r} = \frac{(6.67 \times 10^{-11}) \times (4 \times 10^{44}) \times (2 \times 10^{42})}{1 \text{Mpc} \times (3.1 \times 10^{22} \text{m/kpc})}$$

$$\langle E_k \rangle = \frac{mv^2}{2} = \frac{2 \times 10^{42}}{2} v^2$$

where v is the average velocity of the galaxies in this cluster. For the entire cluster, we can use the virial theorem to solve for v.

$$[200 \langle E_p \rangle] = 2 \times [200 \times \langle E_k \rangle$$

$$200 \times 1.7 \times 10^{52} \simeq 2 \times 200 \times (10^{42} \times v^2)$$

$$v \simeq \sqrt{0.85 \times 10^{10}} \simeq 9.2 \times 10^5 \text{m/s} \simeq 920 \text{ km/s}.$$

18. **M/L.** At $r = 10^5$ light-years from the center of a galaxy the measurement yields $v_{meas} = 225$ km/s while the expected velocity calculated from the luminous mass is of $v_{calc} = 15$ km/s. Calculate the visible and the true galaxy mass, and the ratio M/L between the total and the luminous masses. How high is the average dark matter mass density?

We want to measure the *visible* and the *real* mass of a galaxy. The *visible* mass is the mass of the luminous matter in the galaxy: stars and gas. The *real* mass includes also the non-luminous matter which we can not see directly.

We can consider, for simplicity, a spherical symmetry of the system: we assume that the spiral galaxy (like our Galaxy) is a flattened disk of stars on almost circular orbits.

First of all we derive the velocity $v(R)$ of a body, like a star, that rotates at a distance R from the center of the system. Equating the equations of the centripetal acceleration of a star and that of the gravitational acceleration we have the circular rotation velocity:

$$v(R) = \sqrt{\frac{G \, M(R)}{R}} \tag{8.49}$$

where $M(R)$ is the mass interior to the radius R and G is the gravitational constant. From this formula we obtain the mass enclosed within the radius R:

$$M(R) = \frac{v^2(R) \, R}{G}. \tag{8.50}$$

Now, we can find the *true* and the *visible* mass of the galaxy. The data that we have are:

- $R = 10^5$ ly $\simeq 9.461 \times 10^{20}$ m[1]: distance from the center of the galaxy:
- $v_{meas} = 225$ km/s $= 225 \times 10^3$ m/s: the *true* velocity ($\simeq$ star + gas + dark matter);
- $v_{calc} = 15$ km/s $= 15 \times 10^3$ m/s: the *visible* velocity, calculated from the luminous mass ($\simeq$ star + gas).

True galaxy mass.
Use the values:

$$M(R)_{meas} \simeq \frac{(225 \times 10^3 \text{ m/s})^2 \times 9.461 \times 10^{15} \text{ m}}{6.67 \times 10^{-11} \text{ m}^3\text{kg}^{-1}\text{s}^{-2}} \simeq 7.18 \times 10^{41} \text{ kg}. \quad (8.51)$$

Visible galaxy mass. Use the values:

$$M(R)_{calc} \simeq \frac{(15 \times 10^3 \text{ m/s})^2 \times 9.461 \times 10^{15} \text{ m}}{6.67 \times 10^{-11} \text{ m}^3\text{kg}^{-1}\text{s}^{-2}} \simeq 3.19 \times 10^{39} \text{ kg}. \quad (8.52)$$

Ratio M/L. Now we can calculate the mass-luminosity ratio, that is the ratio between the total and the luminous masses. So, for the total mass we consider the *true* galaxy mass and for the luminous mass we use the *visible* galaxy mass:

$$\frac{M}{L} = \frac{7.18 \times 10^{41} \text{ kg}}{3.19 \times 10^{39} \text{ kg}} \simeq 225. \quad (8.53)$$

We can say that the total mass of the galaxy is about 225 times the mass of the visible mass.

Dark Matter mass density

Now we have all the ingredients to calculate the mass density of the Dark Matter (DM). The formula that we need is:

$$\rho_{DM} = \frac{M_{DM}}{V_{galaxy}} \quad (8.54)$$

where M_{DM} is the mass of the Dark Matter and V_{galaxy} is the volume of the entire galaxy. In our case the mass of the DM is the mass of the non-luminous matter, therefore is the difference between the *true* galaxy mass ($\sim$ total mass of the galaxy) and the mass of the *visible* matter.

$$M_{DM} = M_{meas} - M_{calc} \simeq 7.15 \times 10^{41} \text{ kg} \; (\simeq 1.28 \times 10^{69}\text{GeV}/c^2) \quad (8.55)$$

---

[1] 1 light year $\simeq 9.461 \times 10^{15}$ m $\simeq 0.3066$ pc.

Remembering that $1 \text{ eV}/c^2 = 1.79 \times 10^{-36}$ kg, this is the mass of about 3.6×10^{11} stars like the Sun ($M_{Sun} \simeq 1.89 \times 10^{30}$ kg).

Now we can calculate the volume of the galaxy, always considering a spherical symmetry of the system. The radius of the Milky Way (MW) is about $R_{MW} \simeq 5 \times 10^{20}$ m. Thus:

$$V_{MW} = \frac{4}{3}\pi R_{MW}^3 \simeq 3.93 \times 10^{62} \text{m}^3 (\simeq 3.93 \times 10^{68} \text{cm}^3) \qquad (8.56)$$

And the mass density of the DM is:

$$\rho_{DM} = \frac{M_{DM}}{V_{MW}} \simeq 1.82 \times 10^{-21} \text{kg/m}^3 \ (\simeq 3.26 \times 10^6 \text{ GeV}/c^2 \text{ m}^{-3} \simeq 3.26 \text{ GeV}/c^2 \text{ cm}^{-3})$$

$$(8.57)$$

The average density of dark matter in our Galaxy is about 1 proton-mass for every cubic centimeter considering that the mass of a proton is 1.67×10^{-27} kg or 0.938 GeV/c^2.

19. *WIMP-WIMP annihilation prefers the production of heavy fermions*
Demonstrate that, in the reaction $\chi\chi \to f\bar{f}$ (where χ is a generic WIMP with momentum $p_\chi \ll m_\chi$), if the annihilation state has spin 0 then

$$\mathcal{M} \propto \left(1 - \frac{|p|}{E + m_f}\right),$$

where E, p are respectively the energy and the momentum of the fermion produced in the decay. Use the above relation to compute, for a WIMP of mass $m_\chi = 50$ GeV, the ratios of the branching fractions into $b\bar{b}$ and into $\tau\tau$, in the hypothesis that:

a. the WIMP has spin 0;
b. the WIMP has spin 1/2 (note: you do not know the dynamics, i.e., which particle is the mediator of the annihilation).

In its center-of-mass system, the annihilation state cannot conserve angular momentum unless the helicity of one of the decaying flavors f (τ or b) is reversed – otherwise, angular momentum of the decaying state would be 1. The probability for such a "spin flip" is:

$$\mathcal{M}^2 \propto \left(1 - \frac{|p|}{E + m_f}\right)^2 \simeq \left(\frac{m_f}{2m_\chi + m_f}\right)^2$$

(higher for a higher mass of the particles in the final state).

a. Since the phase space for the two decay states is comparable ($E = 2m_\chi \gg m_b, m_\tau$),

$$\frac{\Gamma_{\chi\chi} \to b\bar{b}}{\Gamma_{\chi\chi} \to \tau^+\tau^-} \simeq \left(\frac{m_b}{m_\tau}\right)^2 \left(\frac{m_\tau + 2m_\chi}{m_b + 2m_\chi}\right)^2 \sim 9 \, .$$

b. If the WIMP has spin 1/2, let us call p the probability to have a final state with spin ± 1, while the probability to have a spin 0 is $(1 - p)$. In the first case, the $\tau\tau$ final state will dominate; in the second case, as we have seen, the probability of the $\tau\tau$ final state will be $\sim 1/9$ of the probability of the $b\bar{b}$ final state.

Since we do not know what is the exchange particle, nor what is the dynamics (V/A) of the interaction, we guess, counting the number of states, $p \sim 0.5$. Thus

$$\frac{\Gamma_{\chi\chi} \to b\bar{b}}{\Gamma_{\chi\chi} \to \tau^+\tau^-} \sim 9 \times (1 - p) \sim 4.5 \, .$$

20. *WIMP "miracle" (17)*
Show that a possible Weak Interacting Massive Particle (WIMP) with a mass of the order of m_χ 100 GeV would have the relic density needed to be the cosmic dark matter (this is the so-called WIMP "miracle").

As we are talking about cold Dark Matter, the appropriate asymptotic form for the equilibrium number density of particles χ with mass $m_\chi \simeq 100\,\text{GeV}$ is the nonrelativistic limit:

$$n \sim (m_\chi T)^{3/2} \exp\left(-m_\chi / T\right) \, . \tag{8.58}$$

Decoupling happens when $\Gamma \sim n\sigma \sim H$, where H is the Hubble constant, which, using the Friedmann equations, can be rewritten as $H \sim T^2/M_{pl}$ with $M_{pl} = (8\pi G)^{-1/2}$. Using this information we can recast the freeze-out condition as:

$$x^{-1/2} e^{-x} \sim \frac{1}{M_{pl} \, \sigma \, m_\chi} \, , \tag{8.59}$$

where $x = m_\chi / T$. Numerical values can be now inserted noting that for a weak interacting particle $\sigma \simeq G_F^2 m_\chi^2 \simeq 10^{-8}\,\text{GeV}^{-2}$. Solving numerically the previous equation in the range $10^{-10} - 10^{-20}$, one finds x in the range $20 - 50$. Now, we know that:

$$\Omega_\chi = \frac{\rho_\chi}{\rho_c} = \frac{m_\chi n}{\rho_c} = \frac{m_\chi}{\rho_c} \frac{n}{T_0^3} T_0^3 \, , \tag{8.60}$$

with $T_0 = 2.75\,\text{K} \simeq 10^{-4}\,\text{eV}$. After decoupling annihilations cease and the density of dark matter particles will just decrease with a^{-3}, exactly like temperature does, so the ratio n/T^3 remains constant and we can take its value at freeze-out. In particular, using $n\sigma \sim H$, we can write $n \simeq T^2/M_{pl}\sigma$ and finally:

$$\Omega_\chi = \left(\frac{T_0^3}{\rho_c M_{pl}}\right)\frac{x}{\sigma} \simeq \frac{x}{40}\left(\frac{10^{-8}\,\text{GeV}^{-2}}{\sigma}\right), \tag{8.61}$$

which gives the right relic density for dark matter $\Omega_\chi \simeq 0.2$, given that $\sigma \simeq 10^{-8}\,\text{GeV}^{-2}$ and $x \simeq 20$.

21. *Optimal mass of the target nucleus in a direct dark matter detection experiment (18)*
    Elastic scattering in a WIMP-nucleon interaction occurs in the non-relativistic limit, being the relative speed between the two particles of the order of 100 km/s.

    a) Justify why this approximation holds.
    b) Using conservation of energy and momentum, derive an expression for the energy of the recoil nucleus, E_R, in terms of the scattering angle θ. Write the equation in terms of the ratio $mM/(m+M)^2$, where m is the mass of the WIMP and M is the mass of the target nucleus.

The estimate of the velocity of a WIMP as $\sim$100 km/s comes from the typical escape velocity of a galaxy. In addition, since we consider cold dark matter, we expect this velocity to be not relativistic.
The recoil energy can be evaluated from the kinematics of the $1+2 \rightarrow 1'+2'$ scattering in the non-relativistic limit. In the LAB system we have:

$$p_1 = (E_1, \vec{p}_1), \quad p_2 = (M, \vec{0}), \quad p_1' = (E_1', \vec{p}_1'), \quad p_2' = (E_2', \vec{p}_2').$$

In the CM system we have:

$$p_1^* = (E_1^*, \vec{p}_1^*), \quad p_2^* = (E_2^*, -\vec{p}_1^*), \quad p_1'^* = (E_1'^*, \vec{p}_1'^*), \quad p_2'^* = (E_2'^*, \vec{p}_1'^*).$$

Energy conservation, combined with the fact that $m_{1,2} = m_{1,2}'$, gives $p^* \equiv |\vec{p}'^*| = |\vec{p}^*|$.
The invariant Mandelstam variable t is defined as $t = (p_1'^* - p_1^*)^2 = (p_1' - p_1)^2$.
In the CM:

$$t_{CM} = (E_1^* - E_1'^*)^2 - (\vec{p}_1'^* - \vec{p}_1^*)^2 = -(\vec{p}_1'^* - \vec{p}_1^*)^2 = -2(p^*)^2 + 2(p^*)^2\cos\theta$$
$$= -2(p^*)^2(1 - \cos\theta).$$

In the LAB:

$$t_{LAB} = (p_1' - p_1)^2 = (p_2' - p_2)^2 = (E_2' - E_2)^2 - (\vec{p}_2' - \vec{p}_2)^2 = (E_2')^2 + (E_2)^2$$
$$- 2E_2'E_2 - |\vec{p}_2'|^2 = 2M^2 - 2E_2'M = 2M(M - E_2') = 2M(E_1' - E_1).$$

Hence:

$$\Delta E_1 = -\frac{2(p^*)^2(1 - \cos\theta)}{2M} = \frac{t}{2M}.$$

In the non-relativistic limit $\Delta E_1 \to -E_R$, where E_R is the recoil energy of the nucleus.

In the CM system and in the non-relativistic limit:

$$p^* = mv^*, \quad v^* = v - V_{CM}, \quad V_{CM} = \frac{mv}{m + M},$$

where v is the velocity of the neutrino in the LAB system, v^* is the velocity in the CM system and V_{CM} is the velocity of the Center of mass in the LAB. Hence:

$$v^* = v - \frac{m}{m + M}v = \frac{mM}{m + M}\frac{v}{m} \Rightarrow p^* = \frac{Mm}{m + M}v \equiv \mu v.$$

Thus we obtain the formula for the recoil energy of the nucleus:

$$E_R = \frac{2\mu^2 v^2(1 - \cos\theta)}{2M} = \frac{mM}{(m + M)^2}mv^2(1 - \cos\theta).$$

22. *Equivalence principle*
A mass m is orbiting around another mass M, with $m \ll M$. For an observer placed at a great distance, m is accelerated by the gravitational field of M and emits gravitational radiation. According to an observer in free fall with m, in a small region around m the gravitational field is canceled: the mass m does not accelerate and therefore does not emit gravitational radiation. Explain this apparent paradox.

We can give a qualitative interpretation of the apparent paradox, though the exact demonstration would prove extremely challenging. The paradox is solved by considering that the free-falling observer solidal with m would necessarily assign an acceleration to M and associate it with the emission of gravitational radiation. From the point of view of the faraway observed, the combined motion of m, M and free-falling observer places the last element in a condition of stationary waves from m.

23. *Age of things in the Universe*
The age of the Universe (that is, the time since the Big Bang) is about 14 billion years. The age of the Solar System is 4.56 billion years. Thus, the Solar System has existed for 1/3 of the age of the Universe. For what fraction of the total age of the Universe have the following things existed?

   a. Helium nuclei;
   b. Neutral atoms;

  c. Galaxies;
  d. Human kind on Earth.

-----

  a. Helium nuclei have been around since the time of primordial nucleosynthesis, at a time $\mathscr{T}_{He} \sim 400$ s after the Big Bang. The age of the Universe, expressed in seconds, is $t_0 \sim 4.4 \times 10^{17}$ s. The fraction of the age of the Universe during which helium nuclei have been around is thus

$$F = 1 - \mathscr{T}_{He}/t_0 \simeq 1 - 10^{-15}.$$

  b. Neutral atoms have been around since the Universe became transparent, at a time $\mathscr{T}_{He} \sim 350\,000$ years after the Big Bang. The fraction of the age of the Universe during which neutral atoms have been around is

$$F = 1 - \mathscr{T}_{trans}/t_0 \simeq 1 - 000025.$$

  Expressed as a percentage, this is about 99.9975% of the age of the Universe.
  c. Galaxies have been around since the Universe had an age $\mathscr{T}_{gal} \sim 5 \times 10^8$ years. The fraction of the age of the Universe during which galaxies have been present is
$$F = 1 - \mathscr{T}_{gal}/t_0 \simeq 1 - 0.04.$$

  Expressed as a percentage, this is about 96% of the age of the Universe.
  d. Human kind on Earth. The species Homo sapiens, which has been called "anatomically modern human", has evolved some $t_{human} \sim 300\,000$ years ago. Expressed as a percentage, $F = t_{human}/t_0$ is about 0.002% of the age of the Universe.

24. *What could we learn from detecting cosmic antideuterium?*
   Why would be the detection of antideuterium nuclei particularly interesting for dark matter search?

   Antideuterons, i.e., antideuterium nuclei, form when an antiproton and an antineutron merge together. The two antinucleons must be at rest with respect to each other in order for fusion to take place successfully, and for kinematic reasons, spallation reactions create very few low-energy particles. The rate of antinuclei in primordial and stellar nucleosynthesis is zero, hence the presence of antideuterons should be the indication of decays of heavier particles – for example, dark matter particles (also more exotic explanations however exist, like galaxies of antimatter).

25. *Curved surfaces and sum of the internal angles of a triangle*
   Show that on a positively curved surface (e.g. sphere) the sum of all angles of a triangle is between 180 and 900°.

By measuring the surface of a spherical triangle in units of square radius, there are obviously two limiting cases that any conceivable spherical triangle cannot attain: 0 and 4π (i.e. no triangle and the whole spherical surface). We can proceed as follows:

Null-surface triangle:

We can consider the limiting case of null-surface spherical triangle by taking the limit of the angle of an arbitrary vertex approaching 0. The geodesics cast on the other two vertices tend to be parallel and, therefore, define the internal angles of the two remaining vertices as the complementary angles of a single geodesic cast on their connecting geodesic. From the arbitrary choice of vertices, it follows that the sum of the internal angles cannot be lower than $\pi = 180°$.

Maximum surface triangle:

We can address this case using Girard's theorem, which states that *the area of a spherical triangle in units of the radius of the sphere equals the spherical excess*: $A = \sum_{i=1}^{3} \alpha_i - \pi$. We have that:

$$\sum_{i=1}^{3} \alpha_i - \pi < 4\pi \implies \sum_{i=1}^{3} \alpha_i < 10\frac{\pi}{2}, \tag{8.62}$$

which is exactly the maximum (unattainable) value of $900°$.

26. *Volume of a hypersphere*

Show that the volume of a 4-sphere with curvature radius r (i.e. the surface "area" of a four dimensional ball of radius r) is $2\pi^2 r^3$.

A n-sphere is a set of points lying at the same "distance" from a chosen reference. The particular symmetry implies two very useful properties:

i. The size of the (n-1) boundary to the n-volume within the sphere can be obtained simply by differentiating the n-volume with respect to its radius. In fact:

$$dV_n = S_n dr \implies S_n = \frac{dV_n}{dr}. \tag{8.63}$$

ii. Given an arbitrary axis along one of the n coordinates, the (n-1)-dimension slices obtained moving an orthogonal (n-1)-plane along the axis, between $-r$ and r, are (n-1)-spheres of radius $r\cos\theta$, where θ is the angle formed on the equatorial plane by the radius pointing to the border of the considered slice.

For the case of the 4-sphere, we can apply the above, starting from (ii) to estimate the 4-volume:

$$V_4 = 2 \int_0^{\pi/2} V_3 r \cos\theta \, d\theta$$
$$= 2 \int_0^{\pi/2} \left(\frac{4\pi}{3} r^3 \cos^3\theta \right) r \cos\theta \, d\theta \qquad (8.64)$$
$$= \frac{8\pi r^4}{3} \int_0^{\pi/2} \cos^4\theta \, d\theta.$$

Since it is:

$$\int_0^{\pi/2} \cos^4\theta \, d\theta = \left[\frac{\sin(4\theta)}{32} + \frac{\sin(2\theta)}{2} + \frac{3\theta}{8} \right]_0^{\pi/2} \qquad (8.65)$$

only the calculation at $\theta = \pi/2$ gives the non-zero contribution of $3\theta/8 = 3\pi/16$. Plugging this result in Eq. (8.64), we obtain:

$$V_4 = \frac{\pi^2}{2} r^4. \qquad (8.66)$$

Therefore, as a consequence of property (i), the requested boundary 4-surface is:

$$S_4 = \frac{dV_4}{dr} = 2\pi^2 r^3. \qquad (8.67)$$

27. *Evolution of the Hubble parameter and density parameters*
    Assume a Universe with only non-relativistic matter. Show that the Friedmann equation can be written as

$$H(z) = H_0 (1+z) \sqrt{1 + \Omega_0 x} \, .$$

If the only energy contribution is from non-relativistic matter, the expansion of the Universe simply dilutes the density of particles as $(a/a_0)^3$; $\Omega_m = \Omega_0 \neq 0$, while all other energy densities are zero.
Starting from the Friedmann equation

$$H^2 = H_0^2 \Omega_m (1+z)^3 - \frac{K}{a^2}$$

and to the fact that

$$\frac{K^2}{H_0^2 a_0^2} = \Omega_0 - 1$$

we obtain, using the equality $(1+z) = a_0/a$:

$$H^2 = H_0^2 (1+z)^2 (1 + \Omega_0 z)$$

and thus:

$$H(z) = H_0(1 + z)\sqrt{1 + \Omega_0 x}\,.$$

28. *Luminosity of galaxies and expansion of the Universe*
    Consider two galaxies, observable at present time, A and B, with the same emitting spectral energy distribution. Suppose at the moment of detection of light signals from them (now) the distances to them are such that $L_{det}^A < L_{det}^B$. In other words, if those galaxies had equal absolute luminosities, the galaxy B would seem to be dimmer. Is it possible for galaxy B (the dimmer one) to be closer to us at the moment of its signal's emission than galaxy A (the brighter one) at the moment of A's signal's emission?

    ───────────────

    Yes, it is possible. The corresponding spacetime diagram is shown on Fig. 8.1. World lines branch out radially in all directions from the big bang. Spatial slices of constant cosmic time are represented by spherical surfaces perpendicular to the world lines, while time is measured along the radial world lines. An arbitrary world line is chosen as the observer and labeled O. At some instant in time – let it be "now" – the observer's lightcone stretches out and back and intersects other world lines such as X and Y. Because of the expansion of space, the lightcone does not stretch out straight as in a static Universe, but contracts back into the big bang. All world lines and all backward lightcones converge into the big bang. Therefore, the extent of the spatial distance of the farthest galaxy B, at its emission time, can have been smaller than the corresponding distance to A, at its later emission time.

29. *The Standard Model of particle physics does not provide good dark matter candidates (10.19)*
    Name all particles which are described by the SM and write down through which force(s) they can interact. Why can we rule out that a dark matter particle does interact through the electromagnetic force? Why can we rule out that a dark matter particle does interact through the strong force? Now mark all particles which pass the above requirements and could account for dark matter, and comment.

    ───────────────

    • Name all particles which are described by the SM and write down through which force(s) they can interact.
    The elementary particles in the Standard Model are 6 quarks, 6 leptons and their corresponding antiparticles, 4 vector bosons and a scalar boson.
    Quarks (up, down, charm, strange, top, bottom) interact via strong interaction. Each quark and antiquark comes in three colors. The antiquarks carry the opposite baryon number (-1/3), electric charge and color charge.
    Leptons (electron, muon, tau, electron neutrino, muon neutrino, tau neutrino) do not interact via the strong interaction. Charged leptons interact via electromagnetic interaction while neutrinos interact weakly. The antileptons carry the opposite electric charge and lepton number.

Fig. 8.1 The reception and emission distances of galaxies X and Y. Although galaxy Y has a greater reception distance, its emission distance is smaller than that of X. Thus Y, which is now farther away than X, was closer to us than X at the time of emission (which is different for X and Y) of the light we now see

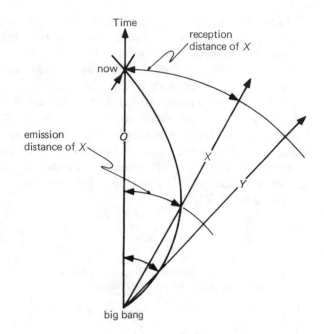

Bosons are either elementary, like photons or composite like mesons. Vector bosons (gluons, photons and the W and Z bosons) are the force mediators and the scalar Higgs boson is responsible for the intrinsic mass of particles.

• Why can we rule out that a dark matter particle does interact through the electromagnetic force? From rotation-curves and galaxy-cluster mass measurements we have the distribution of matter in these objects. The calculated mass exceeds the observed mass, so most of the mass is not observed. Dark matter does not emit, absorb or reflect light (hence dark) and is invisible to the entire electromagnetic spectrum. So we can rule out charged particles as DM candidates.

• Why can we rule out that a dark matter particle does interact through the strong force?

The elementary particles which interact strongly are the quarks. They are all charged (see Fig. 8.1) and interact also via electromagnetic interaction. Furthermore there are many evidences which suggest DM is non-baryonic, such as the fact that baryonic gas being visible if backlit by stars or the anisotropies in CMB which indicates most of total matter interacting with ordinary matter only via gravity. According to Big Bang nucleosynthesis studies, if there were more baryons, then there would be more helium, lithium and heavier elements synthesized during the Big Bang.

• Mark all particles which pass the above requirements and could account for dark matter, and comment.

The weakly interacting particles could account for dark matter. For example the neutrinos are quite abundant and dark. But since they have quite small masses they could not explain the total amount of DM expected.

There are several hypothetical candidates for DM, such as axions, weakly interacting massive particles (WIMPs), gravitationally-interacting massive particles (GIMPs), supersymmetric particles or mirror matter (Table 8.1). A particularly interesting case is the mirror matter, which consists of mirror particles with the same mass and charge as ordinary particles, but with opposite chirality (Table 8.2). The interaction between mirror and ordinary matter occurs mainly through gravity, although kinetic mixing of neutral particles such as photons is allowed and these interactions can only be very weak. Therefore they have been suggested as DM candidates (Table 8.3).

Table 8.1 Three generations of quarks, classified by charge (Q) and the 6 flavor numbers (D, U, S, C, B and T). Antiquarks have opposite signs

quark	Q	D	U	S	C	B	T
d	$-\frac{1}{3}$	-1	0	0	0	0	0
u	$+\frac{2}{3}$	0	1	0	0	0	0
s	$-\frac{1}{3}$	0	0	-1	0	0	0
c	$+\frac{2}{3}$	0	0	0	1	0	0
b	$-\frac{1}{3}$	0	0	0	0	-1	0
t	$+\frac{2}{3}$	0	0	0	0	0	1

Table 8.2 3 generations of leptons, classified by charge (Q), electron number (L_e), muon number (L_μ) and tau number (L_μ). Antileptons have opposite signs

lepton	Q	L_e	L_μ	L_τ
e	-1	1	0	0
ν_e	0	1	0	0
μ	-1	0	1	0
ν_μ	0	0	1	0
τ	-1	0	0	1
ν_τ	0	0	0	1

Table 8.3 Gauge bosons and forces

Gauge bosons		Spin	Charge	Mass	Force
photon	γ	1	0	0	electromagnetic
W−boson	$W^\pm$	1	± 1	80.4 GeV/c^2	weak
Z−boson	Z^0	1	0	91.2 GeV/c^2	weak
gluon	g	1	0	0	strong

Chapter 9
The Properties of Neutrinos

1. *Kinematics of the beta decay (see also Chap. 2).* The study of the electron energy spectrum in the β decays processes was at the origin of the proposal of the neutrino existence and, nowadays, is one of the best ways to measure the neutrino mass. Considering a neutron decaying at rest,

   a) Compute the electron energy spectrum admitting an hypothetical two body neutron decay ($n \to pe^-$).
   b) Supposing now the real beta decay process ($n \to pe^-\bar{\nu}_e$):
      i. compute the minimal and maximal energy that the electron produced in the decay may have in case of a massless neutrino.
      ii. compute again the minimal and maximal energy that the electron in the decay may have, but now considering a neutrino mass of 0.02 eV.

   _____

   a) It can be shown that in the CM frame, the energy of one of the particles emerging from a two-body decay (check the PDG or look into Chap. 2 of the textbook) is given by

   $$E_1^* = \frac{s + m_1^2 - m_2^2}{2\sqrt{s}}. \tag{9.1}$$

   In this particular case, let us call E_1^* the energy carried by the electron, m_1 and m_2 the mass of the electron and the proton, respectively. The center-of-mass energy, $\sqrt{s}$, is the mass of the decaying neutron. Plugging these values into Eq. 9.1, we obtain

   $$E_e^* = \frac{M_n^2 + m_e^2 - m_p^2}{2M_n} \simeq 1.6 \, \text{MeV}. \tag{9.2}$$

© Springer Nature Switzerland AG 2021
A. De Angelis et al., *Particle and Astroparticle Physics*, Undergraduate Lecture Notes in Physics,
https://doi.org/10.1007/978-3-030-73116-8_9

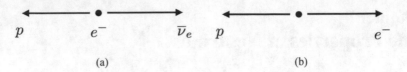

p e^- $\bar{\nu}_e$ p e^-

(a)   (b)

Fig. 9.1 Kinematic configuration for minimum (a) and maximum (b) energy of an electron emerging from a neutron decay. The arrows represent the particle's momentum. The arrow of the neutrino in (b) is not represented as we are taking the limit $P \to 0$

     Notice that the energy spectrum of the electron in the CM reference frame would be monochromatic.

b)   i. For massless neutrinos and in the neutron rest frame it is relatively easy to see that the kinematic configuration that minimizes the electron energy is the one shown in Fig. 9.1a, i.e., when the electron is produced at rest and the proton and the neutrino have the same absolute values of the momentum.

The kinematic configuration for the maximum electron energy is the one shown in Fig. 9.1b. We aim at reaching the highest possible momentum, and the way to achieve this is to give all the energy to the massive particles, i.e., put $E_{\bar{\nu}} \to 0$. In this configuration, we can approximate the whole 3-body system to a 2-body system (consider only the proton and the electron) and Eq. 9.1 can be used to compute the maximum energy that the electron can acquire.

In short, $E_e^*|_{\min} = 511 \, \text{keV}$ and $E_e^*|_{\max} = 1.6 \, \text{MeV}$. The electron maximum energy could also be computed using the so called *Dalitz plot*, discussed in Chap. 2 (problem 2.19) of this book, in particular making use of Eq. 2.73.

ii. Taking the neutrino mass to be $m_{\bar{\nu}} = 20 \, \text{meV}$ we have now that the minimum energy of the electron continues to be when it is produced at rest. The maximum momentum can be computed using Eq. 2.73 giving,

$$|\mathbf{p}_e^*|_{\max} = \frac{\sqrt{(M_n^2 - (m_p + m_{\bar{\nu}} + m_e)^2)(M_n^2 - (m_p + m_{\bar{\nu}} - m_e)^2)}}{2M_n} \simeq 1.515 \, \text{MeV}/c \tag{9.3}$$

which in turn lead to a maximum electron energy of

$$E_e^*|_{\max} = \sqrt{(|\mathbf{p}_e^*|_{\max})^2 + m_e^2} \simeq 1.6 \, \text{MeV}. \tag{9.4}$$

As expected the contribution due to the mass of the neutrino to the maximum energy reach by the electron is extremely small. In fact, $E_e(m_{\bar{\nu}} = 0) - E_e(m_{\bar{\nu}} = 20 \, \text{meV}) \sim 10^{-2} \, \text{eV}$.

2. *Neutrino interaction cross-section (1)*. The cross-section in the (anti)neutrino electron interaction cross section has a peak for an (anti)neutrino energy around 10^{16} eV. Explain why.

The s-channel excitation of a W^- boson on-shell, commonly known as the Glashow resonance, can be initiated by the electron antineutrino hitting an electron at a laboratory energy around 6.3 PeV.

If we call x the direction of motion of the (anti)neutrino, the energy-momentum 4-vector of the system constituted by the (anti)neutrino and by an electron at rest can be written, neglecting the neutrino mass, as

$$(E_\nu^2 + m_e, \ E_\nu^2, 0, 0)$$

(we omitted the bar over the antineutrino symbol). The square of the invariant mass of the system is thus

$$m_{inv}^2 = (E_\nu^2 + m_e)^2 - E_\nu^2 \simeq 2E_\nu m_e .$$

The energy at which one can produce a W boson on shell is

$$E_\nu \simeq \frac{m_W^2}{2m_e} \simeq 6.3\,\text{PeV} ,$$

and this corresponds to a peak in the cross section.

3. *Neutrinos from the Sun (2)*. Neutrinos from the Sun come mostly from reactions which can be simplified into

$$4p \to{}^4 \text{He} + 2e^+ + 2\nu_e .$$

The energy gain per reaction corresponds to the binding energy of He, $\sim$28.3 MeV. The power of the Sun at Earth (nominal solar constant) is $P = 1361\,\text{W/m}^2$. How many solar neutrinos arrive at Earth per square meter per second?

The proton-proton fusion reactions responsible for the production of the neutrinos are also the main energy source of the Sun. Therefore, the number of reactions that corresponds to the energy that arrives at Earth per second are just

$$n = \frac{1361 \times 6.24 \times 10^{12}\,\text{MeV}}{28.3\,\text{MeV}} \sim 3 \times 10^{14},$$

where the factor 6.24×10^{12} comes from the conversion of joule to MeV. The number of neutrinos arriving to Earth is then of the order of

$$n_\nu = 2 \times n \sim 6 \times 10^{14}\,\text{m}^{-2}\,\text{s}^{-1}.$$

4. *Radiation exposure due to solar neutrinos (3).* Assuming a neutrino-nucleon cross section, in the energy range for solar neutrinos, approximately of 10^{-45} cm^2/nucleon, and a human body with a mass of 80 kg:

a. Compute the rate of interactions of solar neutrinos in the human body.
b. Estimate the annual dose received by the human body under the assumption that the radiation damage is caused mainly by electrons produced in the reaction $\nu N \rightarrow e N'$, where on average 50% of the neutrino energy is transferred to the electron, and that the average energy of neutrinos is 100 keV.

---

a. The number of nucleons in a human body with a mass of 80 kg is:

$$n_N = \frac{80 \, \text{kg}}{m_N}, \tag{9.5}$$

where $m_N \sim 1.67 \times 10^{-27}$ kg is the mass of one nucleon. Then

$$n_N = \frac{80 \, \text{kg}}{1.67 \times 10^{-27} \, \text{kg}} \sim 4.8 \times 10^{28}. \tag{9.6}$$

The number of neutrinos arriving to Earth that were produced at the Sun as ν_e is of the order of (see the previous exercise):

$$n_\nu \sim 6 \times 10^{14} \, \text{m}^{-2} \, \text{s}^{-1}.$$

In fact, due to neutrino oscillations, just a fraction of these neutrinos will arrive at Earth as ν_e. This fraction has been experimentally measured in several experiments and is a function of the neutrino energy (see for example Chap. 9 of the textbook). For the purpose of this exercise, where orders of magnitude are just asked, these oscillations will not be considered.

Taking 1 m^2 as the transverse human body area (as the neutrino cross section is very small, there is no shadow effect in the target and the final result will not depend on this assumption), the rate of solar neutrinos interactions in the human body is given by:

$$W_{int} = \sigma \times N_T \times J,$$

where $\sigma = 10^{-49}$ m^2/nucleon, $N_T = n_N$ and $J = n_\nu$. Then:

$$W_{int} \sim 3 \times 10^{-6} \, \text{s}^{-1}.$$

b. The energy absorbed by a human body in one year (annual dose, D) due to the interaction of solar neutrinos is, taking into account the fluxes estimated in the previous item, the data given in the exercise statement, the conversion factor between keV and joule (1 keV $\sim 1.6 \times 10^{-16}$ J) and the number of seconds in one year ($\sim \pi \times 10^7$ s):

$$D \simeq 0.5 \times 100 \times (1.6 \times 10^{-16}) \times 3 \times 10^{-6} \times (\pi\, 10^7)\, \text{J/yr} \simeq 7.5 \times 10^{-13}\, \text{J/yr}.$$

For electrons the radiation weighting factor to convert dose (D) in equivalent dose (H) taking into account the relative biological effectiveness of the radiation, is according to the usual regulations taken as 1. Then in sievert (joule per kg) the equivalent dose is

$$H \sim 10^{-14}\, \text{Sv},$$

which is, by far, much less than the average equivalent dose in one year of the background radiation from natural sources at the Earth surface ($\sim$ few mSv).

5. *Solar neutrino detection in SNO (see also Chap. 2, 4, 6).* The Sudbury Neutrino Observatory (SNO) was able to observe neutrino flavor oscillations by measuring the different possible interactions of the solar neutrinos with heavy water. One of these interactions is the elastic scattering of the neutrinos with the electrons present in the water,

$$\nu_X + e^- \rightarrow \nu_X + e^-$$

where ν_X can be a neutrino of an electron, muon or tau. Assuming an energy for the neutrino of 10 MeV, neglecting its mass and knowing the following properties of the heavy water: $n = 1.33$ and $\rho_{D_2O} = 1.11\,\text{g cm}^{-3}$,

a. Determine the energy of the neutrino and the electron in the center-of-mass (CM) reference frame, before and after the reaction.
b. Compute the velocity, β of the center-of-mass (CM) reference frame.
c. Calculate the maximum angle that the electron can have with respect to the direction of the incoming neutrino.
d. Apart from the neutrino elastic scattering with the electrons, the SNO experiment also uses the interaction of the neutrinos with the deuterium $(D = p\,n)$, present in the heavy water to detect them. Knowing that the area of detection is composed by 1000 tons of heavy water (D_2O), surrounded by PMTs, and that the neutrino interaction cross-section with nucleons is $\sigma_{\nu N} \sim 10^{-42}\,\text{cm}^2$, estimate the expected number of neutrinos to be detected per day. Consider a neutrino solar flux of $\sim 5 \times 10^6\,\text{cm}^{-2}\,\text{s}^{-1}$ and an experimental detection efficiency of $\sim 50\%$.
e. Assuming that the scattered electron travels 5 cm before stopping, estimate the number of photons emitted by Cherenkov radiation.

a. The 4-momentum vector before the interaction in the LAB reference frame can be written as

$$P_\mu^i = (E_\nu + m_e, E_\nu),\tag{9.7}$$

where E_ν is the neutrino energy in the LAB and it was used $E_\nu \simeq P_\nu$ as we are neglecting the neutrino mass. Computing the Lorentz invariant quantity, s, we get

$$s = P_\mu^i P^{\mu,i} = (E_\nu + m_e)^2 - E_\nu^2 = 2E_\nu m_e + m_e^2.\tag{9.8}$$

In the interaction center-of-mass (CM) reference frame, immediately after the collision we can write

$$P_\mu^f = (E_e' + P', \mathbf{0}),\tag{9.9}$$

where again it was considered $E_\nu \simeq P_\nu'$ and $P' \equiv P_e' = P_\nu'$. With this s can be written as

$$s = (E_e' + P')^2 = 2(P')^2 + m_e^2 + 2E_e'P',$$

using $E_e'^2 = P'^2 + m_e^2$. After some algebraic manipulation we can finally reach an expression for the electron (and neutrino) momentum as a function of s and the electron mass:

$$P' = \frac{s - m_e^2}{2\sqrt{s}}.\tag{9.10}$$

Due to energy-momentum conservation and s Lorentz invariance the s in the previous equation is equal to the one computed in 9.8 leading to

$$P' = \frac{E_\nu m_e}{\sqrt{m_e^2 + 2E_\nu m_e}} \simeq 1.6\,\text{MeV}.\tag{9.11}$$

Hence, the energy of the neutrino, E_ν', and of the electron, E_e', in the CM frame, are

$$E_\nu' \quad = P' \simeq 1.6\,\text{MeV}$$
$$E_e' = \sqrt{P'^2 + m_e^2} \simeq 1.7\,\text{MeV}.$$

b. From the Lorentz transformations we have

$$\underbrace{\begin{pmatrix} E^* \\ P^* \end{pmatrix}}_{CM} = \begin{pmatrix} \gamma & -\gamma\beta \\ -\gamma\beta & \gamma \end{pmatrix} \underbrace{\begin{pmatrix} E \\ P \end{pmatrix}}_{LAB}\tag{9.12}$$

where E and E^* is the electron energy in the LAB reference frame work and in the CM, respectively. From this matrix we can derive the following system of equations for the electron, where in the second equation it was used the fact that the electron is at rest in the LAB frame, i.e., $P_e = 0$,

$$\begin{cases} E^* = \gamma(E - \beta P) \\ P^* = \gamma(-\beta E + P) \end{cases} \Leftrightarrow \begin{cases} E^* = \gamma E \\ P^* = -\gamma\beta E \end{cases}. \tag{9.13}$$

As such,

$$\frac{E^*}{P^*} = -\frac{1}{\beta} \Leftrightarrow \beta = -\frac{P^*}{E^*} \simeq 0.95. \tag{9.14}$$

c. The maximum angle that the electron can make with the incoming neutrino direction (boost direction) occurs when, in the CM reference frame, the electron is emitted with $\theta^* = 90°$ with respect to the boost direction. Therefore the angle of emission in the LAB will be given by

$$\tan\theta = \frac{P_T}{P_l}. \tag{9.15}$$

In the LAB the transverse momentum, P_T, is equal to the momentum carried by the electron in the CM, P_e^* and the longitudinal momentum can be obtained through the Lorentz transformation

$$P_l = \gamma\beta E^* + \gamma P^* \underbrace{\cos\theta^*}_{0} = \gamma\beta E^*. \tag{9.16}$$

From the previous problem we know β and consequently we can compute $\gamma = (\sqrt{1 - \beta^2})^{-1} = 3.25$. So

$$\theta = \arctan\left(\frac{1.6}{3.25 \times 0.95 \times 1.7}\right) \simeq 17°. \tag{9.17}$$

d. First, let us start to calculate the number of targets. Taking the mass of protons and neutrons to be nearly the same, $m_p \sim m_n$, then $D_2O \rightarrow 4 + 16 = 20$ nucleons from which 4 are in the deuterium. Hence, only $1/5$ of the total D_2O volume will interact with the neutrino. Moreover, the number of mol which could participate in the interaction is $N_{mol} = 10^9$ g $\times N_A/20$ g, where N_A is the Avogadro's number.
The number of targets is then

$$N_{target} = \frac{1}{5} \times 20 \times N_{mol} = 1.2 \times 10^{32}. \tag{9.18}$$

The number of interactions is

$$N_{\text{int}} = \sigma_{\nu N} \, \phi \, N_{\text{target}} \, \varepsilon$$
$$= 10^{-42} \times 10^6 \times 1.2 \times 10^{32} \times 0.5 = 3 \times 10^{-4} \, \text{s}^{-1}$$

where ϕ is the neutrino flux and ε the detection efficiency; per day we have

$$N_{\text{int}} = 3 \times 10^{-4}(60 \times 60 \times 24) = 25.92 \, /\text{day}. \qquad (9.19)$$

e. The number of Cherenkov photons produced can be approximated by

$$\frac{d^2 N}{d E d X} \simeq 370 \, \sin^2 \theta_C \, \text{photon} \, \text{eV}^{-1} \, \text{cm}^{-1}. \qquad (9.20)$$

Integrating the above equation over the distance crossed by the particle we get

$$\frac{d N}{d E} \simeq 370 \, \sin^2 \theta_C \Delta x \, \text{photon} \, \text{eV}^{-1}, \qquad (9.21)$$

with $\Delta x = 5 \, \text{cm}$. Attending to the fact the minimum emission angle for the electron with the neutrino incoming direction is $0°$ and the maximum angle is $17°$ (see previous problem), one can take as average angle zero. For this configuration ($\theta^* = 0°$) we have that $E_e = \gamma (E_e^* + \beta P_e^* \cos \theta^*) = 10.465 \, \text{MeV}$ and

$$\beta = \sqrt{1 - \left(\frac{m_e}{E_e} \right)^2} \simeq 0.9998, \qquad (9.22)$$

leading to

$$\theta_C = \arccos \left(\frac{1}{\beta n} \right) \simeq 41.23°. \qquad (9.23)$$

Hence, the number of photons as a function of its energy is

$$\frac{d N}{d E} \simeq 804 \, \text{photons} \, \text{eV}^{-1}. \qquad (9.24)$$

6. *Neutrino interaction diagrams (see also Chap. 6).* Solar neutrinos interact with the nucleus of the heavy water ($D_2 O$ with $D = pn$) through the following channels: charged current (CC), neutral current (NC). Additionally, it can undergo elastic scattering (ES) with the electrons present in the heavy water.

a. Draw all the possible diagrams, at the tree level, for the following channels:
    i. Charged current (CC)
    ii. Neutral current (NC)
    iii. Elastic scattering (ES).

b. The charged current interaction is described, at quark level, by the process,

$$\nu_e(p_1) + d(p_2) \rightarrow e^-(p_3) + u(p_4).$$

Draw the possible Feynman diagram(s), at the tree level for this process.

c. Write down the amplitude $\mathcal{M}$ for this process.

d. Assuming that the energy is such that you can neglect the lepton and quark masses, but not the gauge boson masses, simplify the expression for the amplitude. Justify the various steps.

e. In these conditions evaluate $\langle |\mathcal{M}|^2 \rangle$.

f. In the same conditions evaluate the total cross-section.

g. Write the cross-section of the process $\nu_e + D \rightarrow p + p + e^-$ in terms of the parton cross-sections. Consider only valence quarks and leave the result in its integral form.

h. Consider the statement: *The total interaction cross-section of neutrinos with matter is proportional to the square of the CM energy*. Discuss this statement indicating its domain of validity.

i. The results obtained in the SNO experiment led to the Nobel Prize in 2015. Discuss briefly how the neutrino oscillations could be observed using the measurements of the channels described above. What is the importance of this result?

---

a. The solar neutrinos interact in SNO with the deuteron in the heavy water molecule.

  i. As the solar neutrinos are particles, through the charged current, they can only give negatively charged leptons. In the experiment, given the neutrino energies involved, only electrons can be produced. So, we should have the charged current vertex $\nu_e \rightarrow e^-$ with the exchange of a W boson. As charge needs to be conserved the W has to connect to the d-quark and transfer it into a u-quark. Thus, the only possible Feynman diagram at tree-level is the one shown in Fig. 9.2.

Fig. 9.2 Tree-level Feynman dominant diagrams for the process $n + \nu_e \rightarrow p + e^-$

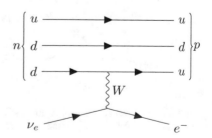

ii. The neutral current is mediated by the Z boson. Now, there is no change of charge and therefore we can have the following diagrams (see Fig. 9.3). In the shown diagrams, ν_X can be any type of neutrino, i.e. $X = e^-$, $\mu^- \tau^-$. Moreover, notice that the Z boson can interact with both u and d valence quarks present in the proton and the neutron.

iii. The elastic scattering is $\nu_X + e^- \rightarrow \nu_X + e^-$, where again ν_X are neutrinos with any possible flavor. Therefore, for the electron neutrino we can have both charged (Fig. 9.4a) and neutral (Fig. 9.4b) weak currents. Contrarily, for ν_μ and ν_τ only neutral weak currents are possible, as shown in Fig. 9.5.

b. The diagram for the process $\nu_e(p_1) + d(p_2) \rightarrow e^-(p_3) + u(p_4)$ is the one shown in Fig. 9.6.

c. The amplitude is

$$i\mathcal{M} = \left(-i\frac{g}{\sqrt{2}}\right)^2 (-i)\bar{u}(p_3)\gamma^\mu P_L u(p_1)\frac{g_{\mu\nu} - \frac{k_\mu k_\nu}{M_W^2}}{t - M_W^2}\bar{u}(p_4)\gamma^\nu P_L u(p_2)$$

(9.25)

where $t = (p_1 - p_3)^2$, $k_\mu = p_1 - p_3$ and $k_\nu = p_4 - p_2$.

d. Neglecting the quark and lepton masses we can discard the terms in the numerator of the W propagator proportional to the momentum. To show this we take, for instance, the neutrino-electron line. Then

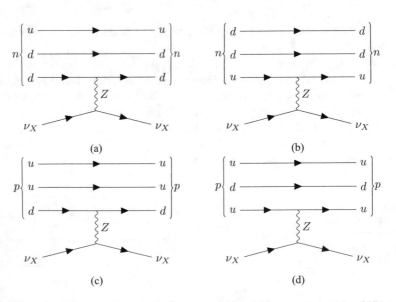

Fig. 9.3 Tree-level Feynman diagrams for the process: (a) and (b) $n\nu_X \rightarrow n\nu_X$; (c) and (d) $p\nu_X \rightarrow p\nu_X$. ν_X is a neutrino of any possible flavor

Fig. 9.4 Tree-level Feynman diagrams for the process $\nu_e e^- \rightarrow \nu_e e^-$

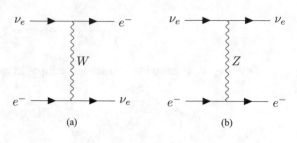

(a)　　　　　(b)

Fig. 9.5 Tree-level Feynman diagrams for the processes: (a) $\nu_\mu e^- \rightarrow \nu_\mu e^-$; (b) $\nu_\tau e^- \rightarrow \nu_\tau e^-$

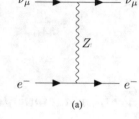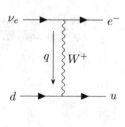

(a)　　　　　(b)

Fig. 9.6 Tree-level Feynman diagrams for the process $\nu_e d \rightarrow e^- u$

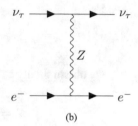

Fig. 9.7 Tree-level Feynman diagrams for the process $\nu_e d \rightarrow e^- u$ with the allowed helicity combinations

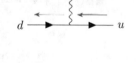

$$k_\mu \bar{u}(p_3)\gamma^\mu P_L u(p_1) = \bar{u}(p_3)(\not{p}_1 - \not{p}_3) P_L u(p_1)$$
$$= -\bar{u}(p_3)\not{p}_3 P_L u(p_1) + \bar{u}(p_3)P_R \not{p}_1 u(p_1) = 0 \qquad (9.26)$$

where it was used the fact that for massless particles the Dirac equation gives

$$\not{p}_1 u(p_1) = 0 \text{ .and } \bar{u}(p_3)\not{p}_3 = 0, \qquad (9.27)$$

showing that this term is zero.

Also, if we assume that the energies are such that $|t| \ll M_W^2$ we can collapse the propagator of the W into

$$\frac{g_{\mu\nu} - \frac{k_\mu k_\nu}{M_W^2}}{t - M_W^2} \rightarrow -\frac{1}{M_W^2} g_{\mu\nu}. \tag{9.28}$$

In this approximation we can re-write Eq. 9.25 as

$$\mathcal{M} = -\frac{g^2}{2M_W^2}\bar{u}(p_3)\gamma^\mu P_L u(p_1)\bar{u}(p_4)\gamma_\mu P_L u(p_2). \tag{9.29}$$

Using

$$\frac{G_F}{\sqrt{2}} = \frac{g^2}{8M_W^2} \tag{9.30}$$

we obtain finally

$$\mathcal{M} = -\frac{4G_F}{\sqrt{2}}\bar{u}(p_3)\gamma^\mu P_L u(p_1)\bar{u}(p_4)\gamma_\mu P_L u(p_2). \tag{9.31}$$

e. Because the charged current is left handed there is only one helicity combination not vanishing. This can be seen because for massless particles we have

$$P_L u(p_1) = u(p_1 \downarrow)$$
$$P_L u(p_2) = u(p_2 \downarrow)$$

and we get solely the diagram shown in Fig. 9.7.
We have then

$$\mathcal{M}(\downarrow\downarrow; \downarrow\downarrow) = -4\frac{G_F}{\sqrt{2}}J_{u1u3}(\downarrow\downarrow) \cdot J_{u2u4}(\downarrow\downarrow)$$

$$= -4\frac{G_F}{\sqrt{2}}s\left(\cos\frac{\theta}{2}, \sin\frac{\theta}{2}, -i\sin\frac{\theta}{2}, \cos\frac{\theta}{2}\right) \cdot \left(\cos\frac{\theta}{2}, -\sin\frac{\theta}{2}, -i\sin\frac{\theta}{2}, -\cos\frac{\theta}{2}\right)$$

$$= -4\frac{G_F}{\sqrt{2}}s\left[\cos^2\frac{\theta}{2} - \left(-\sin^2\frac{\theta}{2} - \sin^2\frac{\theta}{2} - \cos^2\frac{\theta}{2}\right)\right]$$

$$= -\frac{8G_F}{\sqrt{2}}s$$

where the symbol $\cdot$ between two 4-vectors, say a and b, indicates their scalar product $(a \cdot b \equiv a^\mu b_\mu)$.
The average squared amplitude can then be computed as

$$\langle |\mathcal{M}|^2 \rangle = \frac{1}{2}|\mathcal{M}(\downarrow\downarrow; \downarrow\downarrow)||^2 = \frac{32G_F^2 s^2}{2} = 16G_F^2 s^2, \tag{9.32}$$

where the $1/2$ factor appears because the interaction occurs with two left-handed particles.

f. The differential cross-section can be evaluated from

$$\frac{d\sigma}{d\Omega} = \frac{1}{64\pi^2 s} \langle |\mathcal{M}|^2 \rangle = \frac{G_F^2 s}{4\pi^2}. \tag{9.33}$$

Integrating over the solid angle, we get

$$\sigma = \int \frac{d\sigma}{d\Omega} d\Omega = 4\pi \frac{G_F^2 s}{4\pi^2} = \frac{G_F s}{\pi}. \tag{9.34}$$

g. Denoting the momentum of the d quark by $\hat{p}_2$ we can define the elementary cross-section as

$$\hat{\sigma}(\hat{s}) = \frac{G_F \hat{s}}{\pi}, \tag{9.35}$$

where $\hat{s} = (p_1 + \hat{p}_2)^2$, the quark only carries a fraction of the neutron momentum, p_2, so $\hat{p}_2 = x p_2$. We have for the system $\nu + n$,

$$s = (p_1 + p_2)^2 \simeq 2 p_1 \cdot p_2 \tag{9.36}$$

and therefore

$$\hat{s} = (p_1 + x p_2)^2 \simeq 2 x p_1 \cdot p_2 = x s. \tag{9.37}$$

With this we can now compute the neutrino-deuterium interaction total cross-section through

$$\sigma_{\nu D}(s) = \int_0^1 dx f_d^n(x) \hat{\sigma}_{\nu d}(x s), \tag{9.38}$$

where $f_d^n(x)$ is the probability of finding a quark d with fraction of momentum x inside the neutron. Notice that the proton in the D is a spectator in this process.

h. We have seen previously in Eq. 9.34 that $\sigma_{\nu q}$ grows with s, i.e. with the square of the CM energy. However, this is only valid for $m_d, m_\nu, m_e \ll \sqrt{s} \ll M_W$. For $\sqrt{s} \sim M_W$ or $\sqrt{s} \gg M_W$ this equation is not valid anymore. If we go back, we can make the replacement

$$\mathcal{M}(\downarrow\downarrow\downarrow\downarrow) = \frac{8 G_F s}{\sqrt{2}} \frac{M_W^2}{t - M_W^2} \tag{9.39}$$

and we would get for the differential cross-section,

$$\frac{d\sigma}{d\Omega} = \frac{G_F^2 s}{4\pi^2} \frac{M_W^4}{(t - M_W^2)^2}. \tag{9.40}$$

In these conditions $t = (p_1 - p_3)^2 = -2 p_1 \cdot p_3$, and since in the CM reference frame we have $p_1 = \frac{\sqrt{s}}{2}(1, 0, 0, 1)$ and $p_3 = \frac{\sqrt{s}}{2}(1, \sin\theta, 0, \cos\theta)$; then

$$t = -\frac{s}{2}(1 - \cos\theta). \tag{9.41}$$

Therefore, t grows with s biasing Eq. 9.34 for large $\sqrt{s}$.

i. As seen before, the charged weak current is only sensitive to ν_e while the neutral current and elastic scattering are sensitive to all neutrino flavors. The elastic scattering contribution is smaller compared with the charged and weak current processes. In the experiment, the ratio of charged to neutral currents was measured, leading to

$$\frac{\Phi_{\nu e}}{\Phi_{\nu X}} \simeq \frac{1}{3} \tag{9.42}$$

where $\Phi_{\nu e}$ is the flux of the neutrino measured through charged current processes (ν_e), and $\Phi_{\nu X}$ the one measured through neutral processes, with contributions of all neutrino species ($\nu_e + \nu_\mu + \nu_\tau$). Through this measurement it was possible to understand that the total number of neutrinos produced at the Sun was in accordance with the expectations and that the ν_e was oscillating into ν_μ and ν_τ. The elastic scattering was used to confirm the obtained result. This result not only demonstrated that there was no problem with our Solar model, which was thought to be wrong because the number of electron neutrinos was smaller than expected, but it also unveiled a new characteristic of neutrinos, the ability to oscillate between flavors. This is very important because in order to oscillate neutrinos must have nonzero mass.

7. *Neutrino narrow beam spectrum - see chap. 6.* A neutrino narrow band beam is produced by the decay of a π^+ monochromatic beam with an energy of 200 GeV. Determine the energy spectrum and the maximum dispersion of the produced neutrino beam.

Neutrino narrow band beams are produced by the decay of π^{+-} (and k^{+-}) beams with well defined energies. In the present problem the incoming beam is a π^+ beam with an energy of 200 GeV in the laboratory (LAB) reference frame.

The neutrinos are produced in the decay $\pi^+ \rightarrow \mu^+ \nu_\mu$. In the center-of-mass (CM) reference frame the momentum of the μ^+ and of the ν_μ are symmetric. Define p^* as the modulus of those momenta and consider the neutrino as massless ($E_\nu^* = p^*$). Imposing energy conservation:

$$m_\pi = \sqrt{m_\mu^2 + p^{*2}} + p^*$$

and

$$p^* = \frac{m_\pi^2 - m_\mu^2}{2m_\pi}.$$

Let's define θ^* as the scattering angle in the CM (angle between the beam direction and the direction of the μ^+ and the ν_μ). The distribution of $\cos\theta^*$ is uniform (the π is a scalar and then the decay is isotropic) with extreme values of -1 (neutrino emitted in the negative direction $(-z)$ of the beam axis) and 1 (neutrino emitted in the positive direction $(+z)$ of the beam axis). In the LAB frame the energy and the momentum of the neutrino can be computed using the Lorentz transformations:

$$E_\nu = \gamma p^* + \gamma \beta p^* \cos \theta^*$$

$$p_{\nu_z} = \gamma \beta p^* + \gamma p^* \cos \theta^*$$

$$p_{\nu_y} = p^* \sin \theta^*,$$

with

$$\gamma = \frac{E}{m}; \; \beta = \frac{p}{E}; \; \gamma\beta = \frac{p}{m}.$$

For a 200 GeV π beam:

$$\gamma = 1433; \; \beta \sim 1; \; \gamma\beta = 1433.$$

For the extreme values of $\cos\theta^*$ one obtains:

$$E_{\nu_{min}} = 10.4 \, \text{keV}; \; E_{\nu_{max}} = 85.4 \, \text{GeV}.$$

The neutrino energy spectrum will then have a flat distribution between 10.4 keV and 85.4 GeV.

To compute the maximum dispersion of the neutrino beam, one should consider the case of the neutrino transverse emission in the CM frame (p_ν^* emitted along the y axis). In that case $p_{\nu_z} = 42.7 \, \text{GeV}$ and $p_{\nu_y} = 29.8 \, \text{MeV}$, which corresponds to a maximum dispersion angle of $\theta_{max} = 4 \times 10^{-2}$ degrees.

8. *Neutrino oscillation probability (4)*. Given a beam of a pure muon neutrino beam, with fixed energy E, derive the probability of observing another neutrino flavor at a distance L assuming two weak eigenstates related to two mass eigenstates by a simple rotation matrix.

The derivation of the two neutrino oscillation formula is given in Chap. 9 of the textbook for the case of electron neutrino to muon neutrino. It is valid for any combination, changing only the angle and the Δm^2. It reads

$$P(a \to b) = \sin^2(2\theta) \sin^2 \left(1.27 \frac{\Delta m^2 \, L}{E} \right),$$

where m is expressed in eV, L in km and E in GeV.

9. *Tau neutrinos appearance (5)*. OPERA is looking for the appearance of tau neutrinos in the CNGS (CERN neutrinos to Gran Sasso) muon-neutrino beam. The

average neutrino energy is 17 GeV and the baseline is about 730 km. Neglecting matter effects, calculate the oscillation probability $P(\nu_\mu \to \nu_\tau)$ and comment.

The transformation of muon to tau neutrinos is seen in atmospheric oscillations, and the corresponding parameters are (see Chap. 9 of the textbook for a demonstration): $\sin^2(2\theta) \sim 1$ and $\Delta m^2 = 0.0025\,\text{eV}^2$. Simply using the two flavor oscillation formula we get a very small $P(a \to b) < 2\%$.

With the distance fixed by the location of the CERN and Gran Sasso existing laboratories, the probability could be maximized by choosing the beam energy such that

$$1.27 \frac{\Delta m^2\, L}{E} = \frac{\pi}{2}(+k\pi),$$

in which k is any integer; the maximum (positive) energy, $E \simeq 1.5\,\text{GeV}$, is reached for $k = 0$.

However, to prove tau appearance one needs enough energy for the tau mass (1.8 GeV) in a charged current interaction. Moreover, as the interaction cross-section increases with energy, it is the product of the oscillation probability and the cross-section that must be maximized.

In addition, the tau decay products must be clearly distinguishable from the muons produced by the muon neutrino beam. OPERA reconstructs the tau neutrino interactions in events with several (5 to 10) particles in 8 cm long detector bricks, and the resolution in general increases with the particle energy.

10. *Neutrino mass differences (6)*. A neutrino experiment at a distance of 200 m from a nuclear reactor measures that the flux of the 3 MeV anti-neutrino beam is $(90 \pm 10)\%$ of what was expected in case of no oscillation. Assuming a maximal mixing determine the value of Δm_ν^2 at a confidence level of 95%.

We can interpret this as a measurement of a P=0.10 appearance result with a 100% uncertainty. It is consistent with no oscillation, or a $\Delta m^2 = 0$, which sets the lower limit. To get a 95% CL result from one side of a Gaussian measurement, it must be within 1.64 σ, so we look for

$$\sin^2(2\theta) \sin^2\left(1.27 \frac{\Delta m^2\, L}{E}\right) < (0.100 + 0.164 = 0.264)\,.$$

Assuming maximal mixing and the given L and E,

$$\sin^2(85\Delta m^2 + k\pi) < 0.264 \implies \Delta m < 12.6\,\text{eV}\,.$$

Notice that several options of k could be available for the mass, but high values within are already excluded by direct mass measurements in beta decay, which are sensitive to all the mass values with the corresponding electron neutrino weights given by the PNMS components.

11. *Neutrino rotation angles (7)*. There are three neutrino types (electron, muon, tau) and three mass values, related by the 3×3 Pontecorvo-Maki-Nakagawa-Sakata (PMNS) mixing matrix, usually factorized in three rotation matrices. Knowing that the three mass values are such that:

- Δm^2 (solar) $= m_2^2 - m_1^2 \sim 10^{-5} \mathrm{eV}^2$
- Δm^2 (atmospheric) $= \left| m_3^2 - m_2^2 \right| \sim 10^{-3} \mathrm{eV}^2$,

discuss the optimization of reactor and accelerator experiments to measure each of the three rotation angles and to confirm those mass differences. Compare, for example, the pairs of experiments (KamLAND, DayaBay), (T2K, OPERA).

The first thing to notice is that the reactor (electron) anti-neutrinos have energies of the order of 1 MeV, so we will not distinguish if they have oscillated to either muon or tau anti-neutrinos. To measure the two mixing angles one should maximize the distances corresponding to the two mass differences.

$$L \sim \frac{E}{\Delta m^2 (\pi/2 + k\pi)} / 1.27 \, ,$$

with L in km, E in GeV and m in eV. We want to minimize the distance to have a larger flux, so choose $k = 0$; then for mass differences typical of the solar neutrinos we get $L \sim 100$ km (as in KamLAND) and for the atmospheric mass difference we get $L \sim 1$ km (as in DayaBay).

At accelerators, the energies can be tuned and should be large enough to, at least, measure the original muon neutrino beam. The distances should then scale accordingly.

In OPERA the distance was fixed by the locations of CERN and Gran Sasso and the energy was adjusted, not to maximize the oscillation, but to maximize the tau reconstruction efficiency. T2K is designed to measure electron neutrino appearance (a very rare process, additionally to measuring precisely the muon neutrino disappearance into tau neutrino, which is the dominant process). T2K chose in parallel the energy and distance, but within some constraints. The energy is selected to be above the muon production (and detection) threshold but below the GeV, when multiple particle production starts to be important and would decrease the energy resolution and introduce backgrounds (namely π^0 production which can fake electron neutrino appearance in a water Cherenkov detector as the SK). The final energy is centered around 600 MeV, with a small spread. The distance can then be selected as before, keeping $k = 0$ to maximize the flux,

$$L \sim \frac{0.6(\frac{\pi}{2}/1.27)}{2.5 \times 10^{-3}} \sim 295 \, \text{km} \, .$$

The neutrino energy is selected (for the first time in T2K) by using the neutrinos coming at a given angle off-axis from the original proton beam. In fact, pions will be created in the direction of the beam and given their average energy and the relation between the pion and muon mass, the neutrino energy can be calculated from the angle as

$$E_\nu = \frac{m_\pi^2 - m_\mu^2}{2(E_\pi - P_\pi \cos\theta)} = \frac{m_\pi^2 - m_\mu^2}{2E_\pi(1 - \beta_p \cos\theta)},$$

thus, as the angle increases, the maximum neutrino energy is limited.

12. *Neutrino disappearance.* Using the two flavor oscillation formula, determine the value of the muon neutrino energy E_ν which gives a 100% disappearance probability for: (a) atmospheric neutrinos crossing the Earth atmosphere ($L \simeq$ 20 km); (b) atmospheric neutrinos crossing the Earth diameter ($L \simeq 13\,000$ km). Comment the result.

In a two flavor scenario the probability of $\nu_\mu \to \nu_x$ is described by:

$$P\left(\nu_\mu \to \nu_x\right) = \sin^2(2\theta) \sin^2\left(1.27\,\frac{\Delta m^2\,(\text{eV}^2)\,L\,(\text{km})}{E_\nu\,(\text{GeV})}\right).$$

The above equation indeed describes well the observed pattern of the atmospheric muon neutrino disappearance with a maximal mixing $\sin^2(2\theta_{\text{atm}}) \sim 1$, and a squared mass difference $\Delta m_{\text{atm}}^2 \sim 0.0025\,\text{eV}^2$.

Then, muon neutrinos with an energy E_ν will have a disappearance probability of $\sim 100\%$ after travelling a distance L if

$$1.27\,\frac{\Delta m^2\,(\text{eV}^2)\,L\,(\text{km})}{E_\nu\,(\text{GeV})} = \frac{\pi}{2}$$

and thus

$$E_\nu\,(\text{GeV}) \sim 2 \times 10^{-3} L\,(\text{km}).$$

For $L \simeq 20$ km and $L \simeq 13\,000$ km the corresponding neutrino energy will then be 40 MeV and 26 GeV, respectively.

The above relation between E_ν and L was indeed established studying the ratio between the observed number of muon neutrino events and the corresponding predicted number in case of no oscillation. For instance, in Fig. 9.8 (right), coming from the textbook, a dip at $L/E_\nu \sim 500\,(\text{km/GeV})$ is clearly visible. In terms of angular dependence (see the left part of the same figure), this dip corresponds to an observed number of upward-going neutrinos (those which have crossed the Earth) roughly half of the predictions, which means that the energy of the observed neutrinos was of the order of the GeV or tens of GeV.

13. *Neutrinos from Supernova 1987A (8).* In 1987 a supernova explosion was observed in the Large Magellanic Cloud, and neutrinos were measured in three different detectors. The neutrinos, with energies between 10 and 50 MeV, arrived

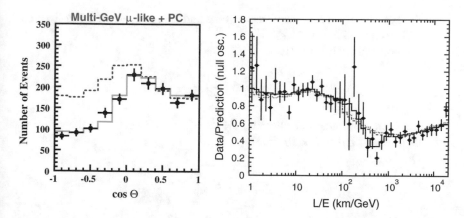

Fig. 9.8 Left: Zenith angle distribution of muon neutrinos in SK. The observed number of upward-going neutrinos was roughly half of the predictions. Right: Survival probability of ν_μ as a function of L/E. Black dots show the observations and the lines shows the prediction based on neutrino oscillation. Data show a dip around $L/E \simeq 500$ km/GeV. The prediction of two-flavor neutrino oscillations agrees well with the position of the dip. From http://www-sk.icrr.u-tokyo.ac.jp/sk/physics/atmnu-e.html and The Super-Kamiokande Collaboration, Y. Ashie et al., "Evidence for an Oscillatory Signature in Atmospheric Neutrino Oscillations," Phys. Rev. Lett. 93 (2004) 101801.

with a time span of 10 s, after a travel distance of about $c(5 \times 10^{12}$ s), and three hours before photons at any wavelength.

a. Can this information be used to determine a neutrino mass? Discuss the quantitative mass limits that could be derived from the SN1987A.
b. This was the only SN observed in neutrinos, up to now, but the same reasoning can be used in pulsed accelerator beams. Derive the needed time and position precision to measure masses ~ 1 eV, given a beam energy $E \sim 1$ GeV and a distance L.

---

a. It is noted that the neutrinos arrive before light; if we would assume that they leave the source at the same time, this would mean that neutrinos are faster than light,

$$\beta = \frac{v}{c} = \frac{p}{E} > 1 \,,$$

which implies $E^2 - m^2 > E^2$ and $m^2 < 0$. In fact, this is not the correct interpretation, but that the light itself is delayed by interaction on the source (more than neutrinos).

Likewise, some neutrinos can be emitted earlier than others, and that explains the time span. However, if they are emitted at the same time but have a large mass, then the most energetic should arrive earlier, which is not reported by the experiments. From the observed time and energy span, an upper limit can still be derived for the mass.

For both calculations we use

$$\Delta\beta = \Delta\left(\frac{v}{c}\right) = \frac{\Delta T}{L}.$$

Between neutrinos and photons, $\Delta\beta \simeq -2 \times 10^{-8}$; between the neutrinos themselves, $\Delta\beta \simeq 2 \times 10^{-12}$.

On the other hand,

$$\beta = \frac{p}{E} = \sqrt{1 - \frac{m^2}{E^2}} \simeq -\frac{1}{2}\frac{m^2}{E^2}$$

$$\Delta\beta \simeq \frac{m^2}{2}\frac{E_1^2 - E_2^2}{E_1^2 E_2^2}$$

$$m = \sqrt{2\Delta\beta\frac{E_1^2 E_2^2}{E_1^2 - E_2^2}}.$$

The comparison between neutrinos and gammas would give an imaginary mass, m = 10 i keV, while the comparison between neutrinos would give an upper limit of 20 eV.

b. Accelerator beams are in general pulsed, and thus we have the time of emission of the neutrino (known within this pulse), and the length until they are detected, as well as their energy. To measure a mass $m \sim 1\,\text{eV}$ with $E \sim 1\,\text{GeV}$, we need a precision $m/E \sim 10^{-9}$. The error on $(\Delta\beta/\beta)$ is about $\sqrt{(\Delta L/L)^2 + (\Delta T/T)^2}$, since the relative uncertainties are summed in quadrature, and thus both the distance and timing precision must be of this order of magnitude or better. Present experiments have accuracies of $L \sim 1\,\text{cm}$ and $T \sim 1\,\text{ns}$.

14. *Double β decay (9).* Double β decay is a rare process, possible only for a small fraction of the nuclear isotopes. The neutrinoless double β decay is only possible if lepton number is not conserved, and is one of the most promising channels to discover lepton number violation. Discuss the optimization (total mass, chosen isotope, backgrounds, energy resolution, . . .) of the experiments looking for $0\nu\beta\beta$. List other possible experimental signals of lepton number violation you can think of.

---

a. In a neutrinoless double beta decay experimental search, the signal is a peak on the sum of the two electron energies at the Q^2-value of the decay, its height depends on the Majorana effective mass. To measure it precisely we need to be able to see it with small statistical and systematic uncertainties.

Because this is a very rare process we want to decrease all backgrounds. Purification of the detector elements and an external shielding is mandatory and the experiment should be located in a clean underground laboratory. Background from the unavoidable $2\nu\beta\beta$ process (at $Q^2 - 2m_{\nu_e}$) will arise due to the finite energy resolution. It can be reduced by choosing a high Q^2 element, in order to increase the energy resolution, and one for which there is a favourable matrix element for 0 neutrino with respect to 2 neutrino decay (which have in general been measured experimentally).

The number of $0\nu\beta\beta$ events depends on the exposure, ie Mass of Isotope x Measurement Time. So it is better to choose an isotope with a large natural abundance, or that can be easily enriched within a given element. In many cases, the element is itself the main detector medium (GERDA uses Ge, EXO uses Xe) or it is loaded in big quantities inside another detector medium without changing too much its properties (the case of large liquid scintillator experiments, KamLAND-Zen or SNO+).

b. A very much studied topic is charged lepton flavor violation, analogue to the neutrino oscillations. For example, in the form of $\mu \to e\gamma$, which can be achieved with a loop with a neutrino and a W, in which the neutrino changes from muon-neutrino to electron-neutrino, and the gamma is emitted by the charged W boson. Neutrinos take time to oscillate, which highly suppresses this process.

In double beta decay, even total lepton number is violated. The same diagram $(dd \to uue^-e^-)$ could give rise, in the LHC, to a topology with two jets and two leptons with the same charge. In this case, the process is not resonant and is again highly suppressed, however, depending on the physics scale for the Majorana mass term, a very high luminosity LHC could be competitive with the double beta decay experiments.

15. *Majorana neutrinos.* Why is neutrinoless double beta decay possible only for massive Majorana neutrinos?

A Majorana neutrino is a neutral spin $1/2$ particle which is identical to its own antiparticle. Therefore two Majorana neutrinos may annihilate as it occurs for any pair of particle and anti-particle. This is the mechanism that may allow the neutrinoless beta decay escaping to a possible lepton number violation that would occur for Dirac neutrinos.

However, the neutrinos produced in a beta decay have a well defined helicity which is a Lorentz invariant in the case of massless particles (for a discussion on helicity and chirality invariance under Lorentz and time transformations see, for example, Chap. 6 of the textbook). On the contrary, for massive fermions this is no longer true since it is always possible to reverse the momentum direction going to a reference frame which would travel at a higher and opposite velocity with respect to the neutrino velocity in the rest frame of the beta decay.

In order that two Majorana neutrinos emitted in a double beta decay may annihilate, they should have also inverse helicity (total spin zero), which would be,
as discussed above, only possible if they were massive.

Chapter 10
Messengers from the High-Energy Universe

1. *Fermi acceleration mechanisms (1).* In the Fermi acceleration mechanism, charged particles increase considerably their energies crossing back and forth many times the border of a magnetic cloud (second-order Fermi mechanism) or of a shock wave (first-order Fermi mechanism). Compute the number of crossings that a particle must do in each of the mechanisms to gain a factor 10 on its initial energy assuming:

   a. $\beta = 10^{-4}$ for the magnetic cloud and $\beta = 10^{-2}$ for the shock wave;
   b. $\beta = 10^{-4}$ for both acceleration mechanisms.

   where β is the velocity of the astrophysical object (shock wave or cloud).

   ---

   a. At each passage through the cloud or shock wave the particle gains some energy that is proportional to its energy. Therefore for n crossings the energy of the particle relatively to its initial energy, E_0 is given by

   $$E = E_0(1 + \varepsilon)^n. \tag{10.1}$$

   where ε is the gain and it is proportional to β for the acceleration in a shock wave (Fermi first order acceleration mechanism) and proportional to β^2 for the *collision* with the magnetic cloud (Fermi second order acceleration mechanism).
   Inverting Eq. (10.1) one can obtain the number of times that a particle should cross in order to increase its energy from E_0 to E,

   $$n = \frac{\ln (E/E_0)}{\ln(1 + \varepsilon)}. \tag{10.2}$$

© Springer Nature Switzerland AG 2021
A. De Angelis et al., *Particle and Astroparticle Physics*, Undergraduate Lecture Notes in Physics,
https://doi.org/10.1007/978-3-030-73116-8_10

In this problem we want to know how many times a particle should cross a cloud or a shock wave to increase its energy by a factor of 10, so $E/E_0 = 10$. Finally, using Eq. (10.2), we have

$$n(\varepsilon \propto \beta; \beta = 10^{-2}) \simeq 2.3 \times 10^2 \text{ cycles} \tag{10.3}$$

$$n(\varepsilon \propto \beta^2; \beta = 10^{-4}) \simeq 2.3 \times 10^8 \text{ cycles.} \tag{10.4}$$

Therefore, in realistic astrophysical conditions the particle needs to cross only 230 times the shock wave to gain a factor of 10 on its energy while it should cross a cloud 2.3×10^8 times to gain the same energy.

b. Here we assume that both the shock wave and the magnetic cloud have the same velocity $\beta = 10^{-4}$.

$$n(\varepsilon \propto \beta; \beta = 10^{-4}) \simeq 2.3 \times 10^4 \text{ cycles} \tag{10.5}$$

$$n(\varepsilon \propto \beta^2; \beta = 10^{-4}) \simeq 2.3 \times 10^8 \text{ cycles.} \tag{10.6}$$

Even considering the same velocity for the two astrophysical phenomena the Fermi first order acceleration needs 10000 times less cycles than the Fermi second order mechanism.

2. *Photon spectrum in hadronic cascades (2).* Demonstrate that in the decay $\pi^0 \rightarrow \gamma\gamma$, the probability to emit a photon of energy E_γ in the laboratory frame is constant over the range of kinematically allowed energies.

In this exercise the symbol γ refers to a photon, while γ_B and β_B refer to the Lorentz factors (where B stands for Boost). All quantities having the apex $'$ are in the rest frame, where the emitting pion π^0 is at rest. In the laboratory frame the emitting pion is moving along the z-axis, so that the azimuth angle φ lies on the xy-plane.

We want to demonstrate that the pdf dN_γ/dE_γ, describing the emission of a photon of energy E_γ, is constant over the range of kinematically allowed energies. Let us consider the system in the rest frame: the emission of the photons is isotropic, with one photon being emitted at θ'_γ and the other photon being emitted at $\theta'_\gamma + \pi$.

We re-write the pdf in the following more convenient way:

$$\frac{dN_\gamma}{dE_\gamma} = \frac{dN_\gamma}{d\cos\theta'_\gamma} \frac{d\cos\theta'_\gamma}{dE_\gamma}. \tag{10.7}$$

The above factorization is helpful, because the first factor can be computed exploiting emission isotropy, while the second factor can be derived by solving the kinematics in the rest frame and then performing a boost into the laboratory frame.

Let us compute the first factor. In the rest frame the gamma emission is isotropic, in formula:

$$\frac{dN_\gamma}{d\Omega'_\gamma} = const \tag{10.8}$$

where $d\Omega'_\gamma$ is the infinitesimal solid angle of gamma emission in the rest frame. Since the system has azimuthal symmetry, the above expression is equivalent to:

$$\frac{dN_\gamma}{d\cos\theta'_\gamma} = const, \tag{10.9}$$

where the constant $const$ can be found by the following normalization condition:

$$1 = \int_{-1}^{+1} \left(\frac{dN_\gamma}{d\cos\theta'_\gamma}\right) d\cos\theta'_\gamma = const \int_{-1}^{+1} d\cos\theta'_\gamma = 2\, const \implies const = \frac{1}{2}. \tag{10.10}$$

It is therefore possible to re-write Eq. 10.7 as:

$$\frac{dN_\gamma}{dE_\gamma} = \frac{1}{2}\frac{d\cos\theta'_\gamma}{dF_\gamma}. \tag{10.11}$$

If we find a way to express E_γ as a function of $\cos\theta'_\gamma$ the problem is solved by computing one derivative. Let us first find an expression for the energy of the gamma in the rest frame E'_γ, and then boost such expression in the laboratory frame. In the rest frame each photon takes half of the rest energy of the pion, so:

$$E'_\gamma = p'_\gamma c = \frac{m_\pi c^2}{2}. \tag{10.12}$$

The 4-momentum of the photon can be written as:

$$P'^{\mu}_\gamma = \begin{pmatrix} E'_\gamma/c \\ p'_{\gamma,x} \\ p'_{\gamma,y} \\ p'_{\gamma,z} \end{pmatrix} = \frac{m_\pi c}{2} \begin{pmatrix} 1 \\ 0 \\ \sin\theta'_\gamma \\ \cos\theta'_\gamma \end{pmatrix} \tag{10.13}$$

where it is assumed that the photons lie on the yz plane. We may also have imagined the emission to take place in the xz plane, or in any intermediate plane, but the result will not change, because there is azimuthal symmetry. We can now perform a boost along the z-axis, in order to compute the gamma energy in the laboratory frame, where $p_\pi > 0$:

$$\begin{cases} \beta_B = p_\pi c / E_\pi \\ \gamma_B = E_\pi / E'_\pi = E_\pi / (m_\pi c^2) \\ E_\gamma = \gamma_B \, (E'_\gamma - \beta_B \, c \, p'_{\gamma,z}). \end{cases} \qquad (10.14)$$

Replacing the first two equations in the third one we obtain:

$$E_\gamma = \frac{E_\pi}{2} \left(1 + \frac{p_\pi c}{E_\pi} \cos \theta'_\gamma \right). \qquad (10.15)$$

Deriving the above equation with respect to $\cos \theta'_\gamma$

$$\frac{dE_\gamma}{d\cos\theta'_\gamma} = \frac{p_\pi c}{2} \implies \frac{d\cos\theta'_\gamma}{dE_\gamma} = \left(\frac{dE_\gamma}{d\cos\theta'_\gamma} \right)^{-1} = \frac{2}{p_\pi c} \qquad (10.16)$$

and inserting the above derivative in Eq. 10.11, we finally find the result:

$$\frac{dN_\gamma}{dE_\gamma} = \frac{1}{p_\pi c} \qquad (10.17)$$

which is indeed constant. For completeness we notice that the interval ends of kinematically allowed energies of the photon are obtained inserting $\theta'_\gamma = 0, \pi$ in Eq. 10.15:

$$E_\gamma^{max,min} = \frac{1}{2}(E_\pi \pm p_\pi c). \qquad (10.18)$$

We can cross-check the obtained result by verifying the normalization condition:

$$\int_{E_\gamma^{min}}^{E_\gamma^{max}} \frac{dN_\gamma}{dE_\gamma} dE_\gamma = \frac{1}{p_\pi c}(E_\gamma^{max} - E_\gamma^{min}) = \frac{1}{p_\pi c}\frac{1}{2}(E_\pi + p_\pi c - E_\pi + p_\pi c) = 1. \quad (10.19)$$

3. *High energy photons and neutrinos fluxes at the sources and on Earth.* Consider the production of high energy photons and neutrinos by Cosmic Ray accelerators by the interaction of protons with matter of the source itself.

   a. Relate the numbers of very high energy γ and each ν flavor at the source.
   b. Assuming no interaction but taking into account neutrino oscillations, estimate on Earth the above relations.

   _____

   a. Due to isospin symmetry in strong interactions, very high energy proton collisions with matter mostly produce pions in a ratio of

$$1\pi^0 : 1\pi^- : 1\pi^+.$$

Pions decay (mostly) to:

$$\pi^0 \to \gamma\gamma$$
$$\pi^- \to \mu^-\bar{\nu}_\mu \to e^-\bar{\nu}_e\nu_\mu\bar{\nu}_\mu$$
$$\pi^+ \to \mu^+\nu_\mu \to e^+\nu_e\bar{\nu}_\mu\nu_\mu$$

leading to a ratio at the source of:

$$2\gamma : 2\nu_\mu : 2\bar{\nu}_\mu : 1\nu_e : 1\bar{\nu}_e : 0\nu_\tau : 0\bar{\nu}_\tau.$$

In fact, while neutral pions have a very short lifetime, charged pions can re-interact producing more pions (of lower energies); so the total ratio of all (anti-) neutrinos to gammas will be below 3. This is the same process used to create neutrino beams in accelerators, or atmospheric neutrinos in cosmic ray showers.

b. The total number of neutrinos and gammas arriving at Earth will keep the above ratio of $6 : 2 \sim 3$, but the fast atmospheric oscillations will change $\sim$ 50% of the ν_μ to ν_τ (and the same happens for the anti-neutrinos). The other oscillation terms are slower, and after all flavors are balanced, there will be no net effect from them. This leads to a ratio at the Earth of:

$$2\gamma : 1\nu_\mu : 1\bar{\nu}_\mu : 1\nu_e : 1\bar{\nu}_e : 1\nu_\tau : 1\bar{\nu}_\tau.$$

4. *Top-down production mechanisms for photons: decay of a WIMP (3).* If a WIMP at rest with a mass $M > M_Z$ decays into γZ, estimate the energy of the photon.

The photon is massless and the Z has a mass $M_Z \simeq 91.2$ GeV. Then imposing also energy-momentum conservation:

$$E_\gamma = |P_\gamma| = |P_Z|$$

and

$$M = E_Z + E_\gamma.$$

Then,

$$M = \sqrt{E_\gamma^2 + M_Z^2} + E_\gamma,$$

and thus

$$E_\gamma = \frac{M^2 - M_Z^2}{2M}.$$

5. *Acceleration and propagation (4).* The transparency of the Universe to a given particle depends critically on its nature and energy. In fact, whenever it is possible to open an inelastic channel of the interaction between the *traveling* particle and

the CMB, its mean free path diminishes drastically. Assuming that the only relevant phenomena that rules the mean free path of the *traveling* particle is the CMB (CνB), estimate the order of magnitude energies at which the transparency of the Universe changes significantly, for:

(a) Photons;
(b) Protons;
(c) Neutrinos.

Assume that for the threshold $E_{\gamma CMB} = E_{\nu C\nu B} \sim 1$ meV. (Note that, $\langle E_{\gamma CMB} \rangle \simeq 0.24$ meV; $\langle E_{\nu C\nu B} \rangle \simeq 0.17$ meV.)

---

a. For photons the dominant process is the interaction with the photons of the cosmic microwave background, γ_{CMB}, through pair creation:

$$\gamma + \gamma_{CMB} \to e^+ + e^-. \tag{10.20}$$

To determine the minimum energy at which this process can occur it is useful to compute the inner product of the four-momentum vector. This quantity is a Lorentz invariant and thus we can easily relate quantities in the laboratory frame with the ones in the center-of-mass frame. For convenience, the calculations for the photons (before the interaction) will be considered in the laboratory while the products of such interaction will be considered in the center-of-mass. Therefore,

$$P_\mu^{LAB} = (E_\gamma + E_b, \mathbf{P_b} + \mathbf{P_\gamma}) \tag{10.21}$$

$$P_\mu^{CM} = (2m_e, \mathbf{0}). \tag{10.22}$$

where it was assumed that $E_b \equiv E_{\gamma CMB}$. Using

$$s = (P_\mu P^\mu)_{LAB} = (P_\mu P^\mu)_{CM} \tag{10.23}$$

$$P_\mu P^\mu = E^2 - \mathbf{P} \cdot \mathbf{P} \tag{10.24}$$

one obtains the following equation:

$$4m_e^2 = (E_b + E_\gamma)^2 - (P_b^2 + P_\gamma^2 + 2\mathbf{P_b} \cdot \mathbf{P_\gamma}). \tag{10.25}$$

Taking into account that for photons $E = |\mathbf{P}|$ and that $\mathbf{P_1} \cdot \mathbf{P_2} = |P_1||P_2| \cos\theta$, one gets that the energy of the incoming photon is given by

$$E_\gamma = \frac{2m_e^2}{E_b(1 - \cos\theta)}. \tag{10.26}$$

Notice that we are looking for the minimal energy that allows for this process to happens, so $\cos\theta = -1$. Inputing the values given in the problem one gets

$$E_\gamma = \frac{m_e^2}{E_b} \simeq 2.5 \times 10^{14}\,\text{eV}. \tag{10.27}$$

b. For protons the dominant inelastic channel is via

$$p + \gamma_{CMB} \rightarrow (\Delta^+) \rightarrow p + \pi^0$$
$$p + \gamma_{CMB} \rightarrow (\Delta^+) \rightarrow n + \pi^+. \tag{10.28}$$

As in the previous problem, we want to find the minimum energy of the proton for which the process is possible. Moreover, we will use the Lorentz invariant s and consider once again that the proton and the gamma are in the Lab frame and the products of the interaction in the center-of-mass frame. Let us then start by defining our kinematics:

$$P_\mu^{LAB} = (E_p + E_b, \mathbf{P_p} + \mathbf{P_b}) \tag{10.29}$$
$$P_\mu^{CM} = (m_p + m_\pi, \mathbf{0}). \tag{10.30}$$

Therefore, using Eq. (10.24), one can write:

$$(E_b + E_p)^2 - P_b^2 - P_p^2 - 2\mathbf{P_b}.\mathbf{P_p} = (m_p + m_\pi)^2. \tag{10.31}$$

Solving the above equation for E_p and recalling that $P_p = \sqrt{E_p^2 - m_p^2}$ one gets:

$$2E_b E_p - 2E_b\sqrt{E_p^2 - m_p^2}\cos(\theta) = m_\pi^2 + 2m_p m_\pi. \tag{10.32}$$

Since $E_p \gg m_p$ the proton momentum can be approximated by its energy $(P_p \simeq E_p)$. Therefore,

$$E_p = \frac{m_\pi^2 + 2m_p m_\pi}{4E_b} \simeq 6 \times 10^{19}\,\text{eV}. \tag{10.33}$$

where $\cos\theta$ was taken to be -1, i.e., the proton and the photon have opposite directions.

c. The Universe becomes opaque to the neutrinos when they the following inelastic interaction channel opens:

$$\nu + \nu_{CvB} \rightarrow Z. \tag{10.34}$$

Again considering that the interaction occurs in the laboratory frame and noticing that the interaction occurs if there is enough energy to produce a Z

at rest, one can write

$$P_\mu^{LAB} = (E_\nu + E_b, \mathbf{P}_\nu + \mathbf{P_b}) \tag{10.35}$$

$$P_\mu^{CM} = (m_Z, \mathbf{0}). \tag{10.36}$$

and similarly as before one can write the following expressions:

$$(E_b + E_\nu)^2 - P_b^2 - P_\nu - 2\mathbf{P_b} \cdot \mathbf{P}_\nu = m_Z^2 \tag{10.37}$$

$$E_\nu = \frac{m_Z^2}{2E_b(1 - \cos\theta))}. \tag{10.38}$$

Therefore, the minimum energy of the neutrino that allows the process in Eq. (10.34) is

$$E_\nu = \frac{m_Z^2}{4E_b} \simeq 2 \times 10^{24} \, \text{eV}. \tag{10.39}$$

6. $\gamma\gamma \to e^+e^-$ *(14)*. Compute the energy threshold for the process as a function of the energy of the target photon for a head-on collision, and compare it to the energy for which the absorption of extragalactic gamma rays is maximal, applying the relation

$$\varepsilon(E) \simeq \left(\frac{900\,\text{GeV}}{E}\right) \text{eV}. \tag{10.40}$$

At the threshold the center-of-mass energy $\sqrt{s}$ should be equal to the sum of the e^+e^- masses. Then, being (E_γ and E_{γ_T} respectively, the energy of the incoming photon and of the target photon:

$$\sqrt{s} = \sqrt{(E_\gamma + E_{\gamma_T})^2 - (E_\gamma - E_{\gamma_T})^2} = 2m_e,$$

and

$$E_\gamma = \frac{m_e^2}{E_{\gamma_T}}.$$

Using GeV as units for the incoming photon and eV for the target photon one obtains:

$$E_\gamma \simeq \left(\frac{250\,\text{GeV}}{E_{\gamma_T}}\right) \text{eV}$$

which, for a given energy of the target photon, the threshold energy of the incoming photon is about one fourth of the energy where the absorption cross section is maximum (see textbook and following exercise).

7. *Photon-photon interactions (5).* Demonstrate that the cross section in Eq. 10.40 (also discussed in the textbook) is maximized for background photons of energy:

$$\varepsilon(E) \simeq \left(\frac{900\,\text{GeV}}{E}\right) \text{eV}$$

when the angular distribution of target photons is isotropical.

Once produced, VHE photons must travel towards the observer. Electron-positron $(e^- e^+)$ pair production in the interaction of VHE photons off extra-galactic background photons is a source of opacity of the Universe to γ rays whenever the corresponding photon mean free path is of the order of the source distance or smaller. The dominant process for the absorption is pair-creation:

$$\gamma + \gamma_{background} \rightarrow e^- e^+.$$

The cross section $\sigma_{\gamma\gamma}$ depends on E, ε and ϕ only through the speed β in natural units of the electron and of the positron in the center-of-mass, given by the equation (discussed in the textbook):

$$\beta(E, \varepsilon, \phi) = \left[1 - \frac{2m_e^2 c^4}{E\varepsilon(1-\cos(\phi))}\right]^{\frac{1}{2}},$$

where ϕ denotes the scattering angle, m_e is the electron mass, E is the energy of the incident photon and ε is the energy of the target (background) photon. The cross section $\sigma_{\gamma\gamma}$ reaches its maximum $\sigma_{\gamma\gamma}^{max} \simeq 1.7 \cdot 10^{-25}$ for $\tilde{\beta} \simeq 0.7$. For the case of the maximized cross section $\sigma_{\gamma\gamma}^{max}$:

$$\cos\phi = \left[1 - \frac{2m_e^2 c^4}{E\varepsilon(1 - \tilde{\beta}^2)}\right].$$

Integrating for an isotropic angle distribution:

$$\varepsilon(E) = \left[\frac{2m_e^2 c^4}{(1 - \tilde{\beta}^2)}\right]\frac{1}{E}.$$

Plugging in the constants, one gets:

$$\varepsilon(E) \simeq \left(\frac{900 GeV}{E}\right) \text{eV}.$$

8. *Neutrinos from SN1987A and neutrino velocity (6).* Neutrinos from SN1987A, at an energy of about 50 MeV, arrived in a bunch lasting 13 s from a distance of 50 kpc, 3 h before the optical detection of the supernova. What can you say on the neutrino mass? What can you say about the neutrino speed (be careful...)?

This exercise is very similar to the following exercise in this chapter, and we refer to it for the first question. About the second question, no relevant conclusions

Fig. 10.1 Timeline of the
SN1987A neutrino
observations

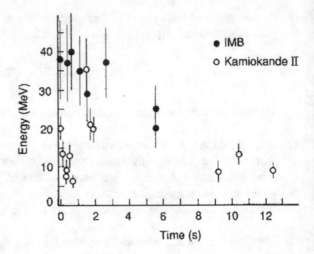

Fig. 10.1 Timeline of the SN1987A neutrino observations

can really be drawn unless a hypothesis is done on the relative time of emission
of neutrinos compared to the time of emission of photons. If you would assume
that the delay between the photon bunch and the neutrino bunch is due to a
superluminal speed c_ν of neutrinos, you would obtain

$$\frac{c_\nu}{c} \simeq 1 + 2 \times 10^{-9}.$$

Notice they the OPERA experiment had claimed that neutrinos were traveling
from CERN to Gran Sasso (a distance of 730 lm) 60 ns before photons; this
would imply $c_\nu/c \simeq 1 + 2 \times 10^{-5}$.

9. *Neutrinos from SN1987A and a bold deduction of neutrino masses (7).* Some
(including one of the authors of this book) saw in Fig. 10.1 a correlation between
the arrival times of some of the neutrinos from the SN1987A, at a distance of
about 168 000 ly, and their energy, and derived a value for the mass of the neutrino
manifesting this correlation, in the hypothesis that neutrinos had been emitted in
a negligible time. What can you say about the mass of this hypothetical neutrino?
How does it compare to the current neutrino mass limits?

In order to give an estimate of the neutrino mass from the "energy vs detection
time" plot shown in the text of the exercise, we start with finding a law relating
energy and time of flight. Let E be the energy of the traveling neutrino, and
TOF the time of flight from the source displaced at distance $L = 168, 000$ ly.
The Lorentz factor of the traveling neutrino is:

$$\gamma = (1 - \beta^2)^{-1/2} = \left[1 - \left(\frac{L}{c\, TOF} \right)^2 \right]^{-1/2} \qquad (10.41)$$

and so its energy is:

$$E = \gamma\, m\, c^2 = \left[1 - \left(\frac{L}{c\, TOF}\right)^2\right]^{-1/2} m\, c^2. \qquad (10.42)$$

The Time of Flight is the difference between the detection time t_{det}, which is shown in the plot, and the emission time t_0:

$$TOF = t_{det} - t_0. \qquad (10.43)$$

Since the emission time is unknown, the E-TOF relation cannot be used to estimate the neutrino mass. Let us obtain the TOF as a function of energy and then get rid of the unknown t_0. By re-arranging Eq. 10.42:

$$TOF = t_{det} - t_0 = \frac{L/c}{\sqrt{1 - (m\, c^2/E)^2}}. \qquad (10.44)$$

Let t_1 (E_1) and t_{last} (E_{last}) being the detection times (energies) of the first and last detected neutrinos: we cancel t_0 (in the hypothesis that neutrinos are emitted in a negligible time), by taking the differences of TOFs:

$$TOF_{last} - TOF_1 = t_{last} - t_0 - t_1 + t_0 = t_{last} - t_1$$
$$= \frac{L/c}{\sqrt{1 - (m\, c^2/E_{last})^2}} - \frac{L/c}{\sqrt{1 - (m\, c^2/E_1)^2}}. \qquad (10.45)$$

The above equation cannot be re-arranged to obtain the mass, but we can simplify it in the very reasonable limit of $(m\, c^2/E)^2 \sim 0$, because the observed energies are tens of MeV, while the neutrino mass is surely smaller. The Taylor series expansion truncated at the first order is:

$$\frac{1}{\sqrt{1 - x}} \sim 1 + \frac{x}{2} \qquad (10.46)$$

and applying this expansion to Eq. 10.45 we obtain:

$$t_{last} - t_1 = L/c \left[1 + \frac{1}{2}(mc^2/E_{last})^2 - 1 - \frac{1}{2}(mc^2/E_1)^2\right]$$
$$= L/c\, \frac{(m\, c^2)^2}{2}\left[\frac{1}{E_{last}^2} - \frac{1}{E_1^2}\right]. \qquad (10.47)$$

By solving the above equation for the mass:

$$m\,c^2 = \sqrt{\frac{2c}{L}\frac{t_{last} - t_1}{(1/E_{last}^2 - 1/E_1^2)}}. \tag{10.48}$$

We can apply this equation to the two datasets, from IMB and Kamiokande-II, to find two estimates of the neutrino mass. This method takes into account only the first and last detected neutrinos, so it is very rough, but we can expect at least to find a reasonable order of magnitude. By eye inspection of the plot the following values are proposed:

Variable	IMB	Kamiokande
t_1 (s)	0	0
E_1 (MeV)	37.5	20
t_{last} (s)	5.5	12.5
E_{last} (MeV)	25	10
m (eV/c^2)	48	25

where it has been used:

$$\frac{L}{2c} \simeq \frac{168\,000\,\mathrm{ly}}{2c} = 84\,000\,\mathrm{years} \simeq 2.65 \times 10^{12}\mathrm{s}. \tag{10.49}$$

The mass estimates found in this exercise are for the electron neutrino, because supernovae like SN1987A produce these kind of neutrinos. The PDG2018 estimate of the upper limit of the electron neutrino is 2 eV, one order of magnitude lower than our rough estimate.

10. *Time lag in the propagation of photons of different energy (8).* Suppose that the speed c of light depends on its energy E in such a way that

$$c(E) \simeq c_0 \left(1 + \xi \frac{E^2}{E_P^2}\right),$$

where E_P is the Planck energy (second-order Lorentz Invariance Violation). Compute the time lag between two VHE photons as a function of the energy difference and of the redshift z.

Let us suppose that the two photons are emitted simultaneously from a source, because of Universe expansion and because of their difference in energy they will arrive on Earth at different times and the distance they cover will be different. The comoving distance, instead, is the same for both photons, so we can derive time lag by imposing that they travel the same comoving distance.

The comoving trajectory of a particle is obtained from the Hamiltonian writing the comoving momentum:

$$\mathscr{H} = \frac{pc_0}{a}\left(1 + \xi\frac{(pc_0)^2)}{a^2 E_P^2}\right)$$

where $a = (1+z)^{-1}$. The velocity can be obtained as $\frac{dx}{dt} = v = d\mathscr{H}/dp$:

$$v = \frac{c_0}{a}\left[1 + \xi\left(\frac{pc_0}{aE_P}\right)^2\right] + \frac{pc_0}{a}\xi\left(2\frac{pc_0^2}{a^2 E_P^2}\right)$$

$$= \frac{c_0}{a}\left[1 + 3\xi\left(\frac{pc_0}{aE_P}\right)^2\right].$$

Indeed the comoving distance is:

$$x(t) = \int_0^t \frac{c_0}{a(t')}\left[1 + 3\xi\left(\frac{E}{a(t')E_P}\right)^2\right]dt'.$$

We can rewrite this distance in term of z, redshift of emission:

$$x(z, E) = \frac{c_0}{H_0}\int_0^z\left[1 + 3\xi\left(\frac{E}{E_P}\right)^2(1+z')^2\right]\frac{dz'}{\sqrt{\Omega_\Lambda + \Omega_M(1+z')^3}}.$$

Let us now analyze the time difference between the photon with energy E_1 emitted from redshift z and arrive at 0 and the other (E_2) emitted at z, but which reaches us at $-\Delta_z$, equating the 2 comoving distance, we obtain:

$$\frac{c_0}{H_0}\int_0^z\left[1 + 3\xi\left(\frac{E_1}{E_P}\right)^2(1+z')^2\right]\frac{dz'}{\sqrt{\Omega_\Lambda + \Omega_M(1+z')^3}}$$

$$= \frac{c_0}{H_0}\int_{-\Delta_z}^z\left[1 + 3\xi\left(\frac{E_2}{E_P}\right)^2(1+z')^2\right]\frac{dz'}{\sqrt{\Omega_\Lambda + \Omega_M(1+z')^3}}. \quad (10.50)$$

We can rewrite the second term as:

$$\frac{c_0}{H_0}\int_{-\Delta_z}^0\left[1 + 3\xi\left(\frac{E_2}{E_P}\right)^2(1+z')^2\right]\frac{dz'}{\sqrt{\Omega_\Lambda + \Omega_M(1+z')^3}}$$

$$+ \frac{c_0}{H_0}\int_0^z\left[1 + 3\xi\left(\frac{E_2}{E_P}\right)^2(1+z')^2\right]\frac{dz'}{\sqrt{\Omega_\Lambda + \Omega_M(1+z')^3}}$$

and approximating the first factor as $\frac{c_0}{H_0}\Delta_z$. Equation (10.50) becomes:

$$\Delta_z = 3\xi \frac{E_1^2 - E_2^2}{E_P^2} \int_0^z \frac{(1+z')^2 dz'}{\sqrt{\Omega_\Lambda + \Omega_M(1+z')^3}}$$

Finally:

$$\Delta t = \frac{\Delta_z}{H_0} = \frac{3\xi}{H_0} \frac{E_1^2 - E_2^2}{E_P^2} \int_0^z \frac{(1+z')^2 dz'}{\sqrt{\Omega_\Lambda + \Omega_M(1+z')^3}}.$$

11. *Difference between the speed of light and the speed of gravitational waves (9).*
    Derive a limit on the relative difference between the speed of light and the speed
    of gravitational waves from the fact that the gamma-ray burst GRB170824A at
    a distance of about 40 Mpc was detected about 1.7 s after the gravitational wave
    GW170817.

Photons should have been emitted later then the gravitational waves and so,
without knowing this time delay, it is not really possible to have an accurate
estimation of the possible relative difference between the speed of light and the
speed of gravitational waves. But a upper limit can of course be derived:
In fact

$$t_\gamma \simeq \frac{d}{c} \;\; ; \;\; t_{GW} \simeq \frac{d}{c_{GW}},$$

then

$$(c_{GW} - c_\gamma) = \frac{\Delta t}{d} c c_{GW} \simeq \frac{\Delta t}{d} c^2,$$

being $d \simeq 40$ Mpc $\simeq 1.2 \times 10^{24}$ m and $\Delta t \simeq 1.7$ s, and thus

$$(c_{GW} - c_\gamma) \simeq 1.3 \times 10^{-7} \mathrm{ms}^{-1}.$$

12. *Flux of photons from Crab (10).* Consider the expression

$$\frac{dN}{dE} \simeq 1.8 \times 10^4 E^{-2.7} \frac{\text{nucleons}}{\text{m}^2 \, \text{s} \, \text{sr} \, \text{GeV}} \tag{10.51}$$

(also discussed in the textbook) and let us assume that the flux of cosmic rays
between 0.05 TeV and 2 PeV follows that expression. On the other hand, the flux
from the most luminous steady (or almost steady) source of, the Crab Nebula,
follows, according to the measurements from MAGIC, a law:

$$N_\gamma(E) \simeq 3.23 \times 10^{-7} \left(\frac{E}{\text{TeV}}\right)^{-2.47-0.24\left(\frac{E}{\text{TeV}}\right)} \text{TeV}^{-1}\text{s}^{-1}\text{m}^{-2}. \tag{10.52}$$

Translate this expression into GeV. Compute the number of photons from Crab
hitting every second a surface of $10000 \, \text{m}^2$ above a threshold of 50 GeV, 100

GeV, 200 GeV, 1 TeV, up to 500 TeV. Compare this number to the background from the flux of cosmic rays in a cone of 1 degree of radius.

First of all we convert from TeV to GeV the above relation for the flux of gamma rays:

$$N_\gamma \simeq 3.23 \times 10^{-14} \left(\frac{E \times 10^{-3}}{\text{GeV}} \right)^{-2.47-0.24\times 10^{-3}\frac{E}{\text{GeV}}} \text{GeV}^{-1}\text{s}^{-1}\text{cm}^{-2}$$

Considering an area of $10000\,\text{m}^2$, the number of photons per second is obtained integrating the expression found above in the appropriate range.

$$\frac{\#\text{photons}}{\text{s}} = \int_x^{500\times 10^3\text{GeV}} 3.23$$

$$\times 10^{-14} \times 10^8 \left(\frac{E \times 10^{-3}}{\text{GeV}} \right)^{-2.47-0.24\times 10^{-3}\frac{E}{\text{GeV}}} dE \quad \text{GeV}^{-1}\text{s}^{-1}$$

For the ranges discussed in the problem, we have found, using Mathematica, the following results:

a. $x = 50$ GeV, so $\frac{\#\text{photons}}{\text{s}} \simeq 1.9 \times 10^{-1}$
b. $x = 100$ GeV, so $\frac{\#\text{photons}}{\text{s}} \simeq 6.9 \times 10^{-2}$
c. $x = 200$ GeV, so $\frac{\#\text{photons}}{\text{s}} \simeq 2.5 \times 10^{-2}$
d. $x = 1000$ GeV, so $\frac{\#\text{photons}}{\text{s}} \simeq 1.6 \times 10^{-3}$.

The expression for the flux of cosmic rays, instead, is:

$$\frac{dN_{CR}}{dE} = 1.8 \times 10^4 E^{-2.7} \frac{\text{nucleons}}{\text{GeV m}^2 \text{ sr}}.$$

For the same area as before and for a cone of 1 degree of radius:

$$\frac{dN_{CR}}{dE} = 1.8 \times 10^8 \times \left(\frac{\pi^3}{180^2} \right) E^{-2.7} \frac{\text{nucleons}}{\text{GeV}}$$

Finally we integrate this expression in the ranges specified above:

a. $x = 50$ GeV, so $\frac{\#CR}{\text{s}} \simeq 130$
b. $x = 100$ GeV, so $\frac{\#CR}{\text{s}} \simeq 40$
c. $x = 200$ GeV, so $\frac{\#CR}{\text{s}} \simeq 12$
d. $x = 1000$ GeV, so $\frac{\#CR}{\text{s}} \simeq 0.80$.

We can compare the two resuts:

a. $\frac{N_\gamma}{N_{CR}} = 1.4 \times 10^{-3}$

Fig. 10.2 Scheme for the travel of protons from the Crab nebula to Earth. Distances and angles are not to scale

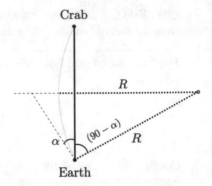

b. $\frac{N_\gamma}{N_{CR}} = 1.70 \times 10^{-3}$

c. $\frac{N_\gamma}{N_{CR}} = 1,98 \times 10^{-3}$

d. $\frac{N_\gamma}{N_{CR}} = 1,94 \times 10^{-3}$.

13. *Astronomy with protons? (11)*. If the average magnetic field in the Milky Way is 1 μG, what is the minimum energy of a proton coming from Crab Nebula (at a distance of 2 kpc from the Earth) we can detect as "pointing" to the source?

   Protons coming from the Crab Nebula at a distance of 2 kpc from the Earth will suffer a deflection due to the Milky Way magnetic field having thus an approximate circular trajectory with a radius of curvature R given by:

   $$\frac{R}{rm1\,kpc} \simeq \frac{E/1\,\mathrm{EeV}}{B/1\,\mu\mathrm{G}}.$$

   Let's now define that "pointing" to the source means that it is possible to determine the position of the source within an angle α. Then the radius of the circular trajectory that connect Crab and Earth should be higher than (see Fig. 10.2):

   $$R \simeq \frac{d}{\cos(\pi/2 - \alpha)}.$$

   where d = 1 kpc is half of the distance between the Earth and the Crab Nebula. Assuming a mean Galactic magnetic field of about $1\mu G$ and an angle $\alpha \simeq 1^o$:

   $$R > 57\mathrm{kpc} \; ; \quad E > 5 \times 10^{16}\,\mathrm{eV} = 50\,\mathrm{PeV}.$$

14. *Maximum acceleration energy for electrons (12)*. The synchrotron loss rate is relatively much more important for electrons than for protons. To find the limit placed by synchrotron losses on shock acceleration of electrons, compare the acceleration rate for electrons with the synchrotron loss rate. The latter is negligible at low energy, but increases quadratically with E. Determine the crossover

energy, and compare it to supernova ages. Is the acceleration of electrons limited by synchrotron radiation?

Considering a supernova explosion and its shock wave, on one hand there will be an acceleration rate given by the Fermi mechanism, which is equal to

$$\frac{dE}{dt} = \frac{\beta E}{T_{cycle}}$$

where T_{cycle} is the time for one back and forth encounter. On the other hand there will be the energy loss rate related to the synchrotron effect

$$-\left(\frac{dE}{dt}\right)_{syncr} \sim 1.6 \times 10^{-3} \frac{erg}{s} \left(\frac{Zm_e}{Am}\right)^4 E^2 B^2$$

for typical values.

To find out what is the value of the crossover energy the modulus of this two equation must be equal:

$$E \sim \frac{\beta}{T_{cycle}} \times \left(1.6 \times 10^{-3} \frac{erg}{s} \left(\frac{Zm_e}{Am}\right)^4 B^2\right)^{-1}$$

where typical values must be substituted.

To compare the result with the supernova ages, maximum energies must be considered:

- integrating the Fermi mechanism on time and considering $T_S = 1000$ yr as the limit of time in which the energy given by the explosion is accessible

$$E_{max} \leq Z \times 3 \times 10^4 \quad GeV.$$

- for the synchrotron loss, comparing its formula with the equation for E_{max} we obtain an upper limit

$$E_{max}^{syncr} \sim 23 \frac{u_1}{cB^{1/2}} \quad TeV.$$

But for typical values of the magnetic field ($B \sim 3\mu G$) and of the velocity of the wave front ($u_1 \sim 5 \times 10^8$ cm^2/s) $E_{max}^{syncr} \sim 220$ TeV, higher than E_{max}. This means that the acceleration of electrons is not limited by synchrotron radiation, making sure that the magnetic field playing is low as much as the one considered. Otherwise for large magnetic fields this is no more valid.

15. *Classification of blazars (13).* Looking to Fig. 10.3, how would you classify Markarian 421, BL Lac and 3C279 within the blazar sequence? Why?

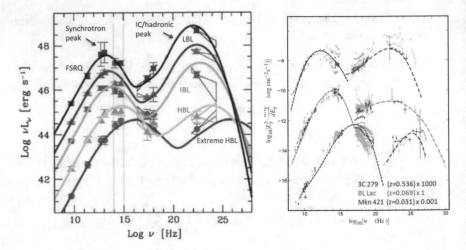

Fig. 10.3 Left: The blazar sequence, from G. Fossati et al., Mon. Not. Roy. Astron. Soc. 299 (1998) 433. Right: The SED of Markarian 421, BL Lac and 3C279

Observationally, blazars are divided into two main subclasses depending on their spectral properties.

- Flat Spectrum Radio Quasars, or FSRQs, show broad emission lines in their optical spectrum.
- BL Lacertae objects (BL Lacs) have no strong, broad lines in their optical spectrum. They are further classified according to the energies of the synchrotron peak $\hat{\nu}_S$ of their SED; they are called accordingly:

  – low-energy peaked BL Lacs (LBLs) if $\hat{\nu}_S \lesssim 10^{14}$ Hz (about 0.4 eV);
  – intermediate-energy peaked BL Lacs (IBL);
  – high-energy peaked BL Lacs (HBL) if $\hat{\nu}_S \gtrsim 10^{15}$ Hz (about 4 eV).

Typically FSRQs have a synchrotron peak at lower energies than LBLs.

Based on the frequency (energy) of the maximum of the synchrotron peak, 3C279 is a LBL or a FSRQ, BL Lac is an IBL and Markarian 421 is a HBL. Spectroscopy shows that 3C279 is a FSRQ.

16. *GZK threshold (15), (2.10).* The Cosmic Microwave Background fills the Universe with photons with a peak energy of 0.37 meV and a number density of $n \sim 400/\text{cm}^3$. Determine:

   a. The minimal energy (known as the GZK threshold) that a proton should have in order that the reaction $p\gamma \rightarrow \Delta$ may occur.
   b. The interaction length of such protons in the Universe considering a mean cross-section above the threshold of 0.6 mb.

c. Pair production $p\gamma \rightarrow e^+e^-$ is a competitive process, with a cross section of about 20 mb. What is its threshold, and the propagation length?

---

a. In order that the reaction $p\gamma \rightarrow \Delta^+$ may occur the center-of-mass energy should be greater than the mass of the Δ particle:

$$\left(P_p + P_{\gamma CMB}\right)^2 \geq m_\Delta^2,$$

$$E_p \geq \frac{m_\Delta^2 - m_p^2}{2\,E_{\gamma CMB}\,(1 - \beta\cos\theta)}.$$

Assuming head-on collisions,

$$E_p \gtrsim \frac{m_\Delta^2 - m_p^2}{4\,E_{\gamma CMB}} \simeq 3.7 \times 10^{20}\text{eV}.$$

In fact the GZK is not a sharp cut-off. Taking into account the full distributions of the energy of the CMB photons (for instance the mean value of this distribution is 0.635 meV) and of the $p\gamma \rightarrow \pi^0 p$ and $p\gamma \rightarrow \pi^+ n$ cross sections, the effect of the GZK suppression should starts to be effective for proton energies of the order of $E_p \sim 6 \times 10^{19}\text{eV}$.

b. The interaction length L_{int} is given by

$$L_{int} = \frac{1}{\sigma\, n}$$

where σ is the total cross section and n the number density of targets (in this case the CMB photons). Replacing the given values one obtains

$$L_{int} \simeq 4 \times 10^{24}\ \text{cm} \simeq 1.3\ \text{Mpc}.$$

Indeed 0.6 mb is the peak photon-pion cross section at the Δ resonance. Taking into account all the relevant GZK energy range, a mean photon-pion cross section of around 0.1 mb should be instead considered and a more realistic value for L_{int} would be around 6 Mpc.

c. Now, the reaction to be considered is: $p\gamma \rightarrow e^+e^-$ and: - the center-of-mass energy should be greater than twice the electron mass:

$$\left(P_p + P_{\gamma CMB}\right)^2 \geq 2\,m_e^2,$$

$$E_p \gtrsim \frac{m_e^2 + m_p m_e}{E_{\gamma CMB}} \simeq 1.3 \times 10^{18}\text{eV};$$

- the interaction length is ($\sigma = \simeq 20$ mb; $n \sim 400/\text{cm}^3$):

$$L_{int} = \frac{1}{\sigma\, n} \simeq 0.04 \text{ Mpc.}$$

17. *Mixing photons with paraphotons (16).* The existence of a neutral particle of tiny mass μ, the paraphoton, coupled to the photon, has been suggested to explain possible anomalies in the CMB spectrum and in photon propagation (the mechanism is similar to the one discussed to the photon-axion mixing, but there are no complications related to spin here). Calling ϕ the mixing angle between the photon and the paraphoton, express the probability of oscillation of a photon to a paraphoton as a function of time (note: the formalism is the same as for neutrino oscillations). Supposing that the paraphoton is sterile, compute a reasonable range of values for ϕ and μ that could explain an enhancement by a factor of 2 for the signal detected at 10 TeV from an AGN at $z \simeq 0.01$.

The probability of a photon not to oscillate into a paraphoton a, as a function of time, can be written as:

$$P(t)_{\gamma \to \gamma} = 1 - \sin^2(2\phi) \sin^2\left(\frac{c^2 \mu^2 t}{4\hbar^2 \omega}\right) \tag{10.53}$$

with where ω is the frequency of the photon, ϕ is the mixing angle and μ is the mass difference between the photon and paraphoton, in this case $m_\gamma = 0$ so μ is just the paraphoton mass (note that the formalism is the same as for a two-flavor neutrino oscillation - see Sect. 9.1.2 of the textbook). Consequently the probability to oscillate is:

$$P(t)_{\gamma \to a} = \sin^2(2\phi) \sin^2\left(\frac{c^2 \mu^2 t}{4\hbar^2 \omega}\right). \tag{10.54}$$

The observed photon intensity at Earth $I_{obs}(z)$ from a source at a redshift z, in the case of no oscillation, is a function of the emitted photon intensity $I(z)$ and of its absorption in the extragalactic medium. Considering an exponential attenuation,

$$I_{obs} = I(z)e^{-\tau(E,z)} \quad , \tag{10.55}$$

where the values for the *optical depth* $\tau(E, z)$ can be found tabulated in www.astro.unipd.it/background/:

$$\tau(E = 10 \text{ TeV}, z = 0.01) \simeq 0.35 \quad . \tag{10.56}$$

The relation between distance d and redshift z can be expressed with the Hubble law:

Fig. 10.4 The parameter $\sin^2(2\phi)$ as a function of μ

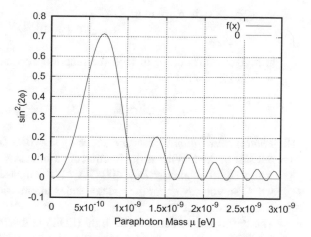

$$d = c\frac{z}{H_0} \quad ; \quad H_0 = 70\,\frac{\text{km}}{\text{s Mpc}}. \tag{10.57}$$

In case of photon-paraphoton oscillations, instead, the attenuation "weighted" by the oscillation probability can be calculated with:

$$
\begin{aligned}
\tau_{osc}(E, z) &= \tau(E, z) \int_0^z \left[1 - P(z')_{\gamma \to a}\right] dz' \\
&= \tau(E, z) \int_0^z \left[1 - \sin^2(2\phi) \sin^2\left(\frac{c^2\mu^2 z'}{4\hbar^2 \omega H_0}\right)\right] dz' \\
&= \tau(E, z) \left[1 - \sin^2(2\phi)\left(\frac{z}{2} - \frac{\hbar^2 H_0 \omega}{c^2\mu^2}\sin^2\left(\frac{c^2\mu^2 z}{2\hbar^2\omega H_0}\right)\right)\right]
\end{aligned}
\tag{10.58}
$$

To obtain a factor of 2 enhancement:

$$2 = \frac{e^{\tau_{osc}(E,z)}}{e^{\tau(E,z)}} \quad ; \quad \tau_{osc}(E, z) - \tau(E, z) = \ln 2 \tag{10.59}$$

$$-\sin^2(2\phi)\left[\frac{z}{2} - \frac{\hbar^2 H_0 \omega}{c^2\mu^2}\sin^2\left(\frac{c^2\mu^2 z}{2\hbar^2\omega H_0}\right)\right] = \frac{\ln 2}{\tau(E, z)}. \tag{10.60}$$

Calling $a = \frac{c^2 z}{2\hbar^2\omega H_0} = 2.4 \times 10^{18}\,\frac{c^2}{eV^2}$:

$$\sin^2(2\phi) = \frac{\ln 2}{\tau(E, z)}\left[\frac{1}{2a\mu^2}\sin^2\left(a\mu^2\right) - \frac{z}{2}\right]. \tag{10.61}$$

The relation between $\sin^2(2\phi)$ and μ, in order to obtain a factor 2 enhancement, is shown in Fig. 10.4; the possible values of μ are limited by the requirement

$\sin^2(2\phi) \geq 0$, in particular:

$$\mu \leq 6.4 \times 10^{-9} \tag{10.62}$$

at higher values of μ, in fact, the function becomes negative. The parameter $\sin^2(2\phi)$ is limited to:

$$\sin^2(2\phi) \leq 0.71 \quad ; \quad \phi \leq 0.497. \tag{10.63}$$

18. *Photon absorption affects the shape of the SED (17).* TXS 0506 +056 has a redshift of 0.34. What is the fraction of gamma rays absorbed due to interaction with EBL at an energy $E = 400$ GeV? If the measured spectral index if of 2.3, what can you say about the spectral index at emission?

The extragalactic background light (EBL) is the integrated intensity of all of the light (mainly ultraviolet, optical, and near-infrared) emitted throughout the history of the Universe across the whole of the electromagnetic spectrum. This light is observed in the visible and near infrared regions and depends on the redshift z, due to the expansion of the Universe.

Photons of energy 100 GeV - 100 TeV (VHE photons) interact with the EBL photons via pair production of electrons and positrons:

$$\gamma_{VHE} + \gamma_{EBL} \rightarrow e^+ + e^- \tag{10.64}$$

resulting in an energy and redshift attenuation of the intrinsic gamma-ray emission. The attenuation of a gamma-ray emitting sources initial photon flux, due to pair production interactions as the photons travel to the observer, can be quantified by the optical depth τ.

In particular, we want to measure the probability for a photon of observed energy E to survive absorption along its path from its source at redshift z to the observer. We can interpret the probability as an attenuation factor for the radiation flux:

$$P = e^{-\tau(E,z)}. \tag{10.65}$$

So, what we need is the optical depth $\tau(E, z)$.

Deriving τ requires integration of the pair production cross section and the number density of EBL photons over energy, as well as the distance between the source and the observer and the angles between the EBL and primary photons. So, the formula is:

$$\tau(E, z) = \int_0^z dl(z) \int_{-1}^1 d\cos\theta \, \frac{1 - \cos\theta}{1} \times \int_{\frac{2(m_e c^2)^2}{E(1-\cos\theta)}}^{\infty} d\varepsilon(z) \, n_\varepsilon\big(\varepsilon(z), z\big) \, \sigma\big(E(z), \varepsilon(z), \theta\big) \tag{10.66}$$

where $l(z) = cdt(z)$ is the distance as a function of the redshift and it depends on the cosmological parameters, θ is the scattering angle, and $n_\varepsilon(\varepsilon(z), z)$ is the density for photons of energy $\varepsilon(z)$ at the redshift z.

We can refer for example to the values of the photon-photon optical depth reported in the web site of the analysis by Franceschini, Rodighiero and Vaccari (www.astro.unipd.it/background/). The value is a function of the energy (E) and redshifts z of the sources.

For this exercise we can consider the values of the photon energy E and of the redshift closer to our:

- $E = 0.381$ TeV
- $z = 0.3$.

Thus, the corresponding value of the optical depth is $\tau = 1.281$.

Now, using the Eq. 10.65, we have the fraction of gamma-rays absorbed by the EBL: $P \simeq 0.28$, which is approximately, the 28% of the initial photon flux of the source.

Considering now the spectral index Γ, it also changes. The observed photon flux (the measured flux) is lower than the initial photon flux of the source because, we have seen, part of the VHE photons are absorbed by the EBL, and so, the measured flux is attenuated.

Then, the spectral index at emission will be lower that the measured spectral index: $\Gamma_{measured} < \Gamma_{intrinsic}$, thus lower than 2.3. Consequently the observed flux results steeper than the emitted one.

19. *Estimating the energy of a cosmic accelerator from the energy of emitted neutrinos (18).* How would you estimate the energy of the proton generating a 300 TeV neutrino in the flare of a blazar (this is the case of the famous source TXS 0506 +056, the first AGN localized as a source of neutrinos)?

There is two way to estimate the proton energy of the blazar by hadronic model of neutrinos and photons production. We can estimate the proton energy via the detected photon energy decayed from the neutral pion produced from the accelerated proton and the photon backgrounds. The approximated formula relating photon energy and proton energy is given by:

$$E_p \sim \frac{350}{E_\gamma/(1eV)} \text{ PeV.} \tag{10.67}$$

For synchrotron photons with a mean energy $\sim 60\,$eV, the estimated proton energy is approximately:

$$E_p \sim \frac{350}{60} \text{ PeV} \sim 5.8 \text{ PeV.} \tag{10.68}$$

For a neutrino energy of 300 TeV, the proton energy can be estimated by the following formula:

$$E_p \sim 20E_\nu. \tag{10.69}$$

Thus:

$$E_p \sim 20(300e^{12}) \sim 6 \text{ PeV.} \tag{10.70}$$

20. *Dwarf Spheroidals and the hunting for Dark Matter (20).* Draco is a dwarf spheroidal galaxy within the Local Group. Its luminosity is $L = (1.8 \pm 0.8) \times 10^5 L_\odot$ and half of it is contained within a sphere of radius of (120 ± 12) pc. The measured velocity dispersion of the red giant stars in Draco is (10.5 ± 2.2) km/s. What is our best estimate for the mass M of the Draco dSph? What about its M/L ratio? Which are our main uncertainties in such determinations?

The Draco dwarf spheroidal galaxy is characterized by these parameters:

$$L = (1.8 \pm 0.8) \times 10^5 L_\odot$$
$$R_{1/2} = (120 \pm 12)\text{pc} = (3.70 \pm 0.37) \times 10^{15}\text{km}$$
$$\sigma = (10.5 \pm 2.2)\text{ km/s}.$$

We defined L as the luminosity, $L_\odot$ as the solar luminosity, $R_{1/2}$ as the radius of the sphere which contains $L/2$ and σ as the velocity dispersion of the red giants stars. The mass of the galaxy can be estimate using the virial theorem:

$$2K + \Omega = 0 \rightarrow \quad \sigma^2 - \frac{MG}{R} = 0$$
$$M = \frac{R\sigma^2}{G}.$$

If we use the values given before and $G = 6.67 \times 10^{-20}$ km^{3}kg^{-1}s^{-2}, we get

$$M_{Draco} = (6.13 \pm 3.18) \times 10^{36}\text{kg} = (3.23 \pm 1.68) \times 10^6 M_\odot$$

inside the $R_{1/2}$ sphere.

We can derive the galaxy mass-luminosity ratio: $M/L = (18 \pm 17) M_\odot/L_\odot$ in the inner part of the galaxy.

The main uncertainties can be found in the σ estimate, derived from Doppler measurements on the stars spectra, and in the way we use the virial theorem. In fact we are assuming that stars are good tracers of the mass distribution without considering Dark Matter.

However it is known that dwarf galaxies are DM dominated: this fact increases the M/L ratio. This is not so evident in this exercise because we focused on the inner part of the galaxy; recent and more accurate studies which include also the external part found M/L over $300 M_\odot/L_\odot$.

21 *Tremaine-Gunn bound (21).* Assume that neutrinos have a mass, large enough that they are non-relativistic today. This neutrino gas would not be homogeneous, but clustered around galaxies. Assume that they dominate the mass of these galaxies (ignore other matter). We know the mass $M(r)$ within a given radius r in a galaxy from the velocity $v(r)$ of stars rotating around it. The mass could be due to a few species of heavy neutrinos or more species of lighter neutrinos.

But the available phase space limits the number of neutrinos with velocities below the escape velocity from the galaxy. This gives a lower limit for the mass of neutrinos. Assume for simplicity that all neutrinos have the same mass. Find a rough estimate for the minimum mass required for neutrinos to dominate the mass of a galaxy. Assume spherical symmetry and that the escape velocity within radius r is the same as at radius r. Compare this value to the result $m_\nu < 0.2\,\mathrm{eV}$, from cosmology. What do you conclude?

Neutrinos are fermions and as such they obey to a Fermi-Dirac statistics. This means that two neutrinos in the same state (spin) cannot overlap. For simplicity, let us assume that the Galaxy is a box with size L. It can be shown, that the total number of states in this box can be written as

$$N_{\mathrm{tot}} = \frac{1}{8}\left[\frac{4}{3}\pi\left(\frac{2m_\nu E_n L^2}{h^2}\right)^{\frac{3}{2}}\right] \tag{10.71}$$

where E_n is the energy in the $n - th$ state, L the size of the box, m_ν the mass of the neutrino and h is the Planck constant. We shall make the approximation that $L \sim r_g$, i.e. the size of the box is equation to the radius of the galaxy.
As a neutrino can have two spin states (up and down) the total mass in the galaxy, M_g, can be written as

$$M_g = 2m_\nu N_{\mathrm{tot}}. \tag{10.72}$$

The energy in the state E_n can be connected to the kinetic energy of the starts at the periphery of the galaxy which is equal to the gravitation potential here such that

$$E_n = \frac{1}{2}mv_g^2 = G\frac{M_g m_\nu}{r_g}, \tag{10.73}$$

where v_g is the stars tangential velocity in the periphery of the galaxy, at a distance r_g from its center. Notice, that above this energy the neutrinos could escape the galaxy. Combining Eqs. 10.71, 10.72 and 10.73 one can finally obtain the following expression for the neutrino mass:

$$m_\nu = 0.76\,h^{\frac{3}{4}} G^{-\frac{3}{8}} M_g^{-\frac{1}{8}} r_g^{-\frac{3}{8}}. \tag{10.74}$$

For our Galaxy $r_g = 12\,\mathrm{kpc}$ and its total mass is $M_g \sim 10^{12}\,M_\odot$. Plugging these values into Eq. 10.74 we obtain

$$m_\nu \simeq 6.51 \times 10^{-35}\,\mathrm{kg} = 36.5\,\mathrm{eV}/c^2. \tag{10.75}$$

From cosmology we know that the mass of neutrinos should be smaller than the obtained value which indicates that neutrinos in our Galaxy are not a candidate for dark matter.

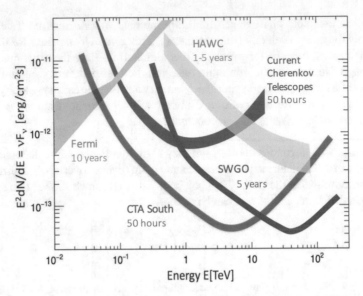

Fig. 10.5 Comparison of the differential flux sensitivity of current and future VHE gamma-ray observatories. From J. Hinton and E. Ruiz-Velasco, https://arxiv.org/abs/1911.06097

22. *Characteristics of synchrotron radiation.* How does the power emitted by a relativistic particle of energy E and mass m moving in a circular orbit (for example in a synchrotron) depend on E/m?

The synchrotron power emitted by a curving charged particle at fixed energy is

$$P \simeq \frac{2}{3}e^2 c R^2 \,,$$

where R is the curvature radius. The emitted power scales as $(E/m)^4$ - electrons radiate much more energy, and it is thus more difficult to accelerate them. Synchrotron radiation is polarized orthogonally to the orbital plane.

23. *Differential sensitivity plot in gamma-ray instruments.* A typical way to express the sensitivity of a gamma-ray instrument is the *differential sensitivity plot* as shown in Fig. 10.5. This plot shows the minimum detectable flux at 5 standard deviations for a source at a given energy, in a collection time of 50h for a Cherenkov detector and of 1 year for extensive air shower detectors and satellite-based telescopes.

   a. Why this time interval of 50h?
   b. How does the sensitivity change as a function of the effective area covered by the detector?
   c. with time and with the number of standard deviations required?

a. The typical observation time, or *duty cycle,* for a Cherenkov telescope during a year (i.e., the time of good weather, dark night or moderate moon) is about 1200 h. As a consequence, the typical time assigned to each observation source is of the order of 50 h. Extensive air shower detectors, instead, are non-pointing instruments, and can observe half of the sky for the full year, day and night. The limit at 5σ is the standard significance to claim the discovery of a new source in astrophysics: the rather conservative number of standard deviations allows compensating for possible background underestimates and for the so-called *look-elsewhere effect*: an apparently statistically significant observation may have actually arisen by chance because of the size of the sky studied.

b. Under the simplification that the sensitivity curve is drawn at

$$\frac{S}{\sqrt{B}} = n = 5$$

we can write, calling Σ the signal per unit area per second and β the background per unit area per second,

$$\frac{\Sigma}{\sqrt{\beta}}\sqrt{\mathscr{A}t} = 5$$

where $\mathscr{A}$ is the effective area.

In the cases in which one can assume that $\mathscr{A} \propto A$, where A is the area covered by the detector, one has simply

$$\Sigma \underset{\sim}{\propto} \frac{5\sqrt{\beta}}{\sqrt{At}}.$$

c. Related to the scaling of the sensitivity curves with time and with the number of standard deviations n, one has

$$\frac{\Sigma}{\sqrt{\beta}}\sqrt{\mathscr{A}t} = n \implies \Sigma_{min} \underset{\sim}{\propto} \frac{n}{\sqrt{At}}.$$

24. *Integral sensitivity plot in gamma-ray instruments.* Another typical way to express the sensitivity of a gamma-ray instrument is the *integral sensitivity plot* as shown in Fig. 10.6. This plot shows the minimum detectable flux at 5 standard deviations for a source integrating all energies above a given energy, in a collection time of 50 h for Cherenkov detectors and of 1 year for extensive air shower detectors, and assuming an energy dependence for the flux of the source.

Fig. 10.6 Comparison of the integral sensitivity of different VHE gamma-ray observatories. From G. Di Sciascio 2019, J. Phys. Conf. Ser. 1263 012003

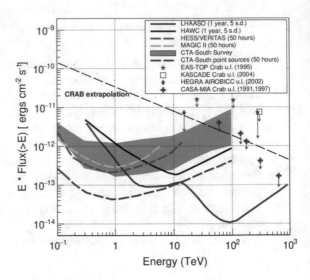

Comment on the difference between the integral and the differential energy plot. What happens if you compare an integral sensitivity plot assuming an energy dependence E^{-2} or E^{-3}?

The differential flux sensitivity plot gives the minimum detectable flux of a source in defined energy intervals. It thus provides a measure of the capability of the detector to assess the energy spectrum from that source. Following the discussion of the previous exercise, the minimum detectable flux in a given energy interval $[E, E + \Delta E]$, $\Phi_{min}(E)$ is

$$\Sigma_{min} = \int_{E}^{E+\Delta E} \Phi_{min}(E')dE',$$

where Σ_{min} is computed as described in the previous exercise, taking the background in the same energy interval.

Considering that ΔE is sufficiently small such that $\Phi(E)$ is constant, one obtains

$$\Phi_{min}(E) = \frac{\Sigma_{min}}{\Delta E}.$$

Assuming that the flux of background events is also constant in ΔE, one can write

$$\Phi_{min}(E) \stackrel{\propto}{\sim} \frac{n}{\sqrt{At}} \frac{\sqrt{\Delta E}}{\Delta E}.$$

Therefore at a given energy and fixed observation conditions one has

$$\Phi_{\min}(E) \underset{\sim}{\propto} \frac{1}{\sqrt{\Delta E}}.$$

There is thus a compromise in choosing ΔE, given that a smaller value allows for a finer measurement of the spectrum but requires a brighter source, as the minimum detectable flux increases.

Considering now the integral sensitivity, it gives the minimum detectable flux above a given energy E_0. In this case one has,

$$\Sigma_{\min} = \int_{E_0}^{\infty} \Phi_{\min}(E')dE'.$$

Assuming a power-law spectrum for the source,

$$\Phi(E) = \Phi_0 \left(\frac{E}{E_0}\right)^{-\alpha} \quad (\alpha > 1),$$

with $\Phi_0 = \Phi(E_0)$, one gets

$$\Phi_{0,min} = \frac{\alpha - 1}{E_0} \Sigma_{\min}.$$

In this case the minimum detectable flux depends on the source's energy spectrum. For a power-law energy dependence, it increases as the power index increases meaning that, for the same observation conditions, a source with a higher flux at high energies is more easily detectable. In particular, when comparing integral sensitivity plots, assuming an energy dependence E^{-2} or E^{-3}, the first will show fluxes two times smaller than the second. When comparing the perfomance of different experiments, the spectrum from the well known Crab nebulae is usually taken as a reference for the integral sensitivity plot.

Chapter 11
Astrobiology and the Relation of Fundamental Physics to Life

1. *Effects of the Sun and of the Moon on tides (1)*
   The mass of the Moon is about 1/81 of the Earth's mass, and the mass of the Sun is 333 000 times the Earth's mass. The average Sun-Earth distance is 150 $\times 10^6$ km, while the average Moon–Earth distance is 0.38 $\times 10^6$ km (computed from center to center).

   (a) What is the ratio between the gravitational forces by the Moon and by the Sun?
   (b) What is the ratio between the tidal forces (i.e., between the differences of the forces at two opposite sides of the Earth along the line joining the two bodies)?

   One has

   $$\left(\frac{M_\odot}{M_{Moon}}\right) \simeq 2.71 \times 10^7 \qquad \left(\frac{D}{d}\right) \simeq 391$$

   and thus:

   a.

   $$\frac{F_{Moon \to Earth}}{F_{\odot \to Earth}} = \left(\frac{M_{Moon}}{M_\odot}\right)\left(\frac{D}{d}\right)^2 \simeq 5.7 \times 10^{-3}.$$

   b. The tidal force on the Earth of a generic celestial body of mass M at distance δ from the Earth's center can be expressed as

   $$F_T = G\frac{M_{Earth}M}{(\delta - R_{Earth})^2} - G\frac{M_{Earth}M}{(\delta + R_{Earth})^2} \simeq 2R_{Earth}G\frac{M_{Earth}M}{\delta^3}.$$

   Thus

© Springer Nature Switzerland AG 2021

A. De Angelis et al., *Particle and Astroparticle Physics*, Undergraduate Lecture Notes in Physics,
https://doi.org/10.1007/978-3-030-73116-8_11

$$\frac{F_{T,Moon \to Earth}}{F_{T,\odot \to Earth}} = \left(\frac{M_{Moon}}{M_\odot}\right)\left(\frac{D}{d}\right)^3 \simeq 2.2 \,.$$

This explains why the Moon is more important than the Sun related to the phenomenon of oceanic tides, despite the smaller gravitational pull.

2. *Temperature of the Earth and Earth's atmosphere (2)*

   What is the maximum temperature for which the Earth could trap an atmosphere containing atomic oxygen O?

   A planet can trap an atom of mass m in its atmosphere when the escape speed is larger than the thermal velocity of the atom itself.

   The average velocity at thermal equilibrium for a molecule or an atom of mass m is given by

$$\frac{1}{2}m <v^2> = \frac{3}{2}k_B T \Rightarrow \sqrt{<v^2>} = \sqrt{\frac{3k_B T}{m}} \,.$$

   where T is the temperature and k_B is the Boltzmann constant.

   The escape speed v_{esc} is instead given by the equation

$$\frac{1}{2}mv_{esc}^2 - G\frac{Mm}{R} = 0 \Rightarrow v_{esc} = \sqrt{\frac{2GM}{R}} \,.$$

   The condition to keep an atom of mass m in the atmosphere is thus

$$\sqrt{<v^2>} = \sqrt{\frac{3k_B T}{m}} \ll v_{esc} = \sqrt{\frac{2GM}{R}} \to T \ll \frac{2GMm}{3k_B R} \,.$$

   Since the mass of the oxygen is 16 atomic mass units (being the atomic mass unit mass about 1.67×10^{-27} kg), the above condition is well verified for the case of oxygen in the atmosphere of the Earth when

$$T \ll 8 \times 10^4 \, \text{K} \,.$$

3. *Equilibrium temperature of the Earth (3)*

   Assuming that the Sun is a blackbody emitting at a temperature of 6000 K (approximately the temperature of the photosphere), what is the temperature of Earth at equilibrium due to the radiation exchange with the Sun? Assume the Sun's radius to be 7000 km, i.e., 110 times the Earth's radius.

   The power absorbed by the planet from the star is equal to the power emitted by the planet: $P_{in} = P_{out}$.

   The power input to the planet is equal to the luminosity (i.e. power emitted) of the star, times the ratio absorbed by the planet (1 minus the albedo, which is the fraction of reflected light), times the area of the planet illuminated by the star, divided by the area of the sphere that all of the star's radiation is cast on at the distance of the planet:

$$P_{in} = L_\odot (1 - a) \left(\frac{\pi R_p^2}{4\pi D^2} \right) .$$

The incoming power to a blackbody is radiated as heat according to the Stefan-Boltzmann law $P = \sigma A T^4$.
The power emitted by the planet is

$$P_{out} = \left(\sigma T_{eq}^4 \right) \left(4\pi R_p^2 \right) .$$

Setting these equal:

$$T_{eq} = \left[\frac{L_\odot (1 - a)}{16\sigma \pi D^2} \right]^{1/4} .$$

The luminosity of the star is equal to the Stefan-Boltzmann constant, times the area of the star, times the fourth power of the temperature of the star:

$$L_\odot = \left(\sigma T_\odot^4 \right) \left(4\pi R_\odot^2 \right) .$$

Inserting this into the previous equation, one has

$$T_{eq} = T_\odot (1 - a)^{1/4} \sqrt{\frac{R_\odot}{2D}}.$$

It is interesting to note that the equilibrium temperature does not depend on the size of the planet, because both the incoming radiation and outgoing radiation are proportional to the area of the planet.
In the case of the Earth, neglecting the albedo (which is about 0.37), the equilibrium temperature is

$$T_{eq,0} \simeq 280 \, \text{K} .$$

The albedo decreases the temperature by 10% (the fourth power in the expression $(1 - a)^{1/4}$ decreases the sensitivity to the albedo effect), so that

$$T_{eq} \simeq 250 \, \text{K} .$$

Fortunately the temperature of the Earth is higher than its equilibrium temperature.
Because of the greenhouse effect, planets with atmospheres can have temperatures higher than the equilibrium temperature – for example, Venus has an equilibrium temperature of approximately 260 K, but a surface temperature of 740 K. Geothermal heating can also play a role.

4. *The Earth will heat up in the future (4)*

In 1 Gyr, the luminosity of the Sun will be 15% higher. By how much will the effective temperature of the Earth change?

The equilibrium temperature of a body orbiting a star of luminosity L can be obtained through the balance between the energy gains from the effective heating by stellar radiation and the corresponding losses of thermal emission:

$$T_{eq} = \left[\frac{L(1-a)}{16\pi\sigma D^2} \right]^{1/4}, \tag{11.1}$$

where a is the planet's albedo, σ is the Stefan-Boltzmann constant and D is the distance between the planet and the star. Assuming that the only change occurring in the lag of 1 Gyr is the luminosity of the Sun, for the Earth this would imply:

$$\frac{T_{1\,Gyr}}{T_{now}} = \left(\frac{L_{1\,Gyr}}{L_{now}} \right)^{1/4}, \tag{11.2}$$

which, for $L_{1\,Gyr} = 1.15\,L_{now}$ leads to:

$$T_{1\,Gyr} = 1.035\,T_{now}. \tag{11.3}$$

With $T_{now} \simeq 250\,K$ (i.e. taking the albedo into account), the net equilibrium temperature increase is $\Delta T_{eq} \simeq 9\,K$.

5. *Moons of giant planets could be habitable (5)*

Although Jupiter is far outside the habitable zone of the Sun, some of its moons, such as Europa, seem possible habitats of life. Where does the energy to sustain such hypothetical life come from? What is the possible role of the other moons?

The main energy production mechanism on the moons of Jupiter is heat from the tidal stress that they suffer through the combination of their rotation and revolution. The presence of several moons with resonant orbital parameters is fundamental in this sense, because the many-body tidal interactions do not allow the tidal stress to dissipate the rotational angular momentum, which, in a two-body system, would lead to the synchronization of the rotation and revolution periods. Well known examples of such effect are the synchronous rotation and revolution of the Moon around the Earth and the almost synchronous orbit of Venus around the Sun. It is remarkable that, without the Moon, even the Earth would show permanently the same portion of surface to the Sun.

6. *Titan (6)*

Why is Titan interesting to study for astrobiology?

Titan is far colder than Earth, and its surface lacks stable liquid water, factors which have led some scientists to consider life there unlikely. On the other hand, its thick atmosphere is chemically active and rich in carbon compounds. On the surface there are bodies of liquid methane and ethane, and it is likely that there is a layer of liquid water under its ice shell; some scientists speculate that these liquid mixes may provide pre-biotic chemistry for living cells different from those on Earth.

Scientists analyzing data from the Cassini-Huygens mission reported in 2010 anomalies in the atmosphere near the surface which could be consistent with the presence of methane-producing organisms, but may alternatively be due to non-living chemical or meteorological processes.

Laboratory simulations have led to the suggestion that enough organic material exists on Titan to start a chemical evolution analogous to what is thought to have started life on Earth. While the analogy assumes the presence of liquid water for longer periods than is currently observable, several hypotheses suggest that liquid water from an impact could be preserved under a frozen isolation layer.

7. *Abundance of elements in the Universe and in living beings (7)*

Look up the average abundance of the chemical elements in the Universe (Chapter 10). Why hydrogen, carbon, oxygen, and nitrogen, the main building blocks for life on Earth, are so abundant? Why is helium not a common element in life?

Concerning the abundance of hydrogen, it is by far the simplest configuration that baryonic matter can achieve, since it requires the simplest nucleus, the proton – a stable constituent of matter. Assembling heavier nuclei is more difficult, due to the increasing amount of neutrons required in the process and their unstable nature, which implies too short a life-time for neutrons to interact in low density environments. However, nuclei with few nucleons (protons and neutrons) can be easily formed in high density environments, such as the cores of stars. The vast majority of the stars in the Universe are able to build nuclei as heavy as oxygen, during their life. Whenever a star is more massive than 2 solar masses, its internal layers are convective and the material processed in the core can be taken to the surface and spread through the stellar wind. Only stars which are more massive than about 8 solar masses can build nuclei as heavy as Fe, while heavier nuclei only form when they eventually explode as supernovae. These explains why such elements are far rarer.

When the environment is cold enough, the low mass complex chemical elements tend to aggregate in molecules and solid particles, rather than sitting in the gas phase, and they generally capture other gas particles, especially hydrogen, to form molecular compounds. Since planetary systems form from the collapse of dusty molecular clouds, they are particularly rich of the main constituents of dust and molecules, like C, N, O and H, of course. Helium, on the other hand, is never captured in compounds due to its complete electron shell configuration, which makes it chemically inert. It therefore tends to be confined in the gas phase and it can only be captured by very massive bodies, which, in the Solar System, are the Sun and the gas giants planets. Life as we know it works best in conditions where water exists in liquid form. Such conditions are met on smaller bodies, where only chemically active light elements can be found in large amounts. These elements are the principal constituents of the simplest DNA-based organisms. A mild availability of heavier elements (Ca, Si, Fe, etc.) allows for the development of more complex life forms, which use them to form highly specialized tissues, such as bones, blood and others.

8. *Detection of exoplanets with astrometry (8)*
   What is the shift of the position of the Sun due to the Earth's orbit? What are the characteristics of an instrument that an alien living near Alpha Centauri would need to detect the Earth using solar astrometry?

In the assumption that we can deal with the Sun-Earth system through the two-body formalism, both the Sun and the Earth will orbit about the common center-of-mass. Using the definition of the center-of-mass position (projected in one dimension):

$$R = \frac{1}{M_\odot + m_\oplus}(M_\odot r_\odot + m_\oplus r_\oplus) \tag{11.4}$$

and considering a system whose origin lies in the center of the Sun ($r_\odot = 0$), we have that:

$$R = \frac{m_\oplus r_\oplus}{M_\odot + m_\oplus}. \tag{11.5}$$

Since $m_\oplus = 5.972 \cdot 10^{24}$ kg, $M_\odot = 1.988 \cdot 10^{30}$ kg and the maximum Earth distance is $r_\oplus = 1.521 \cdot 10^8$ km, we find that the maximum distance between the center of the Sun and the center of mass is:

$$R \simeq 457 \, \text{km}, \tag{11.6}$$

i.e. well within the Sun itself.

Due to the Earth's influence, in a full orbital period the Sun is thus displaced by an offset of $2R = 914$ km. At the distance of Alpha Centauri, which lies 1.44 pc away, this leads to an angular displacement of $2.06 \cdot 10^{-11}$ rad, corresponding to $4.25 \cdot 10^{-6}$ arcsec. This is the angular resolution that should be achieved in order to detect the Earth's influence on the Sun at the distance of Alpha Centauri.

9. *Radial velocity measurement via Doppler spectroscopy (9)*
   What is the Doppler shift of the light emission from the Sun due to the Earth's orbit? What are the characteristics of an instrument that an alien living near Alpha Centauri would need to detect the Earth using Doppler spectroscopy?

Due to the small eccentricity of the Earth's orbit around the Sun ($e \simeq 0.0167$), we can get a reasonable estimate by assuming circular motions with constant velocity. In a two-body approximation, the Sun and Earth move around a center of mass which lies 457 km off the center of the Sun in the period of 1 yr, i.e. $3.156 \cdot 10^7$ s. This corresponds to a circular orbital average speed of $v_{avg} = 9.10 \cdot 10^{-5}$ km s^{-1}.

Alpha Centauri lies at an ecliptic latitude of approximately $i = -43°.5$, meaning that the Doppler effect is obtained by projecting v_{avg} along the radial direction:

$$\frac{\Delta\lambda}{\lambda} = \frac{v_{avg}}{c} \sin i \simeq 2.15 \cdot 10^{-10}. \tag{11.7}$$

At the peak emission wavelength of the Sun, which lies at $\lambda \simeq 509$ nm, this requires the ability to measure $\Delta\lambda = 1.09 \cdot 10^{-7}$ nm. The influence of the Earth on the velocity of the Sun lies at the scale of few cm/s. The solar surface is affected by perturbations which may change its radial velocity measurement on larger scales, yet an extremely sensitive instrument might still be able to point out the existence of a regular modulation of the Sun's radial velocity with a period of 1 year (corresponding to an unknown number of Centaurian time units).

10. *Biosignatures (10)*

    Discuss some of the characteristics of gases in the atmosphere which could be indicators of life.

    _____

    Gases in the atmosphere of an exoplanet provide a means by which one can deduce the possible existence of life. Observers should explore the transmissivity of light from the parent star through the atmosphere of the planet; this carries molecular features from the planet due to the composition, temperature, and pressure structure of its atmosphere.
    In particular one can look for:

    a. Atmospheric gases: Gases formed by metabolic processes, which may be present on a planet-wide scale. For example, large amounts of oxygen and small amounts of methane are generated by life on Earth. Two of the top 14 000 proposed atmospheric biosignatures are dimethyl sulfide and chloromethane (CH_3Cl). An alternative biosignature is the combination of methane and carbon dioxide.

    b. Temporal variability: variations in time of atmospheric gases, reflectivity, or macroscopic appearance that indicate the presence of life. For example, the action of Earth's industrial civilization is changing our atmosphere composition.

    c. Anomalous proportions: proportions of the abundances of gases which are inconsistent with thermodynamical equilibrium. For example, the abundance of methane in the Earth's atmosphere is orders of magnitude above the equilibrium value due to the constant methane flux that life on the surface emits.

    These are of course only part of the possible biosignatures, others being related, for examples, to technological imprints (like radio emissions).

11. *How far can we see humans in space?*

    Moon landing conspiracy theories claim that Moon landings were hoaxes staged by NASA. One of the naïve objections is that men on moon were not detected by telescopes. The most sensitive satellite for this purpose would be WISE (launched in 2009), an infrared telescope whose sensitivity is equal to 1.6×10^{-13} $\mathrm{erg\,cm^{-2}\,s^{-1}}$. How far can a human being be observed with the current astronomical instrumentation?

The problem of detecting human figures in space should be addressed with caution, since some doubt might be cast on the opportunity to use a space suit that radiates at the temperature of the human body. For consistency, the problem is addressed by considering both the cases when the astronaut is the source of radiation, as well as when the astronaut is perfectly reflecting light from the Sun.

Astronaut as radiation source.

This is the IR case. If we assume the astronaut to be a thermal source that radiates at $T = 310\,\mathrm{K}$ ($37\,°\mathrm{C}$) we have a peak emission wavelength of $9.35\,\mu\mathrm{m}$. Using a human body surface of $S = 2 \cdot 10^4\,\mathrm{cm^2}$ and applying the Stefan-Boltzmann's law, we can estimate a radiating luminosity of:

$$L = \sigma T^4 S = 1.05 \cdot 10^{10}\ \mathrm{erg\,s^{-1}} \tag{11.8}$$

and, therefore, a flux at distance D (measured in cm):

$$F = \frac{L}{4\pi D^2} = \frac{8.33 \cdot 10^8}{D^2}\mathrm{erg\,s^{-1}}. \tag{11.9}$$

We can infer the maximum distance at which such flux could be observed by setting F equal to the WISE flux limit and then computing D:

$$D = \sqrt{\frac{8.33 \cdot 10^8\ \mathrm{erg\,s^{-1}}}{1.6 \cdot 10^{-13}\ \mathrm{erg\,cm^{-2}\,s^{-1}}}} = 7.21 \cdot 10^{10}\ \mathrm{cm}, \tag{11.10}$$

that is $D = 7.21 \cdot 10^5$ km, almost twice as much as the distance between the Earth and the Moon.

Externally illuminated astronaut.

In this case, the visible astronaut's surface (half of the total, on average) is reflecting visible Sun light over approximately 2π sr. Assuming perfect reflection of the standard solar flux ($1.361 \cdot 10^6\ \mathrm{erg\,cm^{-2}\,s^{-1}}$) from $10^4\,\mathrm{cm^2}$, the reflected power from the Moon would be:

$$F = \frac{1.361 \cdot 10^{10}\ \mathrm{erg\,s^{-1}}}{2\pi D^2}. \tag{11.11}$$

At the distance of the Moon (approximately $3.85 \cdot 10^{10}$ cm) this would result in a flux of $1.46 \cdot 10^{-12}$ erg cm^{-2} s^{-1}. To have an idea, the Space Telescope Imaging Spectrograph can detect a source at a 5σ-level down to a flux as low as $2.5 \cdot 10^{-13}$ erg cm^{-2} s^{-1} in one second of exposure.

It is therefore expected that we could be able to detect the flux produced by a man wandering on the Moon using the Hubble Space Telescope or the WISE satellite.

Appendix A
Particle Physics Exams

A.1 Discovery of the Positron

Carl D. Anderson was awarded the 1936 Nobel Prize in physics for the discovery of the positron, the antiparticle of the electron. In 1932, during a cloud chamber experiment designed to observe cosmic rays, he detected a track of a positive particle with a mass about equal to that of the electron.

1. Charged particles were deflected in the chamber due to a strong magnetic field of 1.7 T and had to traverse a lead plate. Assume that the particle that caused the track is a positron.

a) What would be the radius of curvature if the positron has an energy of 63 MeV?
b) In the positron discovery paper, Anderson discusses and refutes the possibility of the track in Fig. A.1 to be produced by a proton or an electron with a reverse trajectory. Discuss, succinctly, how one can reach this conclusion with the data above.
c) Knowing that the positron has an energy of 63 MeV before it reached the lead plate and 23 MeV after emerging it, estimate the lead plate thickness in centimeters considering only ionization losses.
d) Accounting for all possible physical phenomena that the positron might suffer while crossing the lead plate, justify if the thickness previously estimated should be smaller, bigger or it is appropriate.
e) One striking feature in Fig. A.1 is that although produced by cosmic rays, the positron seems to be coming from below. Show that a downgoing vertical muon cannot produce a positron with an energy of 63 MeV in the opposite direction. Why? (Remember that: $\mu^+ \to \bar{\nu}_\mu e^+ \nu_e$).
f) Imagine that the thin lead plate is now replaced by a small slab of water, where light can be collected by a photomultiplier (PMT). Could this setup be able to distinguish between a positron and a proton of 63 MeV? Justify.

© Springer Nature Switzerland AG 2021
A. De Angelis et al., *Particle and Astroparticle Physics*, Undergraduate Lecture Notes in Physics,
https://doi.org/10.1007/978-3-030-73116-8

Fig. A.1 Anderson's
photograph of the positron
track

2. Atmospheric positrons may arise from muons which are essentially produced in the decay of charged pions, for instance $\pi^+ \to \mu^+ \nu_\mu$. Assume that the π^+ has an energy of 10 GeV.

a) Compute the mean distance travelled by the pion before decaying.
b) The pion can have also hadronic interactions with the atmosphere atoms. Supposing that the pion was created at 2 km from the ground were the density is relatively constant ($\rho \sim 1.1\,\text{kg}\,\text{m}^{-3}$) and knowing that the pion-Air cross-section is roughly $\sigma_{\pi-Air} \sim 200\,\text{mb}$, compute the interaction length of the pion. With this information discuss if the pion is expected to interact or to decay. Note that the pion interacts with the atoms in the atmosphere and remember that the Earth's atmosphere composition is roughly 80% nitrogen and 20% oxygen.
c) Determine the minimum and maximum energy that the muon can get in the pion reference frame.
d) Determine the minimum and maximum energy that the muon can get in the laboratory (LAB) frame.

3. Cosmic rays interact with the atmosphere's atoms producing cascades of particles that can give origin to positrons at some stage. Verify, from the point of view of quantum numbers, if the following reactions are possible and if not explain why.

a) $pp \to pne^+$
b) $pp \to p\pi^+$

c) $\pi^+ p \to \pi^+ \pi^- \Delta^{++}$
d) $pp \to pnK^+\pi^0$.

A.2 Observation of the Glashow Resonance at IceCube

IceCube is a neutrino observatory built at the South Pole, which uses thousands of detectors distributed over a cubic kilometer of volume under the Antarctic ice. The neutrino is detected when it interacts with the ice atoms producing relativistic charged particles. This experiment is sensitive to neutrinos and anti-neutrinos of all flavors.

1. One interesting channel which was recently reported is the interaction between an anti-neutrino of the electron, $\bar{\nu}_e$, and an electron in the ice, as this is a resonant process via the W boson. This process is known as the Glashow resonance.

a) Consider the interaction between the electron and the anti-neutrino. Knowing that the electron is at rest, what is the energy of the anti-neutrino necessary to produce a W boson?

b) Consider now the process $e^-(p_1) + \bar{\nu}_e(p_2) \rightarrow \mu^-(p_3) + \bar{\nu}_\mu(p_4)$. Draw the corresponding Feynman diagram(s).

c) Write the amplitude $\mathcal{M}$ in its simplest form. Assuming that $m_{\text{fermions}} \ll \sqrt{s} \leq M_Z, M_W$, simplify as much as possible the amplitudes. Justify your answer.

d) Evaluate the spin averaged squared amplitude $\langle|\mathcal{M}|^2\rangle$.

e) For this process evaluate the differential cross section $d\sigma/d\Omega$ in the center-of-mass (CM) frame as a function of the square of the energy in the CM, $s = (p_1 + p_2)^2$, and the scattering angle θ.

f) Determine the expression for the total cross section $\sigma(e^- + \bar{\nu}_e \rightarrow \mu^- + \bar{\nu}_\mu)$ at the energy $\sqrt{s} = M_W$. Calculate its value in nb.

g) Find the value of the total cross section

$$\sigma(e^- + \bar{\nu}_e \rightarrow \text{All})$$

at $\sqrt{s} = M_W$ in nb.

2. Consider the following processes with neutrinos. For each process draw the Feynman diagram(s) and write the amplitude. Assuming that $m_{\text{fermions}} \ll \sqrt{s} \leq M_Z, M_W$ simplify as much as possible the amplitudes. Justify your answer. You do not need to calculate anything.

a) $\nu_e + q \rightarrow e^- + q'$
Identify the possible quarks q and q' assuming that they are valence quarks of nucleons in matter, and not anti-quarks.

b) $\bar{\nu}_\mu + q \rightarrow \mu^+ + q'$
Identify the possible quarks q and q' assuming that they are valence quarks of nucleon in matter, not anti-quarks.

c) $\bar{\nu}_e + e^+ \rightarrow \bar{\nu}_e + e^+$
Indicate what are the non vanishing helicity amplitudes, by drawing in the diagrams the spin arrows.

3. The electron anti-neutrinos measured at IceCube should have an astrophysical origin. Given their extremely high energy, these $\bar{\nu}_e$ are most likely produced in SuperNova Remnants (SNR). Describe the most probable mechanism which can originate such flux of electron anti-neutrinos from the primary matter of a star, namely protons and electrons. (Remember that in SNR high-energy particles are essentially accelerated in electromagnetic shock waves).

A.3   Investigation of the Z Boson at SLAC

The SLAC national accelerator laboratory, originally named Stanford Linear Accelerator Center, hosts the longest linear accelerator ever built (3.2 km). The research done with it produced 3 Nobel prizes (for the discovery of the charm quark, of the tau lepton, and for the studies of deep inelastic scattering of electrons on nucleons, which has been essential for the development of the quark model). In the early to mid 1990s the Stanford Linear Collider (SLC) investigated the properties of the Z boson, using electron-positron beams which could be polarized.

1. Consider the process $e^- + e^+ \to \mu^- + \mu^+$ at the CM energy $\sqrt{s} = M_Z$. Neglect the masses of the leptons.

  a) What is the dominant diagram at these energies and why? Draw the corresponding Feynman diagram.
  b) Considering the conditions of a), write the amplitude $\mathcal{M}$ in its simplest form. Explain all the approximations.
  c) Evaluate the spin averaged squared amplitude $\langle |\mathcal{M}|^2 \rangle$.
  d) For this process evaluate the differential cross section $d\sigma/d\Omega$ in the CM frame as a function of the square of the energy in the CM frame, $s = (p_1 + p_2)^2$.
  e) Determine the expression for total cross section $\sigma(e^- + e^+ \to \mu^- + \mu^+)$ at the energy $\sqrt{s} = M_Z$. Find its value in pb.
  f) Find the value of the total cross section

$$\sigma((e^- + e^+ \to \text{All})$$

at $\sqrt{s} = M_Z$ in pb.

2. One of the most important features of SLC was the use of polarization of the electron beam. A very important quantity measured was the asymmetry A_{LR} defined by

$$A_{LR} \equiv \frac{\sigma_L - \sigma_R}{\sigma_L + \sigma_R},$$

where σ_L (σ_R) are the cross sections for the L and R polarized beams, respectively. Consider that $\sqrt{s} = M_Z$ and, as before, neglect the masses of the leptons.

  a) Assuming 100% polarizations evaluate the spin averaged squared amplitudes $\langle |\mathcal{M}_L|^2 \rangle$ and $\langle |\mathcal{M}_R|^2 \rangle$ for the process $e^- + e^+ \to \mu^- + \mu^+$ for the cases when the electron beam is left-handed polarized (L) and righ-handed polarized (R), respectively. You can re-use the calculations of the helicity amplitudes of Problem 1.
  b) Evaluate the total cross sections σ_L and σ_R for the two polarizations of the electron beam.
  c) Show that A_{LR} can be written as

$$ A_{LR} = \frac{2 g_V^e g_A^e}{(g_V^e)^2 + (g_A^e)^2} . $$

Explain why the result does not depend on g_V^μ and g_A^μ.

d) In practice the electron beam is not 100% polarized, but has some degree of polarization P_e. It is known that in this case

$$ A_{LR} = \frac{1}{P_e} \frac{N_L - N_R}{N_L + N_R} , $$

where N_L (N_R) are the number of events for L (R) polarized beams. In the 1992 run of the experiment the following numbers were obtained (neglecting errors):

$$ P_e = 0.224, \quad N_L = 5226, \quad N_R = 4998 . $$

Determine A_{LR}.

e) Derive an expression for $s_W^2 \equiv \sin^2 \theta_W$, where θ_W is the weak mixing angle, in terms of A_{LR}.

f) Use the previous results to determine the value of s_W^2. Knowing that the experimental error on A_{LR} in the above experiment was ± 0.044, discuss if the determination is reasonable considering the presently measured value

$$ s_W^2 = 0.23129 \pm 0.00005 . $$

A.4 Deep Inelastic Scattering in the ZEUS Experiment

The ZEUS experiment studied the internal structure of the proton through measurements of deep inelastic scattering (DIS) by colliding electrons with protons. The electrons and the protons were accelerated in the HERA particle accelerator.

1. The first run that produced physics results was done accelerating electrons to an energy (in the LAB frame) of 26.7 GeV, and protons to 820 GeV. An integrated luminosity of 2.1 nb^{-1} was reached.

a) Compute the center-of-mass energy for the electron-proton collision.
b) Compute the electron energy in the reference frame were the proton is at rest.
c) To analyse the results, the experimental data are divided in bins of Q^2 and x. Knowing that the cross-section for the bin $Q^2 \sim 8.5 \, \text{GeV}^2$ and $x \sim 4.1 \times 10^{-4}$, is $4.8 \times 10^{-32} \, \text{cm}^2$, how many events in that bin were detected if the detector has an efficiency of 80%?
d) Consider the kinematics of a deep inelastic scattering event. What should be the energy and the scattering angle, θ, with respect to the electron beam, of the out-coming electron so that $Q^2 = 8.5 \, \text{GeV}^2$ and $x = 4.1 \times 10^{-4}$. Perform and present all calculations in the reference frame where the nucleon is at rest.

2. Verify, from the point of view of quantum numbers, if the following reactions are possible and if not explain why.

a) $e^- p \to e^- p \bar{p}$
b) $e^- p \to e^- p K^- \pi^+$
c) $e^- p \to e^- p K^- K^+$
d) $e^- p \to \Delta^- e^+$
e) $e^- p \to \Delta^+ e^- \nu_e \bar{\nu_e}$.

3. In ZEUS the hadrons and electrons, produced in the DIS $e^- p \to e^- h$ were measured with a calorimeter. The calorimeter, which surrounds completely the interaction vertex, consists of alternating layers of 3.3 mm thick uranium and 2.6 mm plastic scintillator.

a) How many radiation lengths has a single layer of uranium?
b) Describe qualitatively the development of an electromagnetic shower initiated by a high-energy electron stating the purpose of each layer type.
c) Consider a charged kaon, K^+, of 20 GeV in the calorimeter. What is the average energy deposited through ionisation by the meson while traversing one uranium layer?

4. The baryon Δ^+ is one of the hadrons that can be produced in a DIS event. Consider that the Δ^+ was produced with a momentum of 4 GeV/c.

a) Compute the mean free path of the Δ^+ in the LAB frame.
b) The Δ^+ decays $\approx 99\%$ of the times in pure hadronic states. Prove that:

$$\frac{\Gamma(\Delta^+ \to p\pi^0)}{\Gamma(\Delta^+ \to n\pi^+)} = 2.$$

c) As seen before, the Δ^+ decays approximately 2/3 of the times into a proton and a π^0. Moreover, the π^0 decays nearly promptly into two photons, such that $\Delta^+ \to p\pi^0 \to p\gamma\gamma$. Knowing that the number of produced Δ^+ is N_0, estimate the number of photons that could be observed by a detector, put 1 m away from the interaction vertex.

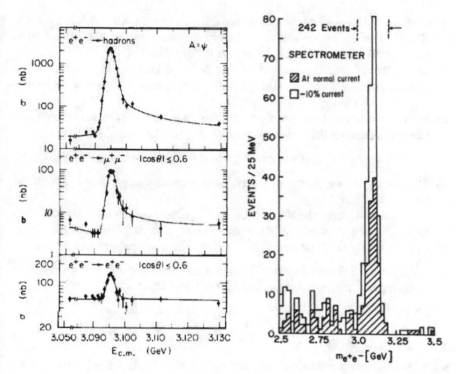

Fig. A.2 Original plots for the discovery of charm at SLAC (left) and at BNL (right)

A.5 Discovery of the J/ψ Meson

The meson, constituted by a charm and an anti-charm quarks, was discovered in 1974 independently by two research groups, one at the Brookhaven National Laboratory (BNL) and the other at the Stanford Linear Accelerator Center (SLAC). For that reason the meson was called J/ψ, the combination of the names given by each group.

1. At BNL, the observation of J/ψ was achieved by colliding protons into a beryllium target.

a) The identification of the J/ψ was done through the identification of an electron-positron pair. Discuss how to evaluate the meson mass in the LAB frame.
b) One of the techniques used to distinguish electrons from protons is to measure the time of flight (ToF) of the particle. Considering an electron and a proton, both with an energy of 1.5 GeV, and knowing that the detector time resolution is ~ 1 ns, determine the minimum distance necessary to distinguish the two types of particles.

c) Consider a J/ψ with a momentum of $10\,\mathrm{GeV/c}$ in the LAB. What is the mean distance that the meson can travel before it decays into an $e^- e^+$ pair? What would be the distance if the J/ψ decays instead into $\rho\pi$?

d) Taking into account the decay $J/\psi \rightarrow e^+ e^-$, and knowing that the meson has an energy of $5\,\mathrm{GeV}$, compute the minimum and maximum angle that the leptons can make with the J/ψ flight direction.

e) In the conditions of the previous question, what are now the minimum and maximum energies that the electron/positron can take in the LAB frame?

2. At SLAC, the J/ψ meson was observed by colliding electrons with positrons.

a) What would be the energy of the positron necessary to produce a J/ψ if the electron is at rest?

b) As stated previously, the time of flight of a particle can be used to identify its mass. Describe another possible experimental apparatus to distinguish different types of highly relativistic particles, namely: electrons, protons and muons.

3. Verify, from the point of view of quantum numbers, if the following reactions involving the J/ψ meson are possible and if not explain why.

a) $e^+ e^- \rightarrow J/\psi\ \mu^-\ \bar{\nu}_\mu$ d) $e^+ e^- \rightarrow J/\psi\ \pi^0$

b) $e^+ e^- \rightarrow J/\psi\ n$ e) $e^+ e^- \rightarrow J/\psi\ \nu_e$

c) $e^+ e^- \rightarrow J/\psi\ K^0$

4. Fig. A.2 (left) was produced accepting events whose collision sub-products are within $|\cos\theta| \leq 0.6$, being θ the angle with the beam direction. How would the plot for $e^+ e^- \rightarrow \mu^+ \mu^-$ and $e^+ e^- \rightarrow e^+ e^-$ change if all events were accepted independently of the angle θ?

5. At SLAC/SPEAR e^- and e^+ beams were collided with a center-of-mass energy in the interval $[2, 8]\,\mathrm{GeV}$. Consider first the process

$$e^-(p_1) + e^+(p_2) \rightarrow \mu^-(p_3) + \mu^+(p_4)$$

in the center-of-mass, for energies in that range. Neglect the masses of all fermions.

a) Draw the corresponding Feynman diagram(s). Identify the most important process at these energies.

b) Considering the conditions of the problem write the amplitude $\mathcal{M}$ in its simplest form for the dominant diagram. Explain all the approximations.

c) Evaluate the spin averaged squared amplitude $\langle|\mathcal{M}|^2\rangle$.

d) For this process, evaluate the differential cross section $d\sigma/d\Omega$ in the CM frame as a function of the square of the energy in the CM frame, s, and scattering angle, θ, defined as the angle between the outgoing μ^- and the incoming e^-.

e) Find the expression for the total cross section. Evaluate it in nb at $\sqrt{s} = 3.13$ GeV.

f) To compare with the experimental result in Fig. A.2 (left-hand side, middle panel) you have to make a cut in the scattering angle θ, as it was done in the experiment. Redo your calculation of the cross section for $|\cos\theta| \leq 0.6$ and compare with the results in the figure. To make this comparison consider the point that is well away from the resonance with $\sqrt{s} = 3.13$ GeV. To extract the values from the plots consider the curve that fits the points and not the data points. Comment on your results.

6. In the experiment at SLAC, Burton Richter and collaborators were scanning the center-of-mass energy $\sqrt{s} \in [2, 8]$ GeV in intervals of 200 MeV. The purpose was to measure the so-called ratio R defined by

$$R = \frac{\sigma(e^- + e^+ \to \text{hadrons})}{\sigma(e^- + e^+ \to \mu^-. + \mu^+)}$$

They found that this ratio *jumped* around $\sqrt{s} \sim 3$ GeV. After this they did a finer scan as shown in Fig. A.2.

a) Determine this ratio for $\sqrt{s} = 2$ GeV.
b) Consider now that $\sqrt{s} \sim 3.5$ GeV. Determine the ratio at this energy.
c) Calculate now R at $\sqrt{s} = 20$ GeV.
d) Make an approximate plot of the ratio R in the interval $\sqrt{s} \in [2, 20]$ GeV.
e) Use Fig. A.2 to extract the value of R below and above the J/ψ resonance. Do not forget that you have to scale up the values for $\sigma(e^- + e^+ \to \mu^- + \mu^+)$ in the middle panel of Fig. A.2 (left) to obtain the cross section for all angles. To extract the values from the plots consider the curve that fits the points and not the data points. Compare with your results.

A.6 Muon/Anti-muon Collider

Muons couple directly to a Standard Model (SM) Higgs particle in s-channel. Thus, a muon collider could be an ideal device for precision measurements of the Higgs boson. Consider a muon collider ($\mu^+ \mu^-$) with a beam energy of 63 GeV. Muons feeding the accelerator come from the decays of pions created through proton-proton interactions.

1.

a) The interactions of the single muon bunch per beam occur with a repetition rate of 30 Hz. Knowing that the experiment is planned to achieve an instantaneous luminosity of 10^{31} cm^{-2}s^{-1}, and that in the interaction point the beam has a transverse area of 0.02 mm^2, compute the average number of muons in each beam.
b) Indicate the number of muons that have to be injected per minute in the collider to maintain a constant luminosity. Consider that the only factor for the disappearing of muons is their decay.

c) Consider the decay of a muon (in its reference frame) into an electron and two neutrinos ($\bar{\nu}_e$ and ν_μ). Indicate the configuration (angles between particles) for which the electron has a minimum and maximum energy, respectively.

d) Determine the minimum and maximum energy that the electron can have in the LAB reference frame.

e) The products of the muon decays are, in these colliders, an important contamination for the selection of the Higgs event candidates. Discuss the angular region where this contamination is more important.

f) Show explicitly, using the appropriate conservation of quantum numbers, which of the following reactions/decays involving muons are possible.

   (i) $\mu^+ \to e^+ \, \nu_e \, \nu_\mu$
   (ii) $\mu^+ \, \mu^- \to e^+ \, e^-$
   (iii) $p \, n \to n \, n \, \mu^+$
   (iv) $\mu^+ \, \mu^- \to \nu_\mu \, \bar{\nu}_\mu$.

g) Estimate the number of Higgs bosons produced in one year using the instantaneous luminosity given in question a, and an Higgs production cross-section $\sigma_H \sim 40\,\text{pb}$.

h) One of the most probable Higgs decays is into $b\bar{b}$. Discuss how one can discriminate experimentally this decay from other hadronic decays of the Higgs, indicating the type of detectors and their typical dimension and position with respect to the interaction point.

2. In the muon collider, the $\mu^\mp$ beams have a contamination of electrons (and positrons) due the μ^- (μ^+) decays. In these context consider the following questions:

a) Draw all Feynman diagrams at lowest order in the SM for the processes:

$$\mu^- \mu^+ \to f\bar{f}$$

   where f is one of the following: $f = e^-, \, \nu_e, \, \nu_\mu, \, b$. Consider neutrinos to be massless, without interaction with the Higgs boson H.

b) Draw all Feynman diagrams, in lowest order in the SM, for the process

$$\mu^- e^+ \to \mu^- e^+.$$

c) Consider now that $\sqrt{s} \simeq M_H$ and the process,

$$\mu^-(p_1) + \mu^+(p_2) \to b(p_3) + \bar{b}(p_4).$$

   Taking only the dominant diagram for this energy, write down the amplitude $\mathcal{M}$ for this process. Hint: See in the formulary below a discussion of the Higgs scalar currents.

d) Neglecting the fermion masses everywhere, except in the Higgs couplings, write down the non-vanishing helicity amplitudes for this process in the CM frame.

e) In these conditions evaluate $\langle|\mathcal{M}|^2\rangle$ and $\dfrac{d\sigma}{d\Omega}$ in the CM frame.

f) In the same conditions evaluate the total cross-section for $\mu^- + \mu^+ \to b + \bar{b}$. For $\sqrt{s} = M_H$ express this cross section in pb.

g) Determine the expression for the ratio

$$R_1 = \frac{\sigma(\mu^-\mu^+ \to b\bar{b})}{\sigma(e^-e^+ \to b\bar{b})}$$

at $\sqrt{s} = M_H$ and give its numerical value.

h) Without making any calculations make an estimate of the value of the ratio

$$R_2 = \frac{\sigma(\mu^-\mu^+ \to b\bar{b})}{\sigma(e^-e^+ \to b\bar{b})}$$

at $\sqrt{s} = M_Z$ and give its numerical value.

i) Comment the following sentence: *A muon collider is a much better Higgs factory than an e^-e^+ collider.* Justify carefully your answer.

A.7 Rare Decays at the LHCb Experiment

With the discovery of the Higgs boson, the search for physics beyond the Standard Model is becoming increasingly important. One important test was performed by the LHCb collaboration[1] using the B_s^0 decay into $\mu^+\mu^-$, a channel with a high sensitivity to possible hidden sectors. The Large Hadron Collider beauty (LHCb) experiment is a single-arm forward spectrometer designed in particular to study particles containing b or c quarks.

1. One of the main physics goals of LHCb is to measure accurately the branching ratio of the $B_s^0 \to \mu^-\mu^+$.

a) Knowing that the search for this decay was done with a total integrated luminosity of 4.4 fb^{-1} and that the production cross-section of particles with at least one b quark in pp collisions is $\sim 100\ \mu$b, compute the number of b-particles produced.

b) Compute the mean free path of a B_s^0 produced in the LAB with a momentum of 100 GeV/c.

c) Consider now the decay of the B_s^0 meson with energy 100 GeV into muons ($B_s^0 \to \mu^+\mu^-$):

  i) What is the minimum and maximum energy that muons can have in the B_s^0 reference frame?

  ii) Compute that minimum and maximum angle between the two muons in the LAB reference frame.

---

[1] The first report of this measurement was done by LHCb and CMS.

iii) What is the minimum energy that the B_s^0 needs to have in the LAB to produce a muon of 8 GeV?

2. Verify, from the point of view of quantum numbers, if the following reactions are possible and if not explain why.

a) $p\,p \to p\,e^+\gamma$ d) $p\,p \to \pi^+\pi^-K^+$

b) $p\,p \to p\,n\,\gamma$ e) $p\,p \to p\,p\,K^0$

c) $p\,p \to p\,n\,e^+\nu_e$ f) $p\,p \to p\,p\,K^+K^-$

3. In order to be able to identify the different kind of particles that emerge from the decays of b-hadrons, LHCb is a multi-detector experiment. In particular, it has two Ring-Imaging Cherenkov (RICH) detectors which are important in the discrimination of pions from kaons. Assume that a π^+ and a K^+ each with momentum 1 GeV/c:

a) What is the minimum (or maximum) medium refractive index to see Cherenkov photons from the pion but not from the kaon?

b) Compute the produced Cherenkov light ratio between pions and kaons. Assume that the particles travel the same distance and that the medium refractive index of the RICH detector is $n = 1.2$.

4. Consider the following interaction possible in proton-proton collisions at the LHC:

$$q(\hat{p}_1) + \bar{q}(\hat{p}_2) \to \mu^-(p_3) + \mu^+(p_4)$$

in the center-of-mass of the $q, \bar{q}$ pair, where $\hat{p}_i$ are the momenta carried by the each quark q. Neglect the masses of all fermions. For simplicity consider only up and down quarks while solving the following exercises.

a) Draw the corresponding first order Feynman diagram(s).

b) Considering the conditions of the problem write the amplitude $\mathcal{M}$ in its simplest form. Explain all the approximations.
Consider for the next questions that $\hat{s} = (\hat{p}_1 + \hat{p}_2)^2 = (M_Z)^2$. Present the results in terms of the general coupling constants.

c) In the above situation evaluate the spin averaged squared amplitude $\langle|\mathcal{M}|^2\rangle$.

d) For this process evaluate the differential cross section $d\sigma/d\Omega$ in the CM frame (of the elementary process) as a function of the square of the energy in the CM, $\hat{s} = (\hat{p}_1 + \hat{p}_2)^2$, and of the scattering angle, $\hat{\theta}$.

e) Evaluate the total elementary cross-section for $\sqrt{\hat{s}} = M_Z$.

f) How can the proton-proton total cross-section for this process be computed? (no calculation is needed).

5. Consider now the decay $B_s^0 \to \mu^-\mu^+$.

a) What is the quark content of the B_s^0 meson?

b) What is (are) the lowest order diagram(s) for this decay? (note that the lowest order has one loop).

c) How could one distinguish the above B_s^0 decay from a Z decaying into muons?

A.8 Neutrino-Nucleon Cross-Section at Very-High-Energies

IceCube is a neutrino observatory build at the South Pole, which uses thousands of detectors distributed over a cubic kilometer of volume under the Antarctic ice. The neutrino is detected when interacts with the ice atoms producing relativistic charged particles. These particles will emit Čerenkov light which can be detected by the surrounding photomultipliers. Although the experiment is sensitive to different flavors of neutrinos one of the most sensitive channels is the muon neutrino.

The neutrino-nucleon charged current cross-section can be given, in first approximation, by,

$$\sigma(\nu N)_{CC} \approx 3.6\,s\,\sqrt{\frac{1.8\,C}{s + 1.8\,C}}\ [\text{fb}] \tag{A.1}$$

where s is the square of center-of-mass energy in GeV^2 and $C = 10^4\,\text{GeV}^2$. Properties of the planet Earth, necessary to solve the following problems:

- Earth's average density $\approx 8\,g\,cm^{-3}$;
- Earth's average molar mass $\sim 30\,g\,mol^{-1}$;
- Earth's average number of nucleons per atom ~ 30;
- Earth's average diameter ≈ 12742 km;
- Antarctic deep ice refraction index ≈ 1.78.

1.

a) Consider a neutrino with an energy of 1 PeV (10^{15} eV) interacting with a proton at rest, and determine the center-of-mass energy of the collision.
b) Compute the probability of a 1 PeV (10^{15} eV) muon neutrino to interact inside the Earth knowing that it crosses it through the center.
c) The interaction of a neutrino with a nucleus can produce a muon. Calculate the Čerenkov critical angle for a muon with an energy of 10 TeV.
d) How many Čerenkov photons arc produced if the muon crosses half the detector (500 m)? Assume that the mean energy of the produced Čerenkov photons is ≈ 1 eV.

2. Verify, from the point of view of quantum numbers, if the following reactions are possible and if not explain why.

a) $\nu_\mu\,p \rightarrow n\,\nu_\mu$
b) $\nu_\mu\,p \rightarrow n\,e^-\,\overline{\nu_e}\,\nu_\mu$
c) $\nu_\mu\,n \rightarrow n\,e^+\,e^-$

d) $\nu_\mu\,p \rightarrow \Delta^+\,\nu_\mu$
e) $\nu_\tau\,n \rightarrow n\,\pi^-\,\tau^+$

3. As the neutrino energy increases the Earth becomes gradually opaque, i.e. the neutrino interacts with its atoms. Assuming that the neutrino crosses through the center of the Earth determine the ν_μ energy for which its survival probability is less than 0.1%.

4. Although the tau neutrino has, similarly to the muon neutrino, an energy for which the Earth is opaque, it can be regenerated via the decay of the tau lepton.

Consider for the next questions the following reactions/decays occurring inside Earth:
$v_\tau\, p \to \tau^- X \to v_\tau\, e^-\, \overline{v}_e\, X$, where X are hadronic particles.

a) What is the interaction mean free path of v_τ and the decay mean free path of the τ lepton, if both of them have an energy of 1 PeV (10^{15} eV)? Compare it with the diameter of Earth.

b) Consider a v_τ with an energy of 1 PeV (10^{15} eV). Knowing that the τ takes 90% of the neutrino energy, compute the maximum and minimum energy that the v_τ, emerging from the τ decay, can get.

5. Discuss how could the IceCube experiment distinguish between different neutrino flavors.

The events studied at IceCube correspond to charged current interactions of the type

$$v_\mu(p_1) + N(p_2) \to \mu^-(p_3) + X(p_4)$$

where N denotes the nucleons in the ice and X the final state excluding the muon.

6. Consider first the interaction of the v_μ with the valence quarks of the nucleon N, that is,

$$v_\mu(p_1) + q(\hat{p}_2) \to \mu^-(p_3) + q'(\hat{p}_4)$$

in the center-of-mass of the v_μ, q pair, where $\hat{p}_2$ is the momenta carried by the quark q. Neglect the masses of all fermions.

a) Identify the valence quarks q and q'.
b) Draw the corresponding Feynman diagram(s).
c) Considering the conditions of the problem write the amplitude $\mathcal{M}$ in its simplest form. Explain all the approximations.
d) Evaluate the spin averaged squared amplitude $\langle|\mathcal{M}|^2\rangle$.
e) For this process evaluate the differential cross section $d\sigma/d\Omega$ in the CM frame (of the elementary process) as a function of the square of the energy in the CM frame, $\hat{s} = (p_1 + \hat{p}_2)^2$ and scattering angle, $\hat{\theta}$, for this elementary process.

7. The neutrinos detected at IceCube have an energy sufficiently large to interact with the anti-quarks inside the nucleons with high probability. Consider then the elementary interaction of the v_μ with the anti-quarks inside the nucleon N, that is,

$$v_\mu(p_1) + \overline{q}(\hat{p}_2) \to \mu^-(p_3) + \overline{q}'(\hat{p}_4)$$

in the center-of-mass of the v_μ, $\overline{q}$ pair, where $\hat{p}_2$ is the momenta carried by the anti-quark $\overline{q}$. Neglect the masses of all fermions.

a) Identify the anti-quarks $\overline{q}$ and $\overline{q}'$.
b) Draw the corresponding Feynman diagram(s).
c) Considering the conditions of the problem write the amplitude $\mathcal{M}$ in its simplest form. Explain all the approximations.

d) Evaluate the spin averaged squared amplitude $\langle |\mathcal{M}|^2 \rangle$.
e) For this process evaluate the differential cross section $d\sigma/d\Omega$ in the CM frame as a function of the square of the energy in the CM frame (of the elementary process), $\hat{s} = (p_1 + \hat{p}_2)^2$ and scattering angle, $\hat{\theta}$, for this elementary process.

8. The expressions for the differential cross sections obtained in the previous problems were given in terms of $\hat{s}$ and $\hat{\theta}$, respectively the energy and scattering angle in the CM frame of the elementary process.

a) Explain why these are not good variables to compare with the experiment.
b) To solve the previous problem, one introduces Lorentz invariant (frame independent) variables. In this case we use the deep inelastic invariant variable

$$y \equiv \frac{p_2 \cdot (p_1 - p_3)}{p_2 \cdot p_1} = \frac{\hat{p}_2 \cdot (p_1 - p_3)}{\hat{p}_2 \cdot p_1}$$

Explain the second equality in the above equation and show that

$$y = \frac{1}{2}(1 - \cos\hat{\theta})$$

c) The final expression for the cross section of the process (ν_μ on proton)

$$\nu_\mu(p_1) + p(p_2) \rightarrow \mu^-(p_3) + X(p_4)$$

is given by

$$\sigma_{\nu p \rightarrow \mu X}(s) = \int_0^1 dx \int_0^1 dy \, \frac{G_F^2 x s}{\pi} \left[d(x) + (1-y)^2 \bar{u}(x) \right] \left[\frac{M_W^2}{Q^2 + M_W^2} \right]^2$$

where $Q^2 = -(p_1 - p_3)^2$. You should be able to derive this expression using the results of the previous problems. But here do not worry about that and just answer the following intermediate necessary steps:

1. Why do you sum the cross sections of problems 1 and 2 and not the amplitudes?
2. What is the meaning of the variable x? What is the relation between s and $\hat{s}$?
3. Find a relation between $d\sigma/d\cos\hat{\theta}$ and $d\sigma/dy$.
4. Explain the integration limits in the above expression.
5. Explain the meaning of the functions $d(x)$ and $\bar{u}(x)$.

A.9   Direct Observation of the Tau Neutrino

The direct observation of the tau neutrino , ν_τ, occurred in the year of 2000 by the DONuT experiment. Such experiment detected the ν_τ through charged current interactions using nuclear emulsion targets.

1.

a) Tau neutrinos were produced colliding protons of 800 GeV against a target of 1 m of tungsten. Knowing that the beam contains 10^{12} protons per second, that there was a total of 10^{17} proton interaction during the operation of the experiment, and that $\sigma_{pp} \sim 40$ mb, compute the effective beam time. Assume that the proton-nucleus cross-section scales linearly with the number of nucleons.

b) In these interactions pions are produced among other particles. Determine what is mean energy of pions, for which decay length is smaller than interaction length. Assume that $\sigma_{\pi p} \sim \frac{2}{3}\sigma_{pp}$.

c) Muons can come from the decay of charged pions. Consider a pion decay into a muon and a muon neutrino. Assuming that the pion as an energy of 10 GeV in the laboratory reference frame, indicate what is the expected muon energy spectrum in the LAB, evaluating its minimum and maximum possible energy.

d) Assuming that the muon has an energy of 5 GeV evaluate the mean energy that the muon loses while crossing the full tungsten target. Assume that the properties of tungsten are similar to the ones of lead.

e) The tau neutrinos measured at DONuT came essentially from the decays of charged mesons, in particular from the D_s^+ state. Knowing that the production cross-section of D_s^+, $\sigma_{D_s^+}$, is ~ 300 nb, compute the number of D_s^+ produced in the experiment.

f) Enumerate the ground state mesons that contain at least one quark c indicating their quark content.

g) Explicitly show, using the conservation of the relevant quantum numbers, which of the following decays of the D_s^+ are possible.

i) $D_s^+ \rightarrow \mu^+ \bar{\nu}_\mu$
ii) $D_s^+ \rightarrow \pi^+ \pi^+ \pi^-$
iii) $D_s^+ \rightarrow p\, K^+ K^-$
iv) $D_s^+ \rightarrow K^+ K^- \pi^+$.

h) Compute the number of tau neutrinos produced in the experiment, assuming that they all come from the D_s^+ decay.

i) The neutrino beam that hits the sensitive region of the DONuT experiment is composed by ν_e, ν_μ and ν_τ. The experimental signature that allowed the detection of the tau neutrino is: *charged track followed by other charged track with a small angle with respect to the first one*, the so-called *kink* (see Fig. A.1). Justify the choice of this signature.

2. The neutrino beam used in the DONuT experiment contains neutrino of all families, ν_e, ν_μ and ν_τ.

Fig. A.3 DONuT event

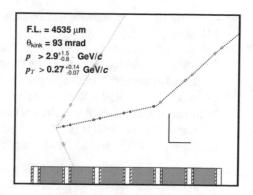

F.L. = 4535 μm
θ_{kink} = 93 mrad
p > 2.9$^{+1.5}_{-0.8}$ GeV/c
p_T > 0.27$^{+0.14}_{-0.07}$ GeV/c

a) Draw the Feynman diagrams for the processes

$$\nu_e e^- \to e^- \nu_e; \quad \nu_\tau e^- \to \tau^- \nu_e$$

b) State what can be the valence quarks, q, q', in the process

$$\nu_\tau q \to \tau^- q',$$

and draw the corresponding Feynman diagram.

c) Consider now the process $\nu_\tau e^- \to \tau^- \nu_e$. Write the amplitude $\mathcal{M}$ without approximations.

d) Neglecting the lepton masses and assuming that the CM energy $\sqrt{s} \ll M_W, M_Z$, write a simplified expression for the amplitude $\mathcal{M}$. Justify the various steps.

e) What are the non-vanishing helicity amplitudes? Evaluate the spin averaged squared amplitude $\langle |\mathcal{M}|^2 \rangle$.

f) For this process evaluate the differential cross section $d\sigma/d\Omega$ in the CM frame as a function of the square of the energy in the CM frame, $s = (p_1 + p_2)^2$.

g) Evaluate the total cross section $\sigma(\nu_\tau e^- \to \tau^- \nu_e)$. For $\sqrt{s} = 1$ GeV, express the result in fb.

h) Is it more likely for the τ^- to be emitted in the front direction (that is with the angle $\theta \in [0, \pi/2]$ where θ is the angle in the center-of-mass reference frame that the emitted τ^- makes with the direction of the ν_τ beam) or on the backwards direction ($\theta \in [\pi/2, \pi]$)?

A.10   Pion-Nucleon Cross-Section

1. In 1951 the group lead by Enrico Fermi in Chicago measured charged pion-nucleon cross sections using pion beams originated in the interaction of primary protons in a fixed target experiment, $p\, A \to \pi^+ X$, where A is the nucleus in the target and X the remaining sub-products of the interaction.

a) Compute the mean distance travelled by a beam of π^+ with a momentum of 145 MeV. ,

b) Discuss a procedure to select the π^+ emitted forward with a momentum of 145 MeV in order to create a pure π^+ beam. Give some quantitative estimation of the main parameters of the proposed selection set-up.

c) Consider that the 145 MeV π^+ beam crossed a target container, filled with liquid hydrogen, and that a set of scintillators were placed before and after the container. Whenever the container is full the number of hydrogen protons projected on the surface of the target was 8.5×10^{23} cm^{-2}.

   i. Compute the π^+ p cross-section at this energy knowing that the ratio between the number of beam particles registered in the scintillators placed after the target container with the container respectively empty and full was 1.116. The error in this ratio was around 0.2%.

   ii. Make an estimation of the mean energy loss and of the angular dispersion suffered by the π^+ beam when crossing the liquid hydrogen target.

d) A fraction of the π^+ decay along the beam line into $\pi^+ \rightarrow \mu^+ \nu_\mu$. Compute the maximum possible angle of the μ^+ with the beam line. Draw the energy spectrum distribution of the μ^+, measured in the laboratory frame, indicating its minimum and maximum energy.

2. In the beginning of the studies of NN and πN scattering (N=nucleon: neutron or proton), a very simple model was used where the spin of the particles was neglected. The model had the interactions

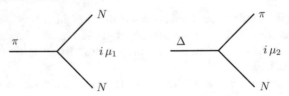

where the constants μ_1 e μ_2 have dimension of a mass in the natural system of units ($\hbar = c = 1$). The charges of the particles are such that the charge is conserved at each vertex.

a) Consider the elastic scattering $\pi^- + p \rightarrow \pi^- + p$. Draw the Feynman diagram(s) for the process.

b) Consider now the process $n(p_1) + p(p_2) \rightarrow p(p_3) + \Delta^0(p_4)$ in the Lab frame where the proton is at rest. What is the minimum energy of the neutron beam for the process to take place?

c) For process b) draw the Feynman diagram(s) and obtain the correspondent amplitude.

d) For process b) evaluate the differential cross section $d\sigma/d\Omega$ in the CM frame as a function of the masses of the particles and the Mandelstam variables, $s = (p_1 + p_2)^2$, $t = (p_1 - p_3)^2$ and $u = (p_1 - p_4)^2$.

e) In the limit that $\sqrt{s} \gg m_\pi, m_N, m_\Delta$ neglect the masses and evaluate the total cross section, σ_T, in the CM frame. Neglect the contribution of the *t-channel* to perform the σ_T calculation. Show that it has the correct dimensions for a cross section.

Appendix B
Particle Physics Exams [Solutions]

B.1 Discovery of the Positron

1.

a) The radius of curvature is 12 cm.
b) If it were an electron traveling in the opposite direction, it would look like the electron would have gained energy when crossing the lead plate, since the curvature radius of the track above is smaller than the one below. A charged particle does not gain energy when crossing a dense material. Actually, the most probable physics behaviour would be to lose energy via inelastic collisions with the atomic (ionization losses). Considering now a proton, one could compute the momentum of such proton as $P \sim 0.3q\,Br$. Taking $r = 0.12\,\text{m}$, $q = +1$ and $B = 1.7\,\text{TeV}$ one would get for the proton a momentum of $P = 0.06\,\text{GeV/c}$. Using this value one could get the mean energy loss by ionization ($\frac{dE}{dX} \gg 10\text{MeV}\,\text{g}^{-1}\,\text{cm}^2$). Given this large value it is expected that the proton deposits all its energy in the lead plate, stopping inside it.
c) $\Delta x = 2.2\,\text{cm}$.
d) The critical energy for the transition between ionization losses and bremsstrahlung radiation occurs for positrons around $\sim 10\,\text{MeV}$. Since the positron has an energy $E \in [23, 63]\,\text{MeV}$, then the most relevant process would be radiation emission by bremsstrahlung. This would lead to a fast degradation in energy, and could inclusively give origin to an electromagnetic shower. So, the thickness computed in the previous question is an upper limit.
e) The most favorable scenario for a e^+ to be produced upwards is if the μ^+ is at rest, i.e., no boost. We want the positron to the the maximum energy, so the best configurations of the decay is when both the neutrinos travel in the opposite direction of the e^+ (see Fig. B.1).

For this configuration the maximum energy of the positron would be $E_e = 52.8\,\text{MeV}$, which is less than the measured one, thus invalidating the possibility that this positron was produced by a vertical muon coming from above.

A. De Angelis et al., *Particle and Astroparticle Physics*, Undergraduate Lecture Notes in Physics, https://doi.org/10.1007/978-3-030-73116-8

Fig. B.1 Scheme of the kinematics of the muon decay

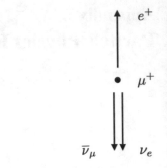

Fig. B.2 Scheme of the kinematics of the pion decay

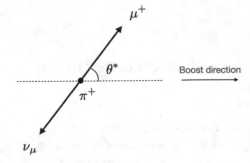

f) Computing the Cherenkov light production threshold one would get for the proton $E_p = 1.4\,\text{GeV}$ and for the positron $E_e = 0.7\,\text{MeV}$. Therefore, this would be a viable option to distinguish a positron from a proton.

2.

a) The distance travelled by the pion in the LAB frame is $x = 557$ m.
b) The interaction length of the pion is $\lambda = 1087$ m. Since this value is larger than the one obtained in the previous question, then it is expected that at this energy the pion decays before it can interact with the atmosphere atoms.
c) $E_\mu^* \simeq 110\,\text{MeV}$.
d) The maximum energy of the muon occurs when the muon is emitted along the boost direction (see Fig. B.2) $\theta^* = 0°$ while the minimum when it goes against the boost, $\theta^* = 180°$. Hence, $E_\mu^{\min} \simeq 5.8\,\text{GeV}$ and $E_\mu^{\max} \simeq 10\,\text{GeV}$.

3.

a) Not possible, violates lepton number. c) Possible.
b) Not possible, violates baryon number. d) Not possible, violates strangeness.

Fig. B.3 Tree-level
Feynman diagram for the
process
$e^- + \bar{\nu}_e \to \mu^- + \bar{\nu}_\mu$

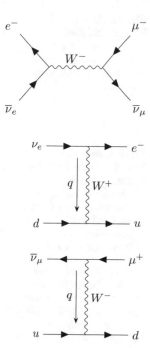

Fig. B.4 Tree-level
Feynman diagram for the
process $\nu_e + d \to e^- + u$

Fig. B.5 Tree-level
Feynman diagram for the
process $\bar{\nu}_\mu + u \to \mu^+ + d$

B.2   Observation of the Glashow Resonance at IceCube

1.

a) $E_\nu = \frac{M_W^2 - m_e^2}{2m_e} = 6.34 \times 10^6 \, \text{GeV} = 6.34 \times 10^{15} \, \text{eV}$.

b) The only diagram for the process $e^- + \bar{\nu}_e \to \mu^-(p_3) + \bar{\nu}_\mu$ is the one shown in Fig. B.3.

c) $\mathcal{M} = \frac{4G_F}{\sqrt{2}} \frac{M_W^2}{s - M_W^2 + iM_W\Gamma_W} \bar{v}(p_2)\gamma_\mu P_L u(p_1)\bar{u}(p_3)\gamma_\mu P_L v(p_4)$

d) $\langle |\mathcal{M}|^2 \rangle = 4G_F^2 s^2 \left(\frac{M_W^2}{s - M_W^2 + iM_W} \right)^2 (1 + \cos\theta)^2$

e) $\frac{d\sigma}{d\Omega} = \frac{G_F^2 s}{16\pi^2} \left(\frac{M_W^2}{s - M_W^2 + iM_W} \right)^2 (1 + \cos\theta)^2$

f) $\sigma = \frac{G_F^2 s}{3\pi} \frac{M_W^4}{(s - M_W)^2 + M_W^2\Gamma_W^2}$

At $\sqrt{s} = M_W$,

$$\sigma = \frac{G_F^2 s}{3\pi} \frac{M_W^4}{\Gamma_W^2} = 53.98 \, \text{nb}.$$

g) $\sigma(e^- + \bar{\nu}_e \to \text{all}) = 9 \times 53.98 = 485.6 \, \text{nb}$.

Fig. B.6 Tree-level
Feynman diagram for the
process $\bar{\nu}_e(p_1) + e^+(p_2) \rightarrow$
$e^+(p_3) + \bar{\nu}_e(p_4)$, (a) has an
amplitude $\mathcal{M}_a$ and (b) an
amplitude $\mathcal{M}_b$

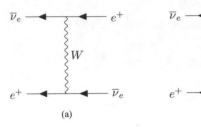

(a)   (b)

Fig. B.7 Non-vanishing
helicities for the Feynman
diagram shown in Fig. B.6a.
The red (grey in black and
white) arrows represent the
helicity and if pointing to the
right it indicated a positive
helicity in case of a fermion

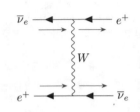

Fig. B.8 Non-vanishing
helicities for the Feynman
diagram shown in Fig. B.6b.
The red (grey in black and
white) arrows represent the
helicity and if pointing to the
right it indicated a positive
helicity in case of a fermion

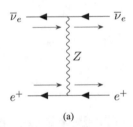
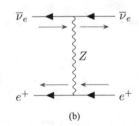

(a)   (b)

2.

a) $q = d$ and $q' = u$ leading to the Feynman diagram depicted in Fig. B.4.

$$\mathcal{M} = \frac{4G_F}{\sqrt{2}} \frac{M_W^2}{t - M_W^2} \bar{u}(p_3)\gamma^\mu P_L u(p_1)\bar{u}(p_4)\gamma_\mu P_L u(p_2).$$

b) $q = u$ and $q' = d$ leading to the Feynman diagram depicted in Fig. B.5.

$$\mathcal{M} = \frac{4G_F}{\sqrt{2}} \frac{M_W^2}{t - M_W^2} \bar{v}(p_1)\gamma^\mu P_L v(p_3)\bar{u}(p_4)\gamma_\mu P_L u(p_2).$$

c) The two diagrams that contribute to this process at the tree level are shown in
Fig. B.6.

$$\mathcal{M}_a = \frac{4G_F}{\sqrt{2}} \frac{M_W}{t - M_W^2} \bar{v}(p_1)\gamma^\mu P_L v(p_3)\bar{v}(P_2)\gamma_\mu P_L v(p_4).$$

$$\mathcal{M}_b = \frac{4G_F}{\sqrt{2}} \frac{M_Z}{t - M_Z^2} \bar{v}(p_1)\gamma^\mu P_L v(p_4)\bar{v}(P_2)\gamma_\mu (g_L^e P_L + g_R^e P_R) v(p_3).$$

The non-vanishing helicity diagrams are shown in Fig. B.7 for the diagram shown
in Fig. B.6a and in Fig. B.8 for the one shown in Fig. B.6b.

3.

Neutrinos are neutral particles so they have to be the interaction/decay products of charged particles, which are accelerated, for instance, in the supernova remnants electromagnetic fields.

High-energy (HE) electrons produce electrons, positrons and gamma rays. Anti-neutrinos of the electron $(\bar{\nu}_e)$ could be produced, for instance, via the following interaction: $e^+ + n \rightarrow \bar{\nu}_e + p$. It should be noted, however, that this is not the main production mechanism: electrons cannot be effectively accelerated. since they tend to rapidly lose energy trough bremsstrahlung emission.

The main production mechanism should then proceed through hadronic interactions of HE protons:

(a) An accelerated proton can interact with other protons in the medium giving origin to hadronic particles (mostly pions). Charged pions decay into: $\pi^+ \rightarrow \mu^+ + \nu_\mu$ and $\pi^- \rightarrow \mu^- + \bar{\nu}_\mu$. The negative muons will then decay giving origin to ν_μ, an e^- and into the *sought* $\bar{\nu}_e$.

(b) Pions can be generated also through photoproduction $p\gamma \rightarrow hadrons$. Although the cross-section for this process is smaller than in (a), typically some 0.5 mb to be compared with ~30 mb, the photon number density in the ambient medium can be orders of magnitude larger than the number density of protons – unless molecular clouds are present.

(c) An alternative to the pion channel would be the creation of high-energy neutrons, resulting from HE proton collisions. Here, the channel to produce $\bar{\nu}_e$ would be the decay $n \rightarrow p + e^- + \bar{\nu}_e$. Note that a PeV neutron would travel in average ~ 30 light years before decaying, which means that there is a high probability that the neutron disappears during its cosmic voyage.

B.3   Investigation of the Z Boson at SLAC

1.

a) At these energies the dominant diagram is the one shown in Fig. B.9.

b) $\mathcal{M} = \dfrac{8G_F}{\sqrt{2}} \dfrac{M_Z^2}{(s - M_Z^2) + i M_Z \Gamma_Z} \bar{v}(p_2) \gamma^\mu (g_V^e - g_A^e \gamma^5) u(p_1) \bar{u}(P_3) \gamma_\mu (g_V^\mu - g_A^\mu \gamma^5) v(p_4).$

c) $\langle |\mathcal{M}|^2 \rangle = \dfrac{32 G_F^2 s^2}{4} \left(\dfrac{M_Z^2}{(s - M_Z^2) + i M_Z \Gamma_Z} \right)^2 [(1 + \cos\theta)^2 ((g_R^e g_R^\mu)^2 + (g_L^e g_L^\mu)^2) + (1 - \cos\theta)^2 ((g_R^e g_L^\mu)^2 + (g_L^e g_R^\mu)^2)]$

d) $\dfrac{d\sigma}{d\Omega} = \dfrac{G_F^2 s}{8\pi^2} \left(\dfrac{M_Z^2}{(s - M_Z^2) + i M_Z \Gamma_Z} \right)^2 [(1 + \cos\theta)^2 ((g_R^e g_R^\mu)^2 + (g_L^e g_L^\mu)^2) + (1 - \cos\theta)^2 ((g_R^e g_L^\mu)^2 + (g_L^e g_R^\mu)^2)]$

e) $\sigma(\sqrt{s} = M_Z) = \dfrac{2}{3} \dfrac{G_F^2}{\pi} \dfrac{M_Z^4}{\Gamma_Z^2} \left[(g_R^e g_R^\mu)^2 + (g_L^e g_L^\mu)^2 + (g_R^e g_L^\mu)^2 + (g_L^e g_R^\mu)^2 \right] = 1969 \, \text{pb}.$

f) $\sigma(e^- e^- \rightarrow \text{All}) = \dfrac{\sigma(e^+ e^- \rightarrow \mu^+ \mu^-)}{BR(Z \rightarrow \mu^+ \mu^-)} = 58.6 \, \text{nb}.$

Fig. B.9 Tree-level
Feynman diagram for the
process
$e^+ + e^- \to \mu^+ + \mu^-$

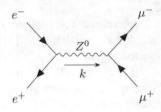

2.

a) $\langle |\mathcal{M}_L|^2 \rangle = 16 G_F^2 s^2 \left(\dfrac{M_Z^2}{(s - M_Z^2) + i M_Z \Gamma_Z} \right)^2 (g_L^e)^2 \left[(1 - \cos\theta)^2 (g_R^\mu)^2 + (1 + \cos\theta)^2 (g_L^\mu)^2 \right]$

$\langle |\mathcal{M}_R|^2 \rangle = 16 G_F^2 s^2 \left(\dfrac{M_Z^2}{(s - M_Z^2) + i M_Z \Gamma_Z} \right)^2 (g_R^e)^2 \left[(1 - \cos\theta)^2 (g_R^\mu)^2 + (1 + \cos\theta)^2 (g_L^\mu)^2 \right]$

b) $\sigma_L = \dfrac{4 G_F^2 s}{3\pi} \left(\dfrac{M_Z^2}{(s - M_Z^2) + i M_Z \Gamma_Z} \right)^2 (g_L^e)^2 \left[(g_R^\mu)^2 + (g_L^\mu)^2 \right]$

$\sigma_L = \dfrac{4 G_F^2 s}{3\pi} \left(\dfrac{M_Z^2}{(s - M_Z^2) + i M_Z \Gamma_Z} \right)^2 (g_R^e)^2 \left[(g_R^\mu)^2 + (g_L^\mu)^2 \right]$

c) $A_{LR} = \dfrac{\sigma_L - \sigma_R}{\sigma_L + \sigma_R} = \dfrac{(g_L^e)^2 - (g_R^e)^2}{(g_L^e)^2 + (g_R^e)^2} = \dfrac{2 g_V^e g_A^e}{(g_V^e)^2 + (g_A^e)^2}$

As we are summing over the polarization of the μ^+ the factor $(g_R^\mu)^2 + (g_L^\mu)^2$
cancels in the ratio.

d) $A_{LR} \simeq 0.099$.

e) $A_{LR} = \dfrac{1 - 4 s_W^2}{1 - 4 s_W^2 + 8 s_W^4}$,

and,

$$s_W^2 = \dfrac{-(1 - A_{LR}) + \sqrt{(1 - A_{LR})^2 + 2(1 - A_{LR}) A_{LR}}}{4 A_{LR}}.$$

f) Taking the error limits and computing s_W^2 one would get,

$$0.2264 \le s_W^2(1992) \le 0.2486$$

which includes the measured value.

Note: this is just a quick way to find if the result is reasonable. A more correct
way would be to propagate the error,

$$\delta s_W^2 = \left| \dfrac{d s_W^2}{d A_{LR}} \right| \delta A_{LR} = 0.0055$$

which would lead to,

$$0.2266 \leq s_W^2(1992) \leq 0.2486.$$

B.4   Deep Inelastic Scattering in the ZEUS Experiment

1.

a) $\sqrt{s} = 295.9\,\text{GeV}.$
b) $E_e = 46.8\,\text{TeV}.$
c) $N \simeq 81$ events.
d) $\theta = 7.1 \times 10^{-5}\,\text{rad} = 0.004°$

2.

a) Not possible, violates baryon number.
b) Not possible, violates strangeness.
c) Possible.
d) Not possible, violates lepton number.
e) Possible.

3.

a) Number of radiation lengths $= 1.04 \simeq 1$.
b) For high-energy electromagnetic particles two processes dominate: pair creation (e^+e^-) and bremsstrahlung radiation. These processes have very similar cross-sections. Hence, a cascade of e^-, e^+ and γ is produced, increasing the number of particles until a critical energy is reached. This critical energy is related with the transition between bremsstrahlung losses and ionization losses. Uranium: *forces* the particles to interact and consequently speeds the shower development. It also is used to absorb low energy photons. Scintillator: the shower low energy particles excite the material producing light that can be guided, through light fibres, and detected by PMTs.
c) $\langle E_{\text{loss}} \rangle = 9.4\,\text{MeV}.$

4.

a) $x = 5.4 \times 10^{-15}\,\text{m}.$
b) $\dfrac{\Gamma(\Delta^+ \to p\pi^0)}{\Gamma(\Delta^+ \to n\pi^+)} = 2.$
c) $N_\gamma = \dfrac{4}{3}N_0.$

Table B.1 Table to assess if a particle of a given type (id) gives signal in each of the detectors: (o) no signal ; (x) signal

Particle Id	a	b	c	d	e
γ	o	x	o	–	o
e	x	x	o	–	o
p	x	x	x	–	o
μ	x	x	x	–	x

B.5   Discovery of the J/ψ Meson

1.

a) $m_{J/\psi} = \sqrt{(E_2 + E_3)^2 - (P_2 + P_3)^2}$.

b) $x \simeq 1\,\mathrm{m}$.

c) $x = 6.8 \times 10^{-13}$ m. The particle width, Γ, or the decay time, τ, do not depend on the decay mode. So the mean distance covered before it decays would be the same.

d) $\theta_{\min} = 0°$ and $\theta_{\max} = 180°$.

e) $E_e^{\min} = 1.53\,\mathrm{GeV}$ and $E_e^{\max} = 6.21\,\mathrm{GeV}$.

2.

a) $E_e \simeq 9.4\,\mathrm{TeV}$.

b) A possible answer would be a set of detectors sequentially deployed along the path of the particle, for instance:

- inner detector sensitive to charged particles (a);
- electromagnetic calorimeter (b);
- hadronic calorimeter (c);
- iron block (d);
- detector sensitive to charged particles (e).

For this apparatus the particles could be identified if they are *seen* in some detector and not in others. The following Table (B.1) can be built for gamma rays (γ), electrons/positrons (e), protons (p) and muons (μ) respectively.

The proton and electron energy can be reconstructed from the hadronic and the electromagnetic calorimeters, respectively. The energy of muons can be obtained by applying a magnetic field and measuring the muon trajectory curvature.

3.

a) Not possible, violates charge conservation.

b) Not possible, violates baryon number.

c) Not possible, violates strangeness.

d) Possible.

e) Not possible, violates lepton number.

Fig. B.10 Tree-level
Feynman diagram for the
process
$e^+ + e^- \rightarrow \mu^+ + \mu^-$ for
$\sqrt{s} \ll M_Z$

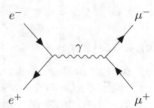

Fig. B.11 R ratio as a
function of the
center-of-mass energy $\sqrt{s}$

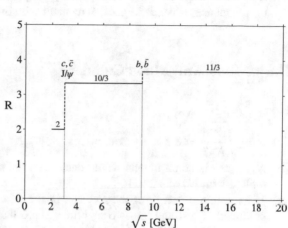

4. For $e^+e^- \rightarrow \mu^+\mu^-$ (a) the only channel contributing is the s-channel while for $e^+e^- \rightarrow e^+e^-$ we would get an s a t and a u-channel. Hence, for (a) the result would be essentially the same but (b) would have an enormous contribution from the forward region that would hide the J/ψ resonance.

5.

a) The dominant diagram taking into account that $\sqrt{s} \ll M_Z$ is the one displayed in Fig. B.10.

b) $\mathcal{M} = \dfrac{e^2}{s}\bar{u}(p_2)\gamma^\mu u(p_3)\gamma_\mu v(p_4)$, with $s = (p_1 + p_2)^2$.

c) $\langle|\mathcal{M}|^2\rangle = (4\pi\alpha)^2 \left[1 + \cos^2\theta\right]$.

d) $\dfrac{\sigma}{d\Omega} = \dfrac{\alpha^2}{4s}\left(1 + \cos^2\theta\right)$.

e) $\sigma = \dfrac{4\pi\alpha^2}{3s}$.

f) $\sigma(|\cos\theta| = 0.6) \simeq 4.5\,\text{nb}$. In Fig. A.2 (left, middle plot) we have $\sigma \sim 5\,\text{nb}$ which is in good agreement with the obtained value since the errors are $\sim 1\,\text{nb}$.

6.

a) $R = 3\left(\dfrac{4}{9} + 2 \times \dfrac{1}{9}\right) = 2$.

b) $R = 3 \left(2 \times \dfrac{4}{9} + 2 \times \dfrac{1}{9} \right) = \dfrac{10}{3}$.

c) $R = 3 \left(2 \times \dfrac{4}{9} + 3 \times \dfrac{1}{9} \right) = \dfrac{11}{3}$.

d) The plot of R as a function of $\sqrt{s}$ is shown in Fig. B.11.

e) $R(\sqrt{s} < 3\,\text{GeV}) \simeq 20/9.9 \simeq 2$ and $R(\sqrt{s} > 3.1\,\text{GeV}) \simeq 35/9.9 \simeq 3.5$. This should be compared with $R = 2$ (below the J/ψ resonance) and $R = 3.3$ above the resonance, which is in good agreement with the previous values.

B.6   Muon/Anti-Muon Collider

1.

a) $N = \sqrt{\dfrac{\mathscr{L} A}{N_b f}} = 8.2 \times 10^{12}$ muons.

b) $N_{\text{decay}} \simeq N_0$, so all muons would decay and the collider has to be completely refilled with $N_0 = 8.2 \times 10^{12}$.

c) The minimum energy of the electron in the CM occurs when the electron is produced at rest and the two neutrinos share the remaining momentum. The maximum energy of the electron in the CM occurs when the neutrinos are produced in the same direction and the electron produced in the opposite direction of the neutrinos.

d) $E_{e,\text{min}} = 1.5\,\text{MeV}$ and $E_{e,\text{max}} = 62.8\,\text{GeV}$.

e) The maximum angle that the electron can make with the muon flight direction in the LAB frame is $\theta_{\text{max}} = 0.1°$. So the forward region of the detector should be avoided.

f) Here it is presented only if the interaction/reaction is possible or not, but the quantum numbers should be evaluated and conserved before and after.

  i) Not possible, violates lepton number.
  ii) Possible.
  iii) Not possible, violates lepton number.
  iv) Possible.

g) $N_{\text{Higgs}} = 12615\,\text{year}^{-1}$.

h) The b quark has a lifetime ~ 1 ps, so any particle containing a b will rapidly decay very near the interaction point. Therefore, one possible strategy would be to have a detector sensitive to charged particles near the interaction point (micro-vertex detector) and look for tracks that emerge from the interaction point that decay rapidly into new tracks.

2.

a) The dominant Feynman diagrams for the process $\mu^+ \mu^- \to f \bar{f}$ are:

  • for $f = e^-$ are displayed in Fig. B.12;
  • for $f = \nu_e$ are displayed in Fig. B.13;

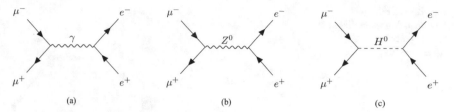

Fig. B.12 Tree-level Feynman dominant diagrams for the process $\mu^+ + \mu^- \to e^- + e^+$

- for $f = \nu_\mu$ are displayed in Fig. B.14;
- for $f = b$ are displayed in Fig. B.15.

b) The dominant channels for the process $\mu^- e^+ \to \mu^- e^+$ are the ones presented in Fig. B.16.

c) For $\sqrt{s} \simeq M_H$ only the diagram with s-channel Higgs exchange is relevant (see Fig. B.15c).

$$\mathcal{M}(h_1 h_2; h_3 h_4) = -g_H^\mu g_H^b \frac{1}{s - M_H^2 + i M_H \Gamma_H} J_{u_1 v_2}(h_1 h_2) J_{u_3 v_4}(h_3 h_4),$$

with $g_H^\mu = \frac{1}{2} g \frac{m_\mu}{M_W}$ and $g_H^\mu = \frac{1}{2} g \frac{m_b}{M_W}$.

d) $\mathcal{M}(\uparrow\uparrow; \uparrow\uparrow) = \mathcal{M}(\downarrow\downarrow; \downarrow\downarrow) = -g_H^\mu g_H^b \frac{s}{s - M_H^2 + i M_H \Gamma_H}$

$\mathcal{M}(\uparrow\uparrow; \downarrow\downarrow) = \mathcal{M}(\downarrow\downarrow; \uparrow\uparrow) = +g_H^\mu g_H^b \frac{s}{s - M_H^2 + i M_H \Gamma_H}$

e) $\langle |\mathcal{M}|^2 \rangle = (g_H^\mu)^2 (g_H^b)^2 \frac{s^2}{(s - M_H^2)^2 + M_H^2 \Gamma_H^2}$

$\frac{d\sigma}{d\Omega} = \frac{(g_H^\mu)^2 (g_H^b)^2}{64 \pi^2} \frac{s}{(s - M_H^2)^2 + M_H^2 \Gamma_H^2}$

f) $\sigma = \frac{(g_H^\mu)^2 (g_H^b)^2}{16 \pi} \frac{1}{\Gamma_H^2} = 2214$ pb.

g) $R_1 = \frac{(g_H^\mu)^2}{(g_H^e)^2} = \frac{m_\mu^2}{m_e^2} = 4.275 \times 10^4$.

h) At $\sqrt{s} = M_Z$ the dominant diagram is the s-channel Z exchange. As the Z couples in the same way to muons and to electrons we will have $R_2 = 1$.

i) A *Higgs factory* means that we want to produce as many H^0 as possible. This is achieved through the Higgs s-channel resonance at $\sqrt{s} = M_H$. As seen before the cross-section is inversely proportional to $1/\Gamma_H^2$ making it very large. Moreover, the Higgs coupling is proportional to the mass of the fermions; hence, muon are favored with respect to electrons. In fact, as seen before, $R_1 = 4.275 \times 10^4$, making the μ-collider much more effective to produce Higgs bosons than a $e^+ e^-$ collider.

Fig. B.13 Tree-level
Feynman dominant diagrams
for the process
$\mu^+\mu^- \to \nu_e\bar{\nu}_e$

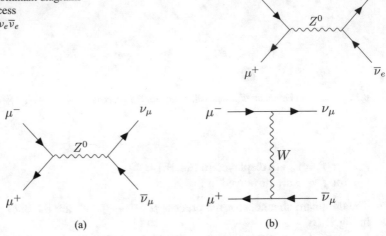

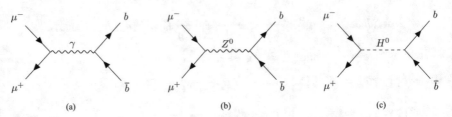

(a)   (b)

Fig. B.14 Tree-level Feynman dominant diagrams for the process $\mu^+\mu^- \to \nu_\mu\bar{\nu}_\mu$

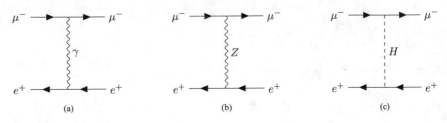

(a)   (b)   (c)

Fig. B.15 Tree-level Feynman dominant diagrams for the process $\mu^+\mu^- \to b\bar{b}$

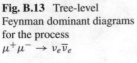

(a)   (b)   (c)

Fig. B.16 Tree-level Feynman dominant diagrams for the process $\mu^- e^+ \to \mu^- e^+$

Fig. B.17 Tree-level Feynman diagram for the process $q(p_1) + \bar{q}(p_2) \to \mu^+(p_3) + \mu^-(p_4)$; (a) has an amplitude $\mathcal{M}_a$ and (b) an amplitude $\mathcal{M}_b$

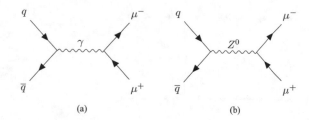

(a)   (b)

B.7 Rare Decays at the LHCb Experiment

1.

a) $N \simeq 4.4 \times 10^{11}$ $b-$particles.
b) $x \simeq 8.4$ mm.
c)   i) $E\mu$, $\min_{-}^{*} E_{\mu,\max}^{*} \simeq 2.68$ GeV.
   ii) $\theta_{\min} = 180°$ and $\theta_{\max} = 0.1$ rad $= 6.15°$.
   iii) $E_{B,\min} \simeq 8.9$ GeV.

2.

a) Not possible, violates baryon and lepton number.
b) Not possible, violates baryon number.
c) Possible.
d) Not possible, violates baryon number, charge and strangeness.
e) Not possible, violates strangeness.
f) Possible.

3.

a) $1.01 \leq n \leq 1.12$.
b) $\dfrac{N_\gamma^\pi}{N_\gamma^k} = 2.15$.

4.

a) The possible Feynman diagrams are the ones depicted in Fig. B.17.
b) $\mathcal{M} = \mathcal{M}_a + \mathcal{M}_b$

$$\mathcal{M}_a = \frac{e^2}{\hat{s}} \bar{v}(\hat{p}_2) \gamma^\mu u(\hat{p}_1) \bar{u}(p_3) \gamma_\mu v(p_4)$$

$$\mathcal{M}_b = \frac{8 G_F}{\sqrt{2}} \frac{M_Z^2}{\hat{s} - M_Z^2 + i M_Z \Gamma_Z^2} \bar{v}(\hat{p}_2) \gamma^\mu (g_L^q P_L + g_R^q P_R) u(\hat{p}_1) \bar{u}(p_3)(g_L^\mu P_L + g_R^\mu P_R) \gamma_\mu v(p_4)$$

c) $\langle |\mathcal{M}|^2 \rangle = 8 G_F^2 \dfrac{M_Z^4}{(\hat{s} - M_Z^2)^2 + M_Z^2 \Gamma_Z^2} s^2 \left[(1 + \cos\theta)^2 \left((g_R^q g_R^\mu)^2 + (g_L^q g_L^\mu)^2 \right) + (1 - \cos\theta)^2 \left((g_R^q g_L^\mu)^2 + (g_L^q g_R^\mu)^2 \right) \right]$

d) $\dfrac{d\sigma_{q\bar{q}}}{d\Omega} = \dfrac{1}{8\pi^2} G_F^2 \dfrac{M_Z^4}{(\hat{s} - M_Z^2)^2 + M_Z^2 \Gamma_Z^2} s \left[(1 + \cos\theta)^2 \left((g_R^q g_R^\mu)^2 + (g_L^q g_L^\mu)^2 \right) + (1 - \cos\theta)^2 \left((g_R^q g_L^\mu)^2 + (g_L^q g_R^\mu)^2 \right) \right]$

e) $\sigma_{q\bar{q}}(\hat{s}) = \dfrac{2}{3\pi} G_F^2 \dfrac{M_Z^4}{(\hat{s} - M_Z^2)^2 + M_Z^2 \Gamma_Z^2} \hat{s} \left[(g_R^q g_R^\mu)^2 + (g_L^q g_L^\mu)^2 + (g_R^q g_L^\mu)^2 + (g_L^q g_R^\mu)^2 \right]$

where $q = u, d$. At $\hat{s} = M_Z^2$ we get

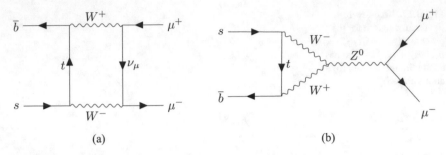

Fig. B.18 Lowest order Feynman diagrams for the decay $B_s^0 \rightarrow \mu^+\mu^-$

$$\sigma_{q\bar{q}}(\hat{s}) = \frac{2}{3\pi} G_F^2 \frac{M_Z^4}{\Gamma_Z^2} \hat{s} \left[(g_R^q g_R^\mu)^2 + (g_L^q g_L^\mu)^2 + (g_R^q g_L^\mu)^2 + (g_L^q g_R^\mu)^2 \right].$$

f) Considering $f_q(x)$ (the quark q parton distribution function with a fraction of the proton momentum x), one could write

$$\sigma_{pp \rightarrow \mu^-\mu^+}(s) = \int_0^1 dx_1 \int_0^1 dx_2 \left[f_u(x_1) f_{\bar{u}}(x_2) \sigma_{u\bar{u}}(x_1 x_2\, s) + f_d(x_1) f_{\bar{d}}(x_2) \sigma_{d\bar{d}}(x_1 x_2\, s) \right].$$

5.

a) $\bar{b}s$
b) The two lowest order possible diagrams are the ones shown in Fig. B.18.
c) By computing the square of the invariant mass, which is a Lorentz invariant, $s = E^2 - P^2 = m_{B_s^0}^2 \neq m_Z^2$. Using muons one would get $s = (E_{\mu^+} + E_{\mu^-})^2 - (P_{\mu^+} + P_{\mu^-})^2 = m_{B_s^0}^2$.

B.8   Neutrino-Nucleon Cross-Section at Very-High-Energies

1.

a) $\sqrt{s} \simeq 1.4\,\text{TeV}$.
b) Probability for the muon neutrino to decay inside the Earth is $\simeq 98\%$.
c) $\theta_C = 0.97\,\text{rad} = 55.8°$.
d) $N_\gamma^{\text{Ch}} \simeq 1.3 \times 10^7$ photons.

2.

a) Not possible, violates charge.   d) Possible.
b) Not possible, violates charge.   e) Not possible, violates lepton number.
c) Not possible, violates lepton number.

Fig. B.19 Tree-level
Feynman diagram for the
process $\nu_\mu d \to \mu^- u$

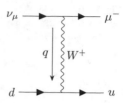

3. $E_\nu \simeq 2.9 \times 10^{15} \, \text{eV} = 2.9 \, \text{PeV}$.

4.

a) The mean interaction free path of the τ_ν is $\lambda \simeq 3.15 \times 10^8$ cm, while the distance travelled in average by the τ-lepton before decaying is $\langle x \rangle \simeq 48.89$ m. Comparing with the Earth diameter, d_T, one gets:

- for τ $\to$ $\langle x \rangle / d_T \simeq 3.8 \times 10^{-6}$;
- for ν_τ $\to$ $\lambda / d_T \simeq 0.25$.

b) The minimum and maximum energy that the ν_τ can get is m_{ν_τ} and $0.9 \times 10^{15} \, \text{eV}$, respectively.

5. To distinguish between different neutrino flavors one could exploit its charged current interation (W-boson t-channel exchange). The leptons arising from this interaction would interact differently with the ice and, as such, neutrinos could be distinguished using the event topology, for instance:

- muons: long-lived with nearly no interaction with matter, consequently producing Cherenkov light along a large path;
- electrons: generate electromagnetic cascades and as a consequence produce shorter and broader tracks;
- tau leptons: would decay after some meters. If decaying into *quarks* (electrons), they will produce hadronic (electromagnetic) showers. Then the signature would be a double-bump structure.

6.

a) $q = d$ and $q' = u$

b) The only Feynman diagram relevant for this process is the one shown in Fig. B.19.

c) $\mathcal{M} = -\dfrac{4G_F}{\sqrt{2}} \dfrac{M_W^2}{Q^2 + M_W^2} \bar{u}(p_3) \gamma^\mu P_L u(p_1) \bar{u}(\hat{p}_4) \gamma_\mu P_L u(\hat{p}_2)$, where $Q^2 = -q^2$.

d) $\langle |\mathcal{M}|^2 \rangle = 16 G_F^2 \hat{s}^2 \left[\dfrac{M_W^2}{Q^2 + M_W^2} \right]^2$.

e) $\dfrac{d\hat{\sigma}}{d\Omega} = \dfrac{G_F^2 \hat{s}}{4\pi^2} \left[\dfrac{M_W^2}{Q^2 + M_W^2} \right]^2$

7.

a) $\bar{q} = \bar{u}$ and $\bar{q}' = \bar{d}$.

Fig. B.20 Tree-level
Feynman diagram for the
process $\nu_\mu \bar{u} \to \mu^- \bar{d}$

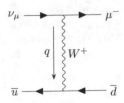

b) The Feynman diagram relevant for this process is the one displayed in Fig. B.20.

c) $\mathcal{M} = -\dfrac{4G_F}{\sqrt{2}} \dfrac{M_W^2}{Q^2 + M_W^2} \bar{u}(p_3)\gamma^\mu P_L u(p_1)\bar{v}(\hat{p}_2)\gamma_\mu P_L v(\hat{p}_4).$

d) $\langle |\mathcal{M}|^2 \rangle = 4G_F^2 \hat{s} \left[\dfrac{M_W^2}{Q^2 + M_W^2} \right]^2 (1 + \cos\hat\theta)^2.$

e) $\dfrac{d\hat\sigma}{d\Omega} = \dfrac{G_F^2 \hat{s}}{16\pi^2} \left[\dfrac{M_W^2}{Q^2 + M_W^2} \right]^2 (1 + \cos\hat\theta)^2.$

8.

a) They are not good variables because the center-of-mass of the elementary process is not the center-of-mass of the overall process and the momenta of the partons are not directly experimentally accessible.

b) If x is the fraction of the proton (nucleon) momentum carried by the quark, we must have

$$\hat{p}_2 = x p_2,$$

and therefore

$$y \equiv \frac{p_2 \cdot (p_1 - p_2)}{p_2 \cdot p_1} = \frac{\hat{p}_2 \cdot (p_1 - p_3)}{\hat{p}_2 \cdot p_1},$$

where the symbol $\cdot$ between two 4-vectors, say a and b, indicates their scalar product $(a \cdot b \equiv a^\mu b_\mu)$.

This quantity is a Lorentz invariant and can be calculated in the CM of the elementary process. In this frame, neglecting the fermion masses, we have

$$p_1 = (E, 0, 0, E)$$
$$\hat{p}_2 = (E, 0, 0, -E)$$
$$p_1 = (E, E\sin\hat\theta, 0, E\cos\hat\theta)$$
$$\hat{p}_4 = (E, -E\sin\hat\theta, 0, E\cos\hat\theta)$$

where the energy $E = \frac{\sqrt{\hat{s}}}{2}$. Substituting in y we get

$$y = \frac{1}{2}(1 - \cos\hat\theta).$$

1. They are distinct processes so we need to sum the probabilities of each, that is the cross-sections.
2. x is the fraction of momentum carried by the quark with respect to the momentum of the proton, that is $\hat{p}_2 = xp_2$. Then it can be shown that $\hat{s} = xs$.
3. $\dfrac{d\sigma}{dy} = 2\dfrac{d\sigma}{d\cos\hat{\theta}}$.
4. From the above relation between $\hat{\theta}$ and $d\hat{y}$ we have $\theta = 0 \Rightarrow y = 0$ and $\theta = \pi \Rightarrow y = 1$. So $y \in [0, 1]$ and $x \in [0, 1]$.
5. The functions $d(x)$ and $\bar{u}(x)$ are the probabilities of finding a quark d and an anti-quark u, respectively, with a fraction x of the momentum of the proton.

B.9   Direct Observation of the Tau Neutrino

1.

a) $\Delta T = 2155\,\text{s}$.
b) $E_\pi \simeq 140\,\text{MeV}$.
c) The muon can have in the LAB reference frame a minimum energy of $E_{\mu,\min} \simeq 5.7\,\text{GeV}$ and a maximum energy of $E_{\mu,\max} \simeq 10\,\text{GeV}$. Its energy spectrum distribution would be constant: $dN/dE \propto 1/\gamma\beta P_\mu^*$.
d) $\langle E_\mu^{\text{loss}} \rangle \simeq 2.9\,\text{GeV}$.
e) $N_{D_s^+} \simeq 7.5 \times 10^{11}$.
f)
  • $c\bar{c} \rightarrow J/\psi$
  • $c\bar{d} \rightarrow D^+$
  • $c\bar{s} \rightarrow D_s^+$
  • $c\bar{b} \rightarrow B_c^+$
  • $c\bar{u} \rightarrow D^0$
g) (i) Not allowed as it violates lepton number conservation. (ii) Allowed. (iii) Not allowed as it violates baryon number conservation. (iv) Allowed.
h) $N_{\nu_\tau} = 4.125 \times 10^{10}$.
i) The interaction of neutrinos with matter produces as charged particles:

  • $\nu_e \rightarrow e^- W^+$;
  • $\nu_\mu \rightarrow \mu^- W^+$;
  • $\nu_\tau \rightarrow \tau^- W^+$.

So the lepton will be the first charged track. The τ^- can be distinguished from the other leptons as it decays very fast (near the neutrino interaction point): it has a $c\tau = 87.03\,\mu\text{m}$. A τ^- can decay into:

  • $\tau^- \rightarrow \nu_\tau \bar{\nu}_e e^-$;
  • $\tau^- \rightarrow \nu_\tau \bar{\nu}_\mu \mu^-$;
  • $\tau^- \rightarrow \nu_\tau$ plus quarks.

Fig. B.21 Lowest order Feynman diagrams for the process $\nu_e e^- \rightarrow e^- \nu_e$

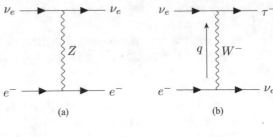

(a)   (b)

Fig. B.22 Tree-level Feynman diagram for the process $\nu_\tau e^- \rightarrow \tau^- \nu_e$

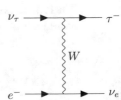

Choosing the leptonic decay channel, which has only a charged particle in the final state, is the simplest choice. The energy of the lepton has to be shared with the remaining two neutrinos, which means that it is improbable that the lepton would follow the direction of the τ^-. Hence, the signature would be for the ν_τ: one charged track (the τ^-) followed by another track, which is the lepton (e^- or μ^-) coming from the τ^- decay, with different direction (kink).

2.

a) The Feynman diagrams for the process $\nu_e e^- \rightarrow e^- \nu_e$ are the ones depicted in Fig. B.21. The only diagram for the process $\nu_\tau e^- \rightarrow \tau^- \nu_e$ is shown in Fig. B.22.

b) $q = d$ and $q' = u$.

c) $i\mathcal{M} = \left(-i\dfrac{g}{\sqrt{2}}\right) \bar{u}(p_3)\gamma^\mu P_L u(p_1) \dfrac{(-i)\left[g_{\mu\nu} - \dfrac{q_\mu q_\nu}{M_W^2}\right]}{q^2 - M_W^2} \bar{u}(p_4)\gamma^\nu P_L u(p_2)$, with $q = p_3 - p_1$.

d) $\mathcal{M} = -\dfrac{4G_F}{\sqrt{2}}\bar{u}(p_3)\gamma^\mu P_L u(p_1)\bar{u}(p_4)\gamma_\mu P_L u(p_2)$.

e) $\mathcal{M}(\downarrow\downarrow; \downarrow\downarrow) = -\dfrac{8G_F}{\sqrt{2}}s$ and

$\langle|\mathcal{M}|^2\rangle = 16G_F^2 s^2$.

f) $\dfrac{d\sigma}{d\Omega} = \dfrac{G_F^2 s}{4\pi^2}$.

g) $\sigma = \dfrac{G_F^2 s}{\pi} = 4.327 \times 10^{-11}\,\text{GeV}^{-2} = 16.85\,\text{fb}$.

h) As $d\sigma/d\Omega$ in the CM does not depend on the CM angle θ, it is likely to be emitted equally in both directions.

i) Neutrinos do not leave tracks on emulsion. So the neutrino beam comes from the left and produces 3 tracks. One of these tracks must be the τ^- because:

Fig. B.23 Tree-level
Feynman diagram for the
process $\nu_\tau d \to \tau^- d$

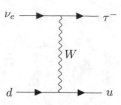

    a. it travels some distance before decaying (note that the τ^- mean decay path
       in its proper reference frame is $\simeq 87.03\,\mu$m while for the muon is 658 m).
    b. the τ^- can decay into a lepton (electron of muon) and two neutrinos.

This last feature would explain the apparent *kink*. Hence, the Feynman diagram
drawn in Fig. B.23, with the subsequent decay of the tau lepton. Notice the it
cannot be $\nu_e e^- \to \tau^- \nu_e$ because the track of the ν_e would not be seen in the
emulsion.

B.10 Pion-Nucleon Cross-Section

1.

a) $\langle x \rangle = 8.08$ m.
b) • Use magnetic field to deflect particles;
    • The deflection depends of the particle mass, charge, velocity and on the of
      the magnetic field $\mathbf{B}$;
    • Use a collimator to stop close-by unwanted particles and create a beam;
    • The collimator has to be placed before the π^+ decays.
    • Hence, making use of the above considerations and taking the pion energy
      and mean decay free path, the radius of curvature should be $r \sim 5$ m and
      consequently $B \sim 0.1$ T.
c) i. $\sigma = 1.29 \times 10^{-25}\,\mathrm{cm}^2 = 129$ mb.
  ii. $\langle E_{\mathrm{loss}} \rangle = 7.05$ MeV and $\theta_0 = 0.017\,\mathrm{rad} \simeq 0.96°$.
d) The maximum angle that the μ^+ can have with the beam line is

$$\theta = \arctan\left(\frac{\sqrt{(E_\mu^*)^2 - m_\mu^2}}{\dfrac{E_\pi P_\pi E_\mu^*}{m_\pi E_\pi}}\right) \simeq 16°.$$

Uniform energy distribution with $E_\mu^{\min} = 126.5$ MeV and $E_\mu^{\min} = 190.3$ MeV.

2.

a) The possible Feynman diagrams at first order are the ones shown in Fig. B.24.
b) $E_n^{\min} = 1.57$ GeV.

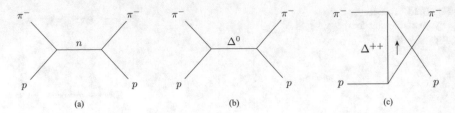

Fig. B.24 Tree-level Feynman dominant diagrams for the process $\pi^- + p \to \pi^- + p$

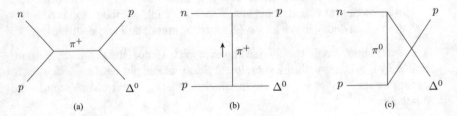

Fig. B.25 Tree-level Feynman dominant diagrams for the process $n(p_1) + p(p_2) \to p(p_3) + \Delta^0(p_4)$

c) For the process $n(p_1) + p(p_2) \to p(p_3) + \Delta^0(p_4)$ we have the following Feynman diagrams shown in Fig. B.25.

The corresponding amplitudes are:

- $\mathscr{M}_a = \dfrac{\mu_1 \mu_2}{s - m_\pi^2}$, with $s = (p_1 + p_2)^2$
- $\mathscr{M}_b = \dfrac{\mu_1 \mu_2}{t - m_\pi^2}$, with $t = (p_1 - p_3)^2$
- $\mathscr{M}_c = \dfrac{\mu_1 \mu_2}{u - m_\pi^2}$, with $t = (p_1 - p_4)^2$.

d) $\dfrac{d\sigma}{d\Omega} = \dfrac{(\mu_1 \mu_2)^2}{64\pi^2 s} \dfrac{\sqrt{s^2 + m_p^4 + m_{\Delta^0}^4 - 2sm_p^2 - 2sm_{\Delta^+}^2 - 2m_p^2 m_{\Delta^0}^2}}{\sqrt{s^2 + m_n^4 + m_p^4 - 2sm_p^2 - 2sm_n^2 - 2m_n^2 m_p^2}} \left[\dfrac{1}{s - m_{\pi^+}^2} + \dfrac{1}{t - m_{\pi^+}^2} + \dfrac{1}{u - m_{\pi^0}^2} \right]$

e) $\sigma = \dfrac{(\mu_1 \mu_2)^2}{16\pi s^3}$.

$[\sigma] = \dfrac{[\text{mass}]^4}{([\text{mass}^2])^3} = \dfrac{1}{[\text{mass}]^2}$

Appendix C
Formulary

Propagators

$$\mu \,\underset{\gamma}{\sim\!\!\sim\!\!\sim\!\!\sim}\, \nu \qquad\qquad -i\frac{g\mu\nu}{k^2} \qquad\qquad (C.1)$$

$$\mu \,\underset{W}{\sim\!\!\sim\!\!\sim\!\!\sim}\, \nu \qquad\qquad -i\frac{g_{\mu\nu} - \frac{k_\mu k_\nu}{M_W^2}}{k^2 - M_W^2 + iM_W\Gamma_W} \qquad\qquad (C.2)$$

$$\mu \,\underset{Z}{\sim\!\!\sim\!\!\sim\!\!\sim}\, \nu \qquad\qquad -i\frac{g_{\mu\nu} - \frac{k_\mu k_\nu}{M_Z^2}}{k^2 - M_Z^2 + iM_Z\Gamma_Z} \qquad\qquad (C.3)$$

$$\xrightarrow[p]{f} \qquad\qquad \frac{i(\not{p} + m_f)}{p^2 - m_f^2} \qquad\qquad (C.4)$$

$$\underset{p}{- - - - H - - - -} \qquad\qquad \frac{i}{p^2 - M_H^2 + iM_H\Gamma_H} \qquad\qquad (C.5)$$

© Springer Nature Switzerland AG 2021
A. De Angelis et al., *Particle and Astroparticle Physics*, Undergraduate Lecture Notes in Physics,
https://doi.org/10.1007/978-3-030-73116-8

Vertices

Charged Current

$$-i\frac{g}{\sqrt{2}}\gamma_\mu \frac{1-\gamma^5}{2}$$

(C.6)

Neutral Current

$$-i\frac{g}{\cos\theta_W}\gamma_\mu \left(g_V^f - g_A^f \gamma_5 \right)$$

$$-ieQ_f\gamma_\mu$$

where

$$g_V^f = \frac{1}{2}T_f^3 - Q_f \sin^2\theta_W, \quad g_A^f = \frac{1}{2}T_f^3 .$$

(C.7)

Higgs Interactions with Fermions

$$-i\frac{g}{2}\frac{m_f}{M_W} \equiv -i g_H^f$$

(C.8)

Results for the Helicity Vector Currents

s-channel

$$J_{u_1 v_2}(\uparrow,\downarrow) = \sqrt{s}\,(0,-1,-i,0)$$

(C.9)

$$J_{u_1v_2}(\downarrow,\uparrow) = \sqrt{s}\,(0,-1,i,0) \tag{C.10}$$

$$J_{u_3v_4}(\uparrow,\downarrow) = \sqrt{s}\,(0,-\cos\theta,i,\sin\theta) \tag{C.11}$$

$$J_{u_3v_4}(\downarrow,\uparrow) = \sqrt{s}\,(0,-\cos\theta,-i,\sin\theta) \tag{C.12}$$

t-channel

$$J_{u_1u_3}(\uparrow,\uparrow) = \sqrt{s}\left(\cos\frac{\theta}{2},\sin\frac{\theta}{2},i\sin\frac{\theta}{2},\cos\frac{\theta}{2}\right) \tag{C.13}$$

$$J_{u_1u_3}(\downarrow,\downarrow) = \sqrt{s}\left(\cos\frac{\theta}{2},\sin\frac{\theta}{2},-i\sin\frac{\theta}{2},\cos\frac{\theta}{2}\right) \tag{C.14}$$

$$J_{v_1v_3}(\uparrow,\uparrow) = \sqrt{s}\left(\cos\frac{\theta}{2},\sin\frac{\theta}{2},i\sin\frac{\theta}{2},\cos\frac{\theta}{2}\right) \tag{C.15}$$

$$J_{v_1v_3}(\downarrow,\downarrow) = \sqrt{s}\left(\cos\frac{\theta}{2},\sin\frac{\theta}{2},-i\sin\frac{\theta}{2},\cos\frac{\theta}{2}\right) \tag{C.16}$$

$$J_{u_2u_4}(\uparrow,\uparrow) = \sqrt{s}\left(\cos\frac{\theta}{2},-\sin\frac{\theta}{2},i\sin\frac{\theta}{2},-\cos\frac{\theta}{2}\right) \tag{C.17}$$

$$J_{u_2u_4}(\downarrow,\downarrow) = \sqrt{s}\left(\cos\frac{\theta}{2},-\sin\frac{\theta}{2},-i\sin\frac{\theta}{2},-\cos\frac{\theta}{2}\right) \tag{C.18}$$

$$J_{v_2v_4}(\uparrow,\uparrow) = \sqrt{s}\left(\cos\frac{\theta}{2},-\sin\frac{\theta}{2},i\sin\frac{\theta}{2},-\cos\frac{\theta}{2}\right) \tag{C.19}$$

$$J_{v_2v_4}(\downarrow,\downarrow) = \sqrt{s}\left(\cos\frac{\theta}{2},-\sin\frac{\theta}{2},-i\sin\frac{\theta}{2},-\cos\frac{\theta}{2}\right) \tag{C.20}$$

Results for Scalar Helicity Currents

The interactions of fermions with the scalar bosons are **scalar** interactions and therefore the scalar currents are numbers, **not four-vectors**! Neglecting the fermion masses they are given below, for the s and t channels.

Important Note:
Notice that the rule for the spin arrows for scalar currents is opposite to that of the vector helicity currents. This means that for s-schannel the only non-vanishing helicity amplitudes are $\uparrow, \uparrow$ and $\downarrow, \downarrow$, while for the t-channel amplitudes they are $\uparrow, \downarrow$ and $\downarrow, \uparrow$.

s-channel

$$J_{u_1 v_2}(\uparrow, \uparrow) = \sqrt{s}$$

$$(C.21)$$

$$J_{u_1 v_2}(\downarrow, \downarrow) = -\sqrt{s}$$

$$(C.22)$$

$$J_{u_3 v_4}(\uparrow, \uparrow) = \sqrt{s}$$

$$(C.23)$$

$$J_{u_3 v_4}(\downarrow, \downarrow) = -\sqrt{s}$$

$$(C.24)$$

t-channel

$$J_{u_1 u_3}(\uparrow,\downarrow) = -\sqrt{s}\sin\frac{\theta}{2} \qquad (C.25)$$

$$J_{u_1 u_3}(\downarrow,\uparrow) = \sqrt{s}\sin\frac{\theta}{2} \qquad (C.26)$$

$$J_{v_1 v_3}(\uparrow,\downarrow) = -\sqrt{s}\sin\frac{\theta}{2} \qquad (C.27)$$

$$J_{v_1 v_3}(\downarrow,\uparrow) = \sqrt{s}\sin\frac{\theta}{2} \qquad (C.28)$$

$$J_{u_2 u_4}(\uparrow,\downarrow) = \sqrt{s}\sin\frac{\theta}{2} \qquad (C.29)$$

$$J_{u_2 u_4}(\downarrow,\uparrow) = -\sqrt{s}\sin\frac{\theta}{2} \qquad (C.30)$$

$$J_{v_2 v_4}(\uparrow,\downarrow) = \sqrt{s}\sin\frac{\theta}{2} \qquad (C.31)$$

$$J_{v_2 v_4}(\downarrow,\uparrow) = -\sqrt{s}\sin\frac{\theta}{2} \qquad (C.32)$$

C.1 Physical and Astrophysical Constants

Quantity	Symbol, Equation	Value
Speed of light in vacuum	c	299 792 458 m s^{-1}
Planck constant	h	6.626 070 040(81)$\times 10^{-34}$ J s
Planck constant, reduced	$\hbar = h/2\pi$	1.054 571 800(13)$\times 10^{-34}$ J s
		$= 6.582\ 119\ 514(40) \times 10^{-22}$ MeV s
electron charge magnitude	e	1.602 176 6208(98) $\times 10^{-19}$ C
conversion constant	$\hbar c$	197.326 9788(12) MeV fm
conversion constant	$(\hbar c)^2$	0.389 379 3656(48) GeV^2 mbarn
electron mass	m_e	0.510 998 9461(31) MeV/c^2
		$= 9.109\ 383\ 56(11) \times 10^{-31}$ kg
proton mass	m_p	938.272 0813(58) MeV/c^2
		1.672 621 898(21)$\times 10^{-27}$ kg
unified atomic mass unit (u)	$m(^{12}C \text{ atom})/12 = (1\text{ g})/(N_A \text{ mol})$	931.494 0954(57) MeV/c^2
		$= 1.660\ 539\ 040(20) \times 10^{-27}$ kg
permittivity of free space	$\epsilon_0 = 1/\mu_0 c^2$	8.854 187 817 ...$\times 10^{-12}$ F m^{-1}
permeability of free space	μ_0	$4\pi \times 10^{-7}$ N A^{-2}
fine-structure constant $(Q^2=0)$	$\alpha = e^2/4\pi\epsilon_0\hbar c$	7.297 352 5664(17)$\times 10^{-3} \simeq 1/137$
classical electron radius	$r_e = e^2/4\pi\epsilon_0 m_e c^2$	2.817 940 3227(19)$\times 10^{-15}$ m
(e^- Compton wavelength)/2	$\lambda_e = \hbar/m_e c = r_e\alpha^{-1}$	3.861 592 6764(18)$\times 10^{-13}$ m
Bohr radius (mnucleus = ∞)	$a_\infty = 4\pi\epsilon_0\hbar^2/m_e e^2$	0.529 177 210 67(12)$\times 10^{-10}$ m
wavelength of 1 eV/c particle	$hc/(1\text{ eV})$	1.239 841 9739(76)$\times 10^{-6}$ m
Rydberg energy	$hcR_\infty = m_e e^4/2(4\pi\epsilon_0)^2\hbar^2 = m_e c^2\alpha^2/2$	3.605 693 009(84) eV
Thomson cross section	$\sigma_T = 8\pi r_e^2/3$	0.665 245 871 58(91) barn
Bohr magneton	$\mu B = e\hbar/2m_e$	5.788 381 8012(26)$\times 10^{-11}$ MeV T^{-1}
nuclear magneton	$\mu N = e\hbar/2m_p$	3.152 451 2550(15)$\times 10^{-14}$ MeV T^{-1}
gravitational constant	G	6.674 08(31)$\times 10^{-11}$ m^{3}kg^{-1}s^{-2}
		$= 6.708\ 61(31) \times 10^{-39}$ $\hbar c$ (GeV/c^2)$^{-2}$
standard gravitational accel.	g_N	9.806 65 m s^{-2}
Avogadro constant	N_A	6.022 140 857(74)$\times 10^{23}$ mol^{-2}
Boltzmann constant	k_B	1.380 648 52(79)$\times 10^{-23}$ J K^{-1}
		$= 8.617\ 3303(50) \times 10^{-5}$ eV K^{-1}
molar volume, ideal gas at STP	$N_A k_B \times 273.15K/101325$ Pa	22.413 962(13)$\times 10^{-3}$ m^{3}mol^{-1}
Wien displacement law constant	$b = \lambda_{max}T$	2.897 7729(17)$\times 10^{-3}$ m K
Stefan-Boltzmann constant	$\sigma = \pi^2 k_B^4/60\hbar^3 c^2$	5.670 367(13)$\times 10^{-3}$ W m^{-2}K^{-4}
Fermi coupling constant	$G_F/(\hbar c)^3$	1.166 378 7(6)$\times 10^{-5}$ GeV^{-2}
weak-mixing angle	$\sin^2 \hat{\theta}(M_Z)$ $(\overline{MS})$	
$W^\pm$ boson mass	m_W	80.385(15) GeV/c^2
Z boson mass	m_Z	91.1876(21) GeV/c^2
strong coupling constant	$\alpha_S(m_Z)$	0.1182(12)

$\pi \simeq 3.141592653589793$	$e \simeq 2.718\ 281\ 828\ 459\ 045$	$\gamma \simeq 0.577215664901532$
1 in $\equiv$ 0.0254 m	1 G $\equiv 10^{-4}$ T	1 eV = 1.602 176 6208(98)$\times 10^{-19}$ J
$k_B T$ at 300 K = $[38.681\ 740(22)]^{-1}$ eV	1 $^\circ$A $\equiv$ 0.1nm	1 dyne $\equiv 10^{-5}$ N
1 eV/c^2 = 1.782 661 907(11) $\times 10^{-36}$ kg	0 $^\circ$C $\equiv$ 273.15 K	1 barn $\equiv 10^{-28}$ m^2
1 erg $\equiv 10^{-7}$ J	2.997 924 58$\times 10^9$ esu = 1C	1 atmosphere $\equiv$ 760 Torr $\equiv$ 101 325 Pa

Quantity	Symbol, Equation	Value
Planck mass	$\sqrt{\hbar c/G}$	$1.220\ 910(29) \times 10^{19}$ GeV/c^2=2.176 $47(5) \times 10^{-8}$kg
Planck length	$\sqrt{\hbar G/c^3}$	$1.616\ 229(38) \times 10^{-35}$ m
tropical year (equinox to equinox) (2011)	yr	$31\ 556\ 925.2$ s $\approx \pi \times 10^7$ s
sidereal year (fixed star to fixed star) (2011)		31558149.8 s $\approx \pi \times 10^7$ s
astronomical unit	au	$149\ 597\ 870\ 700$ m
parsec (1 au/1 arc sec)	pc	$3.08567758149 \times 10^{16}$ m = 3.262 ...ly
light year (deprecated unit)	ly	0.3066 pc=0.946053×10^{16} m
Solar mass	$M_\odot$	$1.988\ 48(9) \times 10^{30}$ kg
Schwarzschild radius of the Sun	$2GM_\odot/c^2$	2.953 km
nominal Solar equatorial radius	$R_\odot$	6.957×10^8 m
nominal Solar constant	$S_\odot$	1361 W m^{-2}
nominal Solar photosphere temperature	$T_\odot$	5772 K
nominal Solar luminosity	$\mathscr{L}_\odot$	3.828×10^{26} W
Earth mass	$M_\oplus$	$5.972\ 4(3) \times 10^{24}$ kg
Schwarzschild radius of the Earth	$2G\ M_\oplus/2c^2$	8.870 mm
nominal Earth equatorial radius	$R_\oplus$	6.3781×10^6 m
jansky (flux density)	Jy	10^{-26} W m^{-2} Hz^{-1}
Solar angular velocity around the GC	Θ_0/R_0	30.3 ± 0.9 km s^{-1} kpc^{-1}
Solar distance from GC	R_0	8.00 ± 0.25 kpc
circular velocity at R_0	v_0 or Θ_0	$254(16)$ km s^{-1}
escape velocity from Galaxy	v_{esc}	498 km/s $< v_{esc} < 608$ km/s
local disk density	ρ_{disk}	312×10^{-24} g cm^{-3} ≈ 27 GeV/c^2 cm^{-3}
local dark matter density	ρ_χ	canonical value 0.4 GeV/c^2 cm^{-3} within factor ~ 2

Quantity	Symbol, Equation	Value
present day CMB temperature	T_0	2.7255(6) K
present day CMB dipole amplitude	d	3.3645(20) mK
Solar velocity with respect to CMB	$v_\odot$	370.09(22) km s^{-1} towards $(\ell, b) =$ (263.00(3)$^\circ$, 48.24(2)$^\circ$)
Local Group velocity with respect to CMB	v_{LG}	627(22) km s^{-1} towards $(\ell, b) =$ (276(3)$^\circ$, 430(3)$^\circ$)
number density of CMB photons	n_γ	410.7(3) (T/2.7255)^3 cm^{-3}
density of CMB photons	ρ_γ	4.645(4)(T/2.7255)$^4 \times 10^{-34}$ g/cm$^3 \approx$ 0.260 eV/cm^3
entropy density/Boltzmann constant	s/k	2 891.2 (T/2.7255)^3 cm^{-3}
present day Hubble expansion rate	H_0	100 h km s^{-1} Mpc^{-1} = $h\times$ (9.777 752 Gyr)$^{-1}$
scale factor for Hubble expansion rate	h	0.678(9)
Hubble length	c/H_0	0.925 0629 $\times 10^{26}$ h^{-1} m=1.374(18) $\times 10^{26}$ m
scale factor for cosmological constant	$c^2/3\,H^2{}_0$	2.85247 $\times 10^{51}$ h^{-2} m^2 = 6.20(17) $\times$ 10^{51} m^2
critical density of the Universe	$\rho_{crit} = 3H_0^2/8\pi G$	1.878 40(9) $\times 10^{-29}$ h^2 g cm^{-3} = 1.053 71(5) $\times 10^5$ h^2 (GeV/c^2) cm^{-3} = 2.775 37(13) $\times 10^{11}$ h^2 M$_\odot$ Mpc^{-3}
baryon-to-photon ratio (from BBN)	$\eta = n_b/n_\gamma$	5.8 $\times 10^{-10} \leq \eta \leq 6.6 \times 10^{-10}$ (95 % CL)
number density of baryons	n_b	2.503(26) $\times 10^{-7}$ cm^{-3} (2.4 $\times$ 10^{-7} <n$_b$ <2.7 $\times 10^{-7}$) cm^{-3}(95% CL)
CMB radiation density of the Universe	$\Omega_\gamma = \rho_\gamma/\rho_{crit}$	2.473 $\times$ 10^{-5} (T/2.7255)^4 $h^{-2}=$ 5.38(15) $\times 10^{-5}$
baryon density of the Universe	$\Omega_b = \rho_b/\rho_{crit}$	0.02226(23)h^{-2} = 0.0484(10)
cold dark matter density of the Universe	$\Omega_c = \rho_c/\rho_{crit}$	0.1186(20) h^{-2} = 0.258(11)
reionization optical depth	τ	0.066(16)
scalar spectral index	n_s	0.968(6)
dark energy density of the Universe	Ω_Λ	0.692 ± 0.012
fluctuation amplitude at 8 h^{-1} Mpc scale	σ_8	0.815 ± 0.009
redshift of matter-radiation equality	z_{eq}	3365 ± 44
redshift at which optical depth equals unity	z_*	1089.9 ± 0.4
comoving size of sound horizon at z_*	r_*	144.9 ± 0.4 Mpc
age when optical depth equals unity	t_*	373 kyr
redshift at half reionization	z_{reion}	8.8$^{+1.7}_{-1.4}$
redshift when acceleration was zero	z_q	≈ 0.65
age of the Universe	t_0	13.80 ±0.04 Gyr
effective number of neutrinos	N_{eff}	3.13 ± 0.32
sum of neutrino masses	$\sum m_\nu$	<0.68 eV (Planck CMB); > 0.06 eV (mixing)
curvature	Ω_K	$-0.005^{+0.016}_{-0.017}$(95 %CL)
primordial helium fraction	Y_p	0.245±0.004

C.2 Particle Properties

Gauge Bosons The gauge bosons all have $J^P = 1^-$.

Particle	Mass	Width	Decay Mode	Fraction (%)
g	0 (assumed)	stable		
γ	0	stable		
$W^\pm$	80.4 GeV/c^2	2.1 GeV/c^2	hadrons	67.41(27)
			$e^+\nu_e$	10.71(16)
			$\mu^+\nu_\mu$	10.63(15)
			$\tau^+\nu_\tau$	11.38(27)
Z	91.2 GeV/c^2	2.5 GeV/c^2	hadrons	69.91(6)
			$\nu_\ell + \bar{\nu}_\ell (all\ \ell)$	20.00(6)
			e^+e^-	3.363(4)
			$\mu^+\mu^-$	3.366(7)
			$\tau^+\tau^-$	3.370(8)

Higgs boson ($J^P = 0^+$)

Particle	Mass	Width
H	125.09(24) GeV/c^2	< 13 MeV/c^2 (~ 4/MeV/c^2?)

Leptons All leptons have $J^P = \frac{1}{2}^+$.

Particle	Mass (MeV/c^2)	Lifetime (s)	Decay Mode	Fraction (%)
ν_e	$< 2 \times 10^{-6}$	Stable		
ν_μ	<0.19	Stable		
ν_τ	<18.2	Stable		
$e^\pm$	0.511	Stable		
$\mu^\pm$	105.66	2.197×10^{-6}	$e^+\nu_e\bar{\nu}_\nu$	≈ 100
$\tau^\pm$	1776.84(12)	$(290.3 \pm 0.5) \times 10^{-15}$	hadrons $+\nu_\tau$	~ 64
			$e^+\nu_e\bar{\nu}_\tau$	17.82(4)
			$\mu^+\nu_\mu\bar{\nu}_\tau$	17.39(4)

Low-Lying Baryon

Particle	I, J^P	Mass (MeV/c^2)	Lifetime or width	Decay Mode	Fraction (%)

Unflavoured states of light quarks ($S = C = B = 0$)

Quark content:

$N = (p, n) : p = uud, n = udd; \Delta^{++} = uuu, \Delta^+ = uud, \Delta^0 = udd, \Delta^- = ddd$

Particle	I, J^P	Mass (MeV/c^2)	Lifetime or width	Decay Mode	Fraction (%)
p	$\frac{1}{2}, \frac{1}{2}^+$	938.272081(6)	$> 2.1 \times 10^{29}$ yr		
n	$\frac{1}{2}, \frac{1}{2}^+$	939.565413(6)	880.2(10) s	$pe^- \bar{\nu}_e$	100
Δ	$\frac{3}{2}, \frac{3}{2}^+$	1232(1)	117(2) MeV	$N\pi$	99.4

Strange baryons ($S = -1, C = B = 0$)

Quark content: $\Lambda = uds : \Sigma^+ = uus, \Sigma^0 = uds, \Sigma^- = dds$, similarly for $\Sigma^* s$.

Particle	I, J^P	Mass (MeV/c^2)	Lifetime or width	Decay Mode	Fraction (%)
Λ	$0, \frac{1}{2}^+$	1115.683(6)	$2.632(20) \times 10^{-10}$ s	$p\pi^-$	63.9(5)
				$n\pi^0$	35.8(5)
Σ^+	$1, \frac{1}{2}^+$	1189.37(7)	$8.018(26) \times 10^{-11}$ s	$p\pi^0$	51.57(30)
				$n\pi^+$	48.31(30)
Σ^0	$1, \frac{1}{2}^+$	1192.642(24)	$7.4(7) \times 10^{-20}$ s	$\Lambda\gamma$	100
Σ^-	$1, \frac{1}{2}^+$	1197.449(30)	$1.479(11) \times 10^{-10}$ s	$n\pi^-$	99.848(5)
Σ^{*+}	$1, \frac{3}{2}^+$	1382.8(4)	37.0(7) MeV	$\Lambda\pi$	87.0(15)
				$\Sigma\pi$	11.7(15)
Σ^{*0}	$1, \frac{3}{2}^+$	1383.7(10)	36(5) MeV	as above	
Σ^+	$1, \frac{3}{2}^+$	1387.2(5)	39.4(21) MeV	as above	

Strange baryons ($S = -2, C = B = 0$)

Quark content: $\Xi^0 = uss, \Xi^- = dss$, similarly for $\Xi^* s$.

Particle	I, J^P	Mass (MeV/c^2)	Lifetime or width	Decay Mode	Fraction (%)
Ξ^0	$\frac{1}{2}, \frac{1}{2}^+$	1314.86(20)	$2.90(9) \times 10^{-10}$ s	$\Lambda\pi^0$	99.524(12)
Ξ^-	$\frac{1}{2}, \frac{1}{2}^+$	1321.71(7)	$1.639(15) \times 10^{-10}$ s	$\Lambda\pi^-$	99.887(35)
Ξ^{*0}	$\frac{1}{2}, \frac{3}{2}^+$	1531.80(32)	9.1(5) MeV	$\Xi\pi$	100
Ξ^{*-}	$\frac{1}{2}, \frac{3}{2}^+$	1535.0(6)	9.9(18) MeV	as above	

Low-Lying Baryon

Particle	I, J^P	Mass (MeV/c^2)	Lifetime or width	Decay Mode	Fraction (%)

Strange baryons ($S = -3$, $C = B = 0$)
Quark content: $\Omega^- = sss$

Particle	I, J^P	Mass (MeV/c^2)	Lifetime or width	Decay Mode	Fraction (%)
Ω^-	$0, \frac{3}{2}^+$	1672.45(29)	$8.21(11) \times 10^{-11}$ s	ΛK^-	67.8(7)
				$\Xi^0 \pi^-$	23.6(7)
				$\Xi^- \pi^0$	8.6(4)

Charmed baryons ($S = 0$, $C = +1$, $B = 0$)
Quark content: $\Lambda_c^+ = udc$: $\Sigma^{++} = uuc$, $\Sigma^+ = udc$, $\Sigma^- = ddc$, similarly for $\Sigma_c^* s$.

Particle	I, J^P	Mass (MeV/c^2)	Lifetime or width	Decay Mode	Fraction (%)
Λ_c^+	$0, \frac{1}{2}^+$	2286.46(14)	$2.00(6) \times 10^{-13}$ s	$n + X$	50(16)
				$p + X$	50(16)
				$\Lambda + X$	35(11)
				$\Sigma^\pm + X$	10(5)
				$e^+ + X$	4.5(17)
Σ_c^{++}	$1, \frac{1}{2}^+$	2453.97(14)	1.89(14) MeV	$\Lambda_c^+ \pi^+$	≈ 100
Σ_c^+	$1, \frac{1}{2}^+$	2452.9(4)	< 4.6 MeV		
Σ_c^0	$1, \frac{1}{2}^+$	2453.75(14)	1.83(15) MeV		
Σ_c^{*++}	$1, \frac{3}{2}^+$	2518.41(20)	14.78(35) MeV	$\Lambda_c^+ \pi^+$	≈ 100
Σ_c^{*+}	$1, \frac{3}{2}^+$	2517.5(23)	< 17 MeV		
Σ_c^{*0}	$1, \frac{3}{2}^+$	2518.48(20)	15.3(4) MeV		

Charmed strange baryons ($S = -1, -2$, $C = +1$, $B = 0$)
Quark content: $\Xi_c^+ = usc$, $\Xi_c^0 = dsc$, similarly for Ξ_c^* s; $\Omega_c^0 = ssc$

Particle	I, J^P	Mass (MeV/c^2)	Lifetime or width	Decay Mode	Fraction (%)
Ξ_c^+	$\frac{1}{2}, \frac{1}{2}^+$	2467.87(30)	$4.42(26) \times 10^{-13}$ s		
Ξ_c^0	$\frac{1}{2}, \frac{1}{2}^+$	2470.87(29)	$1.12(11) \times 10^{-13}$ s		
Ω_c^0	$\frac{1}{2}, \frac{1}{2}^+$	2695.2(17)	$6.9(1.2) \times 10^{-14}$ s		
Ξ_c^{*+}	$\frac{1}{2}, \frac{3}{2}^+$	2645.53(31)	2.14(19) MeV		
Ξ_c^{*0}	$\frac{1}{2}, \frac{3}{2}^+$	2646.32(31)	2.35(22) MeV		

Bottom baryons ($S = C = 0$, $B = -1$)
Quark content: $\Lambda_b^0 = udb$, $\Xi_b^0 = usb$, $\Xi_b^- = dsb$

Particle	I, J^P	Mass (MeV/c^2)	Lifetime or width	Decay Mode	Fraction (%)
Λ_b^0	$0, \frac{1}{2}^+$	5619.58(17)	1.47(01) ps	$\Lambda_c^+ + X$	$\sim 11.5(2)$
Ξ_b^0	$\frac{1}{2}, \frac{1}{2}^+$	5791.9(5)	1.479(31) ps		
Ξ_b^-	$\frac{1}{2}, \frac{1}{2}^+$	5794.5(14)	1.571(40) ps		

Low-Lying Mesons

Particle	I, J^{PC}	Mass (MeV/c^2)	Lifetime or width	Decay Mode	Fraction (%)

Unflavored states of light quarks ($S = C = B = 0$)

Quark content:

$I = 1$ states, $u\bar{d}$, $\frac{1}{\sqrt{2}}(u\bar{u} - d\bar{d})$, $d\bar{u}$; $I = 0$ states, $c_1(u\bar{u} - d\bar{d}) + c_2 s\bar{s}$ ($c_{1,2}$ are constants)

Particle	I, J^{PC}	Mass (MeV/c^2)	Lifetime or width	Decay Mode	Fraction (%)
$\pi^\pm$	$1, 0^-$	139.57061(24)	$2.6033(5) \times 10^{-8}$ s	$\mu^+ \nu_\mu$	99.98770(4)
π^0	$1, 0^{-+}$	134.9770(5)	$8.52(18) \times 10^{-17}$ s	$\gamma\gamma$	98.823(34)
η	$0, 0^{-+}$	547.862(17)	1.31(5) keV	$\gamma\gamma$	39.41(20)
				$\pi^0\pi^0\pi^0$	32.68(23)
				$\pi^+\pi^-\pi^0$	22.92(28)
				$\pi^+\pi^-\gamma$	4.22(8)
ρ	$1, 1^{--}$	775.26(25)	149.1(8) MeV	$\pi\pi$	≈ 100
ω^0	$0, 1^{--}$	782.65(12)	8.49(8) MeV	$\pi^+\pi^-\pi^0$	89.2(7)
				$\pi^0\gamma$	8.40(22)
η'	$0, 0^{-+}$	957.78(6)	0.196(9) MeV	$\pi^+\pi^-\eta$	42.6(7)
				$\rho^0\gamma$	28.9(5)
				$\pi^0\pi^0\eta$	22.8(8)
				$\omega\gamma$	2.62(13)
ϕ	$0, 1^{--}$	1019.460(16)	4.247(16) MeV	K^+K^-	48.9(5)
				$K_L^0 + K_S^0$	34.2(4)
				$\rho\pi + \pi^+\pi^-\pi^0$	15.32(32)

Strange mesons ($S = \pm 1, C = B = 0$)

Quark content: $K^+ = u\bar{s}$, $K^0 = d\bar{s}$, $\bar{K}^0 = s\bar{d}$, $K^- = s\bar{u}$, similarly for K^*s

Particle	I, J^{PC}	Mass (MeV/c^2)	Lifetime or width	Decay Mode	Fraction (%)
$K^\pm$	$\frac{1}{2}, 0^-$	493.677(16)	$1.2380(20) \times 10^{-8}$ s	$\mu^+\nu_\mu$	63.56 (11)
				$\pi + \pi^0$	20.67(8)
				$\pi + \pi^+\pi^-$	5.583(24)
				$\pi^0 e^+ \nu_e$	5.07(4)
				$\pi^0 \mu^+ \nu_\mu$	3.352(33)
$K^0, \bar{K}^0$	$\frac{1}{2}, 0^-$	497.611(13)			
K_S^0			$8.954(4) \times 10^{-11}$ s	$\pi^+\pi^-$	69.20(5)
				$\pi^0\pi^0$	30.69(5)
K_L^0			$5.116(21) \times 10^{-8}$ s	$\pi^\pm e^\mp \nu_e (\bar{\nu}_e)$	40.55(11)
				$\pi^\pm \mu^\mp \bar{\nu}_\mu (\bar{\nu}_\mu)$	27.04(7)
				$\pi^0\pi^0\pi^0$	19.52(12)
				$\pi^+\pi^-\pi^0$	12.54(5)
$K^{*\pm}$	$\frac{1}{2}, 1^-$	891.76(25)	50.3(8) MeV	$K\pi$	~ 100
K^{*0}	$\frac{1}{2}, 1^-$	895.55(20)	47.3(5) MeV	$K\pi$	~ 100

Particle	I, J^{PC}	Mass (MeV/c^2)	Lifetime or width	Decay Mode	Fraction (%)

Charmed mesons ($S = 0, C = \pm 1, B = 0$)
Quark content: $D^+ = c\bar{d}$, $D^0 = c\bar{u}$, $\bar{D}^0 = u\bar{c}$, $D^- = d\bar{c}$, similarly for D^*s

Particle	I, J^{PC}	Mass (MeV/c^2)	Lifetime or width	Decay Mode	Fraction (%)
$D^\pm$	$\frac{1}{2}, 0^-$	1869.59(9)	1.040(7) ps	$\bar{K}^0 + X$	61(5)
				$K^- + X$	25.7(14)
				$\bar{K}^{*0} + X$	23(5)
				$e^+ + X$	16.07(30)
				$\bar{K}^+ + X$	5.9(8)
$D^0, \bar{D}^0$	$\frac{1}{2}, 0^-$	1864.83(5)	$4.101(15) \times 10^{-13}$ s	$K^- + X$	54.7(28)
				$\bar{K}^0 + X$	47(4)
				$\bar{K}^{*0} + X$	9(4)
				$e^+ + X$	6.49(11)
				$K^+ + X$	3.4(4)
$D^{*\pm}$,	$\frac{1}{2}, 1^-$	2010.26(5)	83.4(18) keV	$D^0 \pi^+$	67.7(5)
				$D^+ \pi^0$	30.7(5)
$D^{*0}, \bar{D}^{*0}$	$\frac{1}{2}, 1^-$	2006.85(5)	< 2.1 MeV	$D^0 \pi^0$	64.7(9)
				$D^0 \gamma$	35.3(9)

Charmed strange mesons ($S = C = \pm 1 B = 0$)
Quark content: $D_s^+ = c\bar{s}$, $D_s^- = s\bar{c}$, similarly for $D_s^* s$

Particle	I, J^{PC}	Mass (MeV/c^2)	Lifetime or width	Decay Mode	Fraction (%)
$D_s^\pm$,	$0, 0^-$	1968.28(10)	$5.00(7) \times 10^{-13}$ s	$K^+ + X$	28.9(07)
				$K_s^0 + X$	19.0(11)
				$\phi + X$	15.07(10)
				$K^- + X$	18.7(5)
				$e^+ + X$	6.5(4)
				$\tau \nu_\tau$	5.48(23)
$D^{*\pm}$	$0, 1^-$	2112.1(4)	< 1.9 MeV	$D_s^+ \gamma$	93.5(7)
				$D_s^+ \pi^0$	5.8(7)

Particle	I, J^{PC}	Mass (MeV/c^2)	Lifetime or width	Decay Mode	Fraction (%)
Bottom strange mesons $(S = \pm 1, C = 0, B = \pm 1)$					
Quark content: $B_s^0 = s\bar{b}, \bar{B}_s^0 = b\bar{s}$					
$B_s^0, \bar{B}_s^0$	$0, 0^-$	5366.89(19)	1.505(05) ps	$D_s^- + X$	93(25)
				$D_s^- \ell^+ \nu_\ell + X$	8.1(13)
Bottom charmed mesons $(S = 0, B = C = \pm 1)$					
Quark content: $B_c^+ = c\bar{b}, B_c^- = b\bar{c}$					
$B_c^\pm$	$0, 0^-$	6274.9(8)	$5.07(12) \times 10^{-13}$ s		
$c\bar{c}$ mesons					
$J/\psi(1S)$	$0, 1^{--}$	3096.900(6)	92.9(28) keV	hadrons	87.7(5)
				$e^+ e^-$	5.971(32)
				$\mu^+ \mu^-$	5.961(33)
$b\bar{b}$ mesons					
$\Upsilon(1S)$	$0, 1^{--}$	9460.30(26)	54.02(125) keV	$\eta' + X$	2.94(24)
				$\tau^+ \tau^-$	2.60(10)
				$e^+ e^-$	2.38(11)
				$\mu^+ \mu^-$	2.48(5)

C.3 Periodic Table of the Elements

PERIODIC TABLE OF THE ELEMENTS

1 IA	2 IIA	3 IIIB	4 IVB	5 VB	6 VIB	7 VIIB	8	9 VIII	10	11 IB	12 IIB	13 IIIA	14 IVA	15 VA	16 VIA	17 VIIA	18 VIIIA
1 H hydrogen 1.008																	2 He helium 4.002602
3 Li lithium 6.94	4 Be beryllium 9.012182											5 B boron 10.81	6 C carbon 12.0107	7 N nitrogen 14.007	8 O oxygen 15.999	9 F fluorine 18.998403163	10 Ne neon 20.1797
11 Na sodium 22.98976928	12 Mg magnesium 24.305											13 Al aluminum 26.9815385	14 Si silicon 28.085	15 P phosphorus 30.973761998	16 S sulfur 32.06	17 Cl chlorine 35.45	18 Ar argon 39.948
19 K potassium 39.0983	20 Ca calcium 40.078	21 Sc scandium 44.955908	22 Ti titanium 47.867	23 V vanadium 50.9415	24 Cr chromium 51.9961	25 Mn manganese 54.938044	26 Fe iron 55.845	27 Co cobalt 58.933195	28 Ni nickel 58.6934	29 Cu copper 63.546	30 Zn zinc 65.38	31 Ga gallium 69.723	32 Ge germanium 72.630	33 As arsenic 74.921595	34 Se selenium 78.971	35 Br bromine 79.904	36 Kr krypton 83.798
37 Rb rubidium 85.4678	38 Sr strontium 87.62	39 Y yttrium 88.90584	40 Zr zirconium 91.224	41 Nb niobium 92.90637	42 Mo molybdenum 95.95	43 Tc technetium (97.907212)	44 Ru ruthenium 101.07	45 Rh rhodium 102.90550	46 Pd palladium 106.42	47 Ag silver 107.8682	48 Cd cadmium 112.414	49 In indium 114.818	50 Sn tin 118.710	51 Sb antimony 121.760	52 Te tellurium 127.60	53 I iodine 126.90447	54 Xe xenon 131.293
55 Cs caesium 132.90545196	56 Ba barium 137.327	57–71 LANTHA- NIDES	72 Hf hafnium 178.49	73 Ta tantalum 180.94788	74 W tungsten 183.84	75 Re rhenium 186.207	76 Os osmium 190.23	77 Ir iridium 192.217	78 Pt platinum 195.084	79 Au gold 196.966569	80 Hg mercury 200.592	81 Tl thallium 204.38	82 Pb lead 207.2	83 Bi bismuth 208.98040	84 Po polonium (208.98243)	85 At astatine (209.98715)	86 Rn radon (222.01758)
87 Fr francium (223.01974)	88 Ra radium (226.02541)	89–103 ACTINIDES	104 Rf rutherford. (267.12169)	105 Db dubnium (268.12567)	106 Sg seaborgium (271.13393)	107 Bh bohrium (272.13826)	108 Hs hassium (270.13429)	109 Mt meitnerium (276.15159)	110 Ds darmstadt. (281.16451)	111 Rg roentgen. (280.16514)	112 Cn copernicium (285.17712)	113 Nh (nihonium) (284.17873)	114 Fl flerovium (289.19042)	115 Mc (moscovium) (288.19274)	116 Lv livermorium (293.20449)	117 Ts (tennessine) (292.20746)	118 Og (oganesson) (294.21392)

Lanthanide series	57 La lanthanum 138.90547	58 Ce cerium 140.116	59 Pr praseodym. 140.90766	60 Nd neodymium 144.242	61 Pm promethium (144.91276)	62 Sm samarium 150.36	63 Eu europium 151.964	64 Gd gadolinium 157.25	65 Tb terbium 158.92535	66 Dy dysprosium 162.500	67 Ho holmium 164.93033	68 Er erbium 167.259	69 Tm thulium 168.93422	70 Yb ytterbium 173.054	71 Lu lutetium 174.9668
Actinide series	89 Ac actinium (227.02775)	90 Th thorium 232.0377	91 Pa protactinium 231.03588	92 U uranium 238.02891	93 Np neptunium (237.04817)	94 Pu plutonium (244.06420)	95 Am americium (243.06138)	96 Cm curium (247.07035)	97 Bk berkelium (247.07031)	98 Cf californium (251.07959)	99 Es einsteinium (252.08298)	100 Fm fermium (257.09511)	101 Md mendelevium (258.09844)	102 No nobelium (259.10103)	103 Lr lawrencium (262.10961)

Fig. C.1 Chemical elements periodic table. Taken from P.A. Zyla et al. (Particle Data Group) 2020, Prog. Theor. Exp. Phys. 2020, 083C01

C.4 Clebsch-Gordan Coefficients

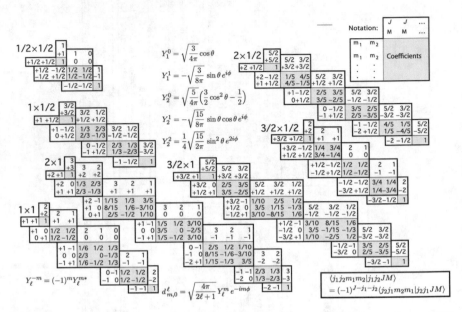

Fig. C.2 Clebsch-Gordan coefficients. Taken from P.A. Zyla et al. (Particle Data Group) 2020, Prog. Theor. Exp. Phys. 2020, 083C01

C.5 Material Properties

Table C.1 Table with some of the material properties. Taken from P.A. Zyla et al. (Particle Data Group) 2020, Prog. Theor. Exp. Phys. 2020, 083C01

Material	Z	A	⟨Z/A⟩	Nucl.coll. length λ_T {g cm⁻²}	Nucl.inter. length λ_I {g cm⁻²}	Rad.len. X_0 {g cm⁻²}	dE/dx|min {MeV cm⁻²}	Density g⁻¹ {g cm⁻³} {g l⁻¹}	Melting point (K)	Boiling point (K)	Refract. index 589.2 nm
H₂	1	1.008(7)	0.99212	42.8	52.0	63.05	(4.103)	0.071(0.084)	13.81	20.28	1.11
D₂	1	2.014101764(8)	0.49650	51.3	71.8	125.97	(2.053)	0.169(0.168)	18.7	23.65	1.11
He	2	4.002602(2)	0.49967	51.8	71.0	94.32	(1.937)	0.125(0.166)		4.220	1.02
Li	3	6.94(2)	0.43221	52.2	71.3	82.78	1.639	0.534	453.6	1615.	
Be	4	9.0121831(5)	0.44384	55.3	77.8	65.19	1.595	1.848	1560.	2744.	
C diamond	6	12.0107(8)	0.49955	59.2	85.8	42.70	1.725	3.520			2.419
C graphite	6	12.0107(8)	0.49955	59.2	85.8	42.70	1.742	2.210	Sublimes at 4098. K		
N₂	7	14.007(2)	0.49976	61.1	89.7	37.99	(1.825)	0.807(1.165)	63.15	77.29	1.20
O₂	8	15.999(3)	0.50002	61.3	90.2	34.24	(1.801)	1.141(1.332)	54.36	90.20	1.22
F₂	9	18.998403163(6)	0.47372	65.0	97.4	32.93	(1.676)	1.507(1.580)	53.53	85.03	
Ne	10	20.1797(6)	0.49555	65.7	99.0	28.93	(1.724)	1.204(0.839)	24.56	27.07	1.09
N	13	26.9815385(7)	0.48181	69.7	107.2	24.01	1.615	2.699	933.5	2792.	
Al	13	26.9815385(7)	0.48181	69.7	107.2	24.01	1.615	2.699	933.5	2792.	
Si	14	28.0855(3)	0.49848	70.2	108.4	21.82	1.664	2.329	1687.	3538.	3.95
Cl2	17	35.45s(2)	0.47951	73.8	115.7	19.28	(1.630)	1.574(2.980)	171.6	239.1	
Ar	18	39.948(1)	0.45059	75.7	119.7	19.55	(1.519)	1.396(1.662)	83.81	87.26	1.23
Ti	22	47.867(1)	0.45961	78.8	126.2	16.16	1.477	4.540	1941.	3560.	
Fe	26	55.845(2)	0.46557	81.7	132.1	13.84	1.451	7.874	1811.	3134.	
Cu	29	63.546(3)	0.45636	84.2	137.3	12.86	1.403	8.960	1358.	2835.	
Ge	32	72.630(1)	0.44053	86.9	143.0	12.25	1.370	5.323	1211.	3106.	
Sn	50	118.710(7)	0.42119	98.2	166.7	8.82	1.263	7.310	505.1	2875.	
Xe	54	131.293(6)	0.41129	100.8	172.1	8.48	(1.255)	2.953(5.483)	161.4	165.1	1.39
W	74	183.84(1)	0.40252	110.4	191.9	6.76	1.145	19.300	3695.	5828.	
Pt	78	195.084(9)	0.39983	112.2	195.7	6.54	1.128	21.450	2042.	4098.	
Au	79	196.966569(5)	0.40108	112.5	196.3	6.46	1.134	19.320	1337.	3129.	
Pb	82	207.2(1)	0.39575	114.1	199.6	6.37	1.122	11.350	600.6	2022.	
U	92	[238.02891(3)]	0.38651	118.6	209.0	6.00	1.081	18.950	1408.	4404.	
Air (dry, 1 atm)			0.49919	61.3	90.1	36.62	(1.815)	(1.205)		78.80	1.0003
Shielding concrete			0.50274	65.1	97.5	26.57	1.711	2.300			
Borosilicate glass (Pyrex)			0.49707	64.6	96.5	28.17	1.696	2.230			
Lead glass			0.42101	95.9	158.0	7.87	1.255	6.220			
Standard rock			0.50000	66.8	101.3	26.54	1.688	2.650			
Methane (CH₄)			0.62334	54.0	73.8	46.47	(2.417)	(0.667)	90.68	111.7	
Ethane (C₂H₆)			0.59861	55.0	75.9	45.66	(2.304)	(1.263)	90.36	184.5	
Propane (C₃H₈)			0.58962	55.3	76.7	45.37	(2.262)	0.493(1.868)	85.52	231.0	
Butane (C₄H₁₀)			0.59497	55.5	77.1	45.23	(2.278)	(2.489)	134.9	272.6	
Octane (C₈H₁₈)			0.57778	55.8	77.8	45.00	2.123	0.703	214.4	398.8	
Paraffin (CH₃(CH₂)ₙ.₂₃CH₃)			0.57275	56.0	78.3	44.85	2.088	0.930			
Nylon (type 6, 6/6)			0.54790	57.5	81.6	41.92	1.973	1.18			
Polycarbonate (Lexan)			0.52697	58.3	83.6	41.50	1.886	1.20			
Polyethylene ([CH₂CH₂]ₙ)			0.57034	56.1	78.5	44.77	2.079	0.89			
Polyethylene terephthalate (Mylar)			0.52037	58.9	84.9	39.95	1.848	1.40			
Polyimide film (Kapton)			0.51264	59.2	85.5	40.58	1.820	1.42			
Polymethylmethacrylate (acrylic)			0.53937	58.1	82.8	40.55	1.929	1.19			1.49
Polypropylene			0.55998	56.1	78.5	44.77	2.041	0.90			
Polystyrene ([C₆H₅CHCH₂]ₙ)			0.53768	57.5	81.7	43.79	1.936	1.06			1.59
Polytetrafluoroethylene (Teflon)			0.47992	63.5	94.4	34.84	1.671	2.20			
Polyvinyltoluene			0.54141	57.3	81.3	43.90	1.956	1.03			1.58
Aluminum oxide (sapphire)			0.49038	65.5	98.4	27.94	1.647	3.970	2327.	3273.	1.77
Barium flouride (BaF₂)			0.42207	90.8	149.0	9.91	1.303	4.893	1641.	2533.	1.47
Bismuth germanate (BGO)			0.42065	96.2	159.1	7.97	1.251	7.130	1317.		2.15
Carbon dioxide gas (CO₂)			0.49989	60.7	88.9	36.20	1.819	(1.842)			
Solid carbon dioxide (dry ice)			0.49989	60.7	88.9	36.20	1.787	1.563	S	ublimes at 194.7K	
Cesium iodide (CsI)			0.41569	100.6	171.5	8.39	1.243	4.510	894.2	1553.	1.79
Lithium fluoride (LiF)			0.46262	61.0	88.7	39.26	1.614	2.635	1121.	1946.	1.39
Lithium hydride (LiH)			0.50321	50.8	68.1	79.62	1.897	0.820	965.		
Lead tungstate (PbWO₄)			0.41315	100.6	168.3	7.39		8.300	1403.		2.20
Silicon dioxide (SiO₂, fused quartz)			0.49930	65.2	97.8	27.05	1.699	2.200	1986.	3223.	1.46
Sodium chloride (NaCl)			0.47910	71.2	110.1	21.91	1.847	2.170	1075.	1738.	1.54
Sodium iodide (NaI)			0.42697	93.1	154.6	9.49	1.305	3.667	933.2	1577.	1.77
Water (H₂O)			0.55509	58.5	83.3	36.08	1.992	1.000	273.1	373.1	1.33
Silica aerogel			0.50093	65.0	97.3	27.25	1.740	0.200	(0.03 H2O, 0.97 SiO2)		

Printed in the United States
by Baker & Taylor Publisher Services